All-Russia Scientific and Practical Conference on Materials Treatment

Current Problems and Solutions

IOP Conference Series: Materials Science and Engineering Volume 125

Yurga, Russia
26 – 28 November 2015

ISBN: 978-1-5108-2413-3
ISSN: 1757-8981

Printed from e-media with permission by:

Curran Associates, Inc.
57 Morehouse Lane
Red Hook, NY 12571

Some format issues inherent in the e-media version may also appear in this print version.

Printed by Curran Associates, Inc. (2016)

Institute of Physics
Dirac House, Temple Back
Bristol BS1 6BE UK

Phone: 44 1 17 929 7481
Fax: 44 1 17 920 0979

techtracking@iop.org

Additional copies of this publication are available from:

Curran Associates, Inc.
57 Morehouse Lane
Red Hook, NY 12571 USA
Phone: 845-758-0400
Fax: 845-758-2633
Email: curran@proceedings.com
Web: www.proceedings.com

TABLE OF CONTENTS

IOP Conf. Series: Materials Science and Engineering **125** (2016) 012001 doi:10.1088/1757-899X/125/1/012001

Evaluation of the Use of Complex Mineral Concentrate as a Modifier Steel

R A Gizatulin, S N Fedoseev, R S Dariev

Yurga Institute of Technology, National Research Tomsk Polytechnic University affiliate Kemerovo region, Yurga, Leningradskaya str. 26, RUS

e-mail: sfedoseev@tpu.ru

Abstract. Increasing customer demands for quality of the resulting metal, and in the first place, the impurities, metallurgists dictate need to develop new and improved technologies. Thus, a significant reduction in metal losses can be achieved by developing new complex alloy steels, special purpose, improving technology of their production and developing new technology of smelting to improve the physical, mechanical, foundry and operational characteristics by influencing the structure of the steel by modifying the liquid melt, change more favorable morphology of nonmetallic inclusions. For complex-alloyed steels expensive and scarce alloying elements Ti, Nb, Zr, etc., are used, which are inaccessible to conventional structural steels. In this regard, the paper also presents the results of applying of innovative modifiers containing alloying elements (Ti, Nb, Zr, etc.) based on mineral concentrates in the Tomsk region.

1. Introduction

Contemporary mechanical engineering requires the use of materials having not only improved strength, but also a series of special properties that provide a long and reliable performance parts in a variety of operating conditions. Thus one and the same steel alloying element imparts simultaneously some special properties. These are determined primarily chemical composition.

Requirements for quality, performance and service characteristics of steel require continuous improvement of their composition and production technology. It depends on the increase in the life of modern equipment and machinery. In this context, an important issue is to increase the wear resistance, corrosion resistance, heat resistance, sealing et al., Which is relevant to many industries (mining, metallurgy, machine building, construction, etc.).

Resolve this issue is possible by means micro-alloying steel small additions of soluble and insoluble impurities, vacuuming and refining, exposure to ultrasound and low-frequency vibrations, external fields (electric, magnetic), speed control and other heat removal.

Most cost effective way to improve the quality of the steel melt should be considered modifying soluble and insoluble additives because is not required any additional expensive equipment. Added to the melt small additives of soluble and insoluble impurities influence the crystallization parameters, dislocation structure, degassing, formation of nonmetallic inclusions and secondary phases, liquation, change of shrinkage, deformation and solidification rate strand shell, as well as recrystallization and grain growth.

A feature of the complex modification of steel alloys is that in parallel with the changes nature of the refinement of the structure and the shape of of nonmetallic inclusions is reduced to 1.5-2.0 times the pollution of the austenite grain boundaries of the oxide, sulfide, and nitride inclusions, increased uniformity of structural components for enhanced ductility and the toughness of the steel.

But the high cost used in the modification of the alloying elements and their deficiency reduces the spread of method for improving steel. This problem can be solved by using the modification steels of natural mineral concentrates affluent necessary alloying elements (Ti, Nb, Zr, and others.). In this paper the use of complex mineral concentrates fields of Tomsk region.

2. Experimental methods

A study was carried out over the length of the metal and the volume of blanks and included the following tests and studies:

- Tensile testing;
- Impact test at a temperature of + 20 °C – 50 °C;
- Determination of crack resistance and impact toughness on samples with a crack at a temperature of – 50 °C;
- Control of macrostructure and fracture of metal;
- Control of the content of gases (oxygen and nitrogen) in the metal;
- Study of the nature and control of the content of nonmetallic inclusions;
- Study of the microstructure of metal;
- Control of the grain size and distribution.

All testing and research (including sample preparation) carried by standardized procedures, which ensures the objectivity and comparability of results, including the ability to compare the results of the previously conducted research and testing, including tests of serial production.

Control of macrostructure and bend metal, and tensile testing and impact strength at a temperature of + 20 °C – 50 °C, carried directly on the manufacturer; testing and fracture toughness of the on samples with a crack at a temperature of – 50 °C and a durability of metal research conducted under conditions of Nano-Center" TPU, Tomsk on samples prepared by enterprises-producers of blanks after finishing their required by the specialists of "Novosibirsk Switch Plant".

Tensile tests were conducted in accordance the requirements of National State Standard 1497-84 samples of type III № 4 $l_0 = 5\ d_0$ latching of the following indicators:

- proportional limit σ_p;
- yield strength $\sigma_{0,2}$;
- the ultimate tensile stress (tensile strength), R_m;
- elongation δ;
- Reduction ψ.

Control was carried out on metal macrostructure Macro section prepared in accordance the requirements of National State Standard 10243-75. Etching micro sections was performed in 50% solution of hydrochloric acid at a temperature of 60 °C for 10-15 minutes.

Fractures were produced from the same plate, which was held control of the macrostructure of the metal. To do so to plates plotted incision Macro section the opposite side, and then conducted its destruction on copra or press.

Kind of macrostructures and fractures was fixed with the help of photography.

Evaluation macrostructures and fractures metal was carried out in accordance with the requirements of National State Standard 5192-78.

Control the content of gases (oxygen and nitrogen) in the metal was carried out using an automated a gas analyzer Eltra ON900.

The principle of operation of the a gas analyzer is in a molten sample of known mass in a stream of neutral gas, which captures and transports oxygen and nitrogen released from the molten metal to the spectral detector. The detector registers the integral content of oxygen and nitrogen in the transmitted stream and calculates the weight and content of the gases in the metal micro sample.

To improve the accuracy of the determination of gases gas analysis procedure conducted repeatedly for each sample and the results were subjected to statistical analysis.

Micro sections to study of nonmetallic inclusions were made of halves of the samples after tensile testing, toughness and crack resistance.

Before the metallographic studies samples passed several stages polishing, which are selected accordance with the features of the material samples. Stages of polishing included:

1. Tenderloin samples cutting machine BUEHLER Delta Abrasimet Cutter using abrasive wheel type HH for metallic materials and steel hardness 50-60 HRC at a constant water cooling;
2. Hot Mounting Presses in the phenolic resin in an automatic press BUEHLER SimpliMet 1000;
3. The grinding and polishing of machine tool to BUEHLER Phoenix 4000 using the following consumables: grinding paper Buehler CarbiMet 180 grit, Buehler CarbiMet 240 grit, Buehler CarbiMet

IOP Conf. Series: Materials Science and Engineering **125** (2016) 012001 doi:10.1088/1757-899X/125/1/012001

600 grit; polishing with diamond pastes tissue Metadi particle size 6 microns and 3 microns and 0.05 microns MasterPrep suspension.

Analysis of the microstructure of the metal was carried out using motorized optical microscopes Zeiss Axiovert 200 MAT and Nikon Epiphot TME, which are part of the complex analytical Thixomet.Pro.

Analysis of the contamination steel non-metallic inclusions was carried out in sections of the standard ASTM E1245-03 and according to National State Standard 1778-70 with appropriate image analyzer modules Thixomet .Pro.

Micro x-ray spectroscopic studies of the chemical composition of non-metallic inclusions were performed on a scanning electron microscopic ZEISS SUPRA 55 VP. The microscope is equipped with a computer control of a system an electron beam scanning and digital recording signals and images, X-ray microanalyses INCA WAVE and INCA X-MAX.

In order to identify the grain structure of micro sections etching performed after determining the composition and the content of nonmetallic inclusions 4% alcoholic solution of nitric acid and of Marshall's reagent, consisting of two parts which are mixed in a ratio of A: B = 1:1 just before use:

- Part A: 5 ml of concentrated H_2SO_4, 8 grams of oxalic acid and 100 mL H_2O;
- Part B: 30% solution of hydrogen peroxide H_2O_2.

Metallographic studies were carried out with the help of a motorized light microscope Axiovert 200 MAT at magnifications from 50 to 1000, which is part of the complex analytical Thixomet Pro.

The procedure for conducting the analysis is to build a panoramic image etched structure with an increase 1000-fold and the quantitative analysis of metal structure by intersecting concentric circles, in accordance with National State Standard 5639-82 and the appointment of grain size, by comparison with the standard scales in accordance with National State Standard 8233-56 (scale 3) in the corresponding module Thixomet.Pro an image analyzer.

3. Results and Discussion

The test results blanks volume acceptance tests after thermal treatment in various models are shown in Table 1. The test results of periodic blanks №№ R6942A2 and R6986A1, previous final thermal treatment for two different modes, are presented in Table 2.

In order to achieve sustained and higher levels of mechanical properties, especially on indicators of percentage reduction and impact toughness at subzero temperatures, additional studies were conducted and testing of modes quenching and tempering.

Table 1. The results of acceptance tests of blanks products 2A82-1M

Mode of heat treatment	№	σ_p, MPa	$\sigma_{0.2}$, MPa	$\sigma_в$, MPa	δ, %	ψ, %	$KCU^{+20°C}$ MJ/m^2	$KCU_{-50°C}$, MJ/m^2	$KCT_{-50°C}$, MJ/m^2	K_{1C} $_{-50°C}$, H/mm$^{3/2}$
blank № R6790A1-1										
Mode № 1	K	1167	1324	1442	8.0	37.0	0.33	0.27	0.10	2099
	D	1197	1324	1452	8.6	30.0	0.29	0.22	0.10	2276
Mode № 2	K	1187	1305	1432	13.0	38.0	0.30	0.20	0.10	2001
	D	1187	1315	1462	8.4	32.0	0.33	0.24	0.11	2551
blank № R6942A2-1										
Mode № 1	K	1187	1295	1413	11.5	35.0	0.29	0.20	0.12	2815
	D	1197	1344	1472	9.6	33.0	0.29	0.20	0.10	2443
Mode № 2	K	1167	1285	1403	10.0	28.0	0.25	0.20	0.11	2443
	D	1177	1295	1432	8.6	31.0	0.27	0.22	0.11	2521
blank № R6986A1-1										
Mode № 1	K	1216	1344	1481	10.0	26.0	0.27	0.24	0.09	2021
	D	1197	1334	1501	9.6	30.0	0.27	0.20	0.08	2325
Mode № 2	K	1158	1305	1452	10.0	23.5	0.27	0.16	0.10	2090
	D	1167	1295	1472	10.5	37.0	0.29	0.20	0,09	2403

IOP Conf. Series: Materials Science and Engineering **125** (2016) 012001 doi:10.1088/1757-899X/125/1/012001

Table 2. The results of periodic testing blank products 2A82-1M

№№	σ_p, MPa	$\sigma_{0.2}$, MPa	$\sigma_в$, MPa	δ, %	ψ, %	KCU $^{+20\,°C}$, MJ/m^2	KCU $^{-50\,°C}$, MJ/m^2	KCT $^{-50\,°C}$, MJ/m^2	$K_{1C}^{-50\,°C}$, H/mm$^{3/2}$
1, 8	1177	1285	1393	9.2	25.0	0.35	0.22	0.13	2462
	1167	1285	1393	10.5	23.0	0.31	0.20	0.11	2629
2, 9	1177	1305	1413	8.0	26.0	0.29	0.20	0.10	2099
	1177	1305	1413	10.0	26.0	0.29	0.20	0.09	2207
3, 10	1158	1275	1383	8.4	29.0	0.29	0.20	0.11	2825
	1187	1295	1413	8.0	28.0	0.27	0.20	0.11	2639
4, 11	1167	1295	1422	10.0	33.0	0.29	0.20	–	–
	1167	1295	1422	8.4	24.5	0.29	0.20	–	–
5, 12	1177	1285	1413	9.6	32.0	0.27	0.20	0.11	2551
	1187	1295	1422	7.5	30.0	0.29	0.20	0.10	2443
6, 13	1187	1305	1422	10.0	30.0	0.31	0.22	–	–
	1177	1285	1413	9.2	29.0	0.29	0.22	–	–
7, 14	1187	1315	1432	9.4	29.0	0.29	0.22	0.10	1972
	1177	1305	1432	9.6	33.0	0.29	0.20	0.10	2580

From the above it is evident that all the blanks in all modes tested final heat treatment meet the requirements of National State Standard V5192-78 established for the category of strength of O-120, on indicators of mechanical properties and crack resistance, except number R6986A1-1blanks having a value of percentage reduction below the established standard (ψ = 20%). When smelting steel billet R6986A1-1 number was excluded additive of titanium in steel for the comparative assessment of its impact on the nitrogen content and the mechanical properties of the metal blanks.

The test results of blanks №№ R6942A2-1 and R6986A1-1 by cutting along the length shows that fluctuations in the properties of the metal are observed not only at the ends of the of blanks, but also along the length. When this billet number R6986A1-1, thermally treated for mode number 2, has a uniform refractive properties throughout.

Results of control of the chemical composition of the metal of blanks are presented in table 3. The inspection results the impurity content of non-ferrous metals is presented in table 4.

Table 3. The inspection results of the chemical composition of the metal products 2A82-1M

№№ blanks	Chemical composition, %									
	C	Mn	Si	P	S	Cr	Ni	Cu	Mo	V
R6942A2-1	0.42	0.26	0.25	0.012	0.003	1.04	3.42	0.22	0.57	0.16
R6790A1-1	0.40	0.31	0.28	0.006	0.009	1.10	3.38	0.17	0.57	0.16
R6986A1-1	0.39	0.30	0.32	0.009	0.007	1.05	3.39	0.21	0.58	0.15
R7186A1-1	0.40	0.30	0.26	0.005	0.004	1.09	3.40	0.24	0.59	0.13

IOP Conf. Series: Materials Science and Engineering **125** (2016) 012001 doi:10.1088/1757-899X/125/1/012001

Table 4. The results of the control the content of impurities in the material non-ferrous metal products 2A82-1M

No blanks	Chemical composition, %							
	Sn	As	Pb	Sb	Zn	Bi	Al	Ti
R6942A2-1	0.009	0.012	< 0.001	0.003	< 0.001	0.002	< 0.005	0.020
R6986A1-1	0.006	0.012	< 0.001	0.002	< 0.001	0.001	0.010	0.022

As can be seen from the data on the chemical composition of the metal of blanks fully meets the requirements of National State Standard 5192-78. It should be noted that the metal blank number R6942A2-1 observed higher (than other metal of blanks) phosphorus content and the metal blank number R7186A1-1 is the purest on the content of sulfur and phosphorus.

Was a detailed study inclusions detected in fracture specimens for mechanical testing of selected of blanks from №№ R6986A1-1 and R7186A1-1, which had negative results on indicators of the plastic properties of metal (relative narrowing and impact toughness at room and low temperature). Type characteristic inclusions encountered in metal of blanks 2A82-1M product is shown in Figure.

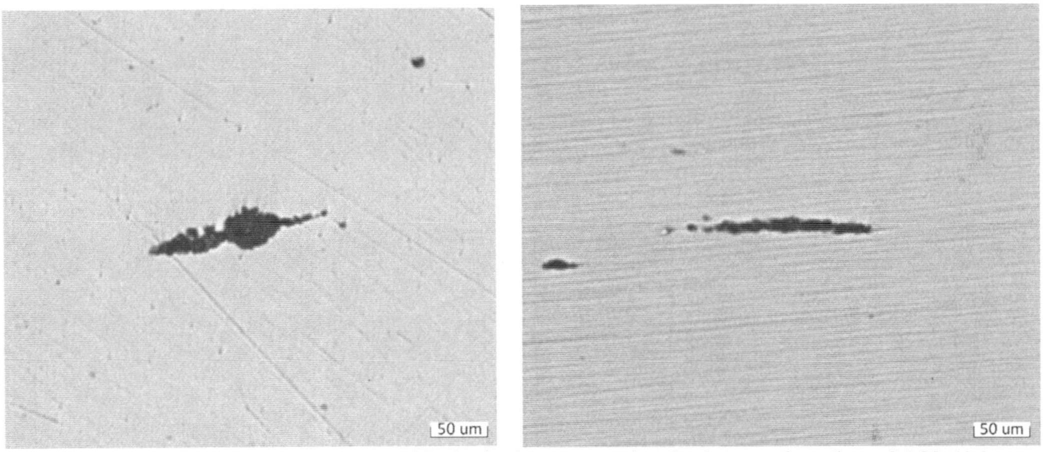

Figure 1. Kind of typical non-metallic inclusions, occurring in the metal products 2A82-1M

As a result, studies have shown that the cause of lowering properties of blanks is large concentrations of nonmetallic inclusions onto the surface of the fracture. View kinks samples presented in Figures 2.

As can be seen from the photos, presented in Figure 2, on the samples taken from the work piece number R6986A1-1, inclusions were golden color, which suggests their nature nitride.

Figure 2. Kind of fracture of the sample with lowered plastic properties, selected products 2A82-1M

IOP Conf. Series: Materials Science and Engineering **125** (2016) 012001 doi:10.1088/1757-899X/125/1/012001

During metallographic studies micro sections inclusion of both types had the appearance of lines. To determine the composition of inclusions was conducted microprobe analysis, which resulted in the maps of the distribution of chemical elements contained in the inclusions.
Photos inclusions micro sections and map the distribution of elements in the inclusions are shown in Figure 3 and Figure 4.

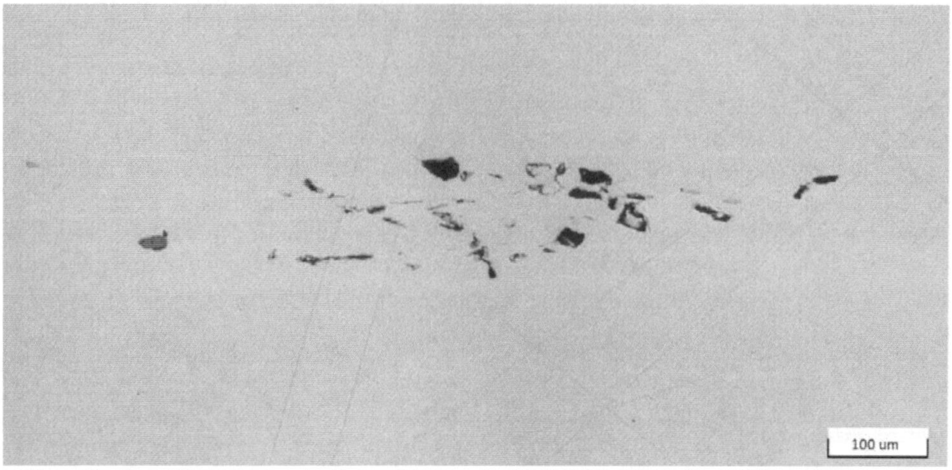

Figure 3. Specific inclusion in the metal sample, № R6986A1-1 view of the micro section

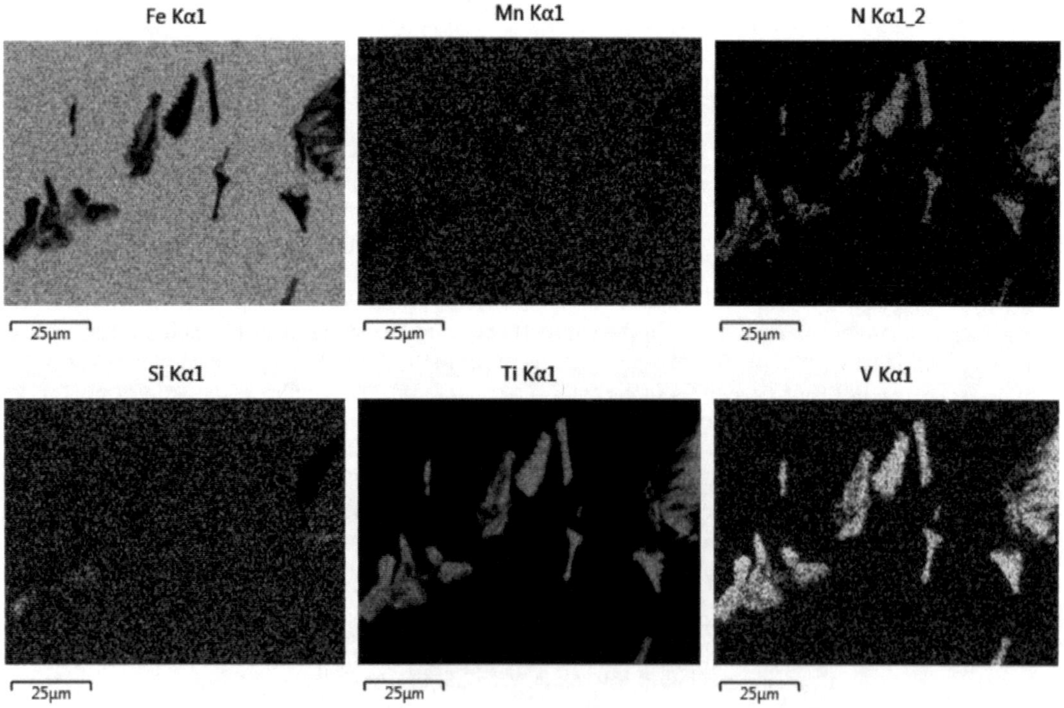

Figure 4. Typical inclusions in the metal sample № R6986A1-1 map the distribution of elements

The presented data show that in the absence of the titanium metal formed in the inclusions containing large amounts of calcium aluminates, and calcium and manganese sulphides. A characteristic of such impurities is that they are arranged in clusters and the layers on the primary grain boundary and lead to a decrease in the material properties, primarily plastic.

The excess of titanium contrary, leads to formation of the metal nitride and carbonitride inclusions, which are located in the metal in the form of clusters and chains extending along the direction of deformation.

IOP Conf. Series: Materials Science and Engineering **125** (2016) 012001 doi:10.1088/1757-899X/125/1/012001

As can be seen from the data, the average content of non-metallic inclusions in the metal of blanks №№ R6942A2-1 and R6986A1-1 is in the range of up to 3 points, but there were some areas (volumes of metal) of blanks, which was an increase in the content of some types of non-metallic inclusions to 4 points. The main types of inclusions - non-deformable and brittle silicates and nitrides stitch.

4. Conclusion

The results of research of mechanical properties and chemical composition comply with the requirements of National State Standard. All the blanks for all modes tested final heat treatment meet the requirements of National State Standard V5192-78. As a result, the modification does not change the basic chemical composition of the steel, but reduces the number and size of nonmetallic inclusions in the grain boundaries and also a decrease in grain size.

From the presented data it can be concluded that the modification of natural concentrates of deposits of Tomsk region has an influence on the mechanical characteristics of the resulting of blanks. The resulting research findings suggest a favorable impact of natural concentrates data on the mechanical characteristics of products.

The developed technologies allow expanding opportunities for the production of inoculants containing precious metals (Ti, Nb, Zr) by engaging in the production of natural concentrates of the Tomsk region.

References

[1] Cherepanov A N, Poluboyarov V A, Kalinin A P, Korotaeva Z A 2000 J. Materials 10 45-53

[2] Gizatulin R A, Valuev D V, Serikbol A Improving the Process Parameters Regenerative Processes at Steel Processing Ladle Furnace 2014 J. Modern Problems of Science and Education 3 1-8

[3] Fedoseev S N, Mukhtar Z M. Increasing the Operational Stability of Railway Core by Modifying 2015 J. IOP Conference Series: Materials Science and Engineering 91 (012050) 1-7

[4] Kalinin V T, Hrychikov V E, Krivosheev V A 2004 J. Metallurgical and Mining Industry 6 38-42

[5] Brover, A V Effect of Surface Heat Treatment Using of Concentrated Energy Flows on the Structural Strength of Steel 2005 J. Strengthening Technology and Coatings 4 22-25

[6] Gizatulin R A, Nokhrina O I, Kozyrev N A Steel microalloying with nitrogen by ladle bottom and top submerged lance blowing 2010 J Proceedings of Higher School Ferrous Metallurgy 6 30-33

[7] Volynova T F 1988 Metallurgy (Moscow)

[8] Fedoseev S N, Lychagin D V, Sharafutdinova A S 2014 J. Advanced Materials Research 1040 236-240

[9] Kuznetsov M A, Zernin E A, Danilov V I, Kolmogorov D E, Zubenko L N Corrosion and mechanical properties of austenic steel weld joints 2015 J. IOP Conference Series: Materials Science and Engineering 91 (012010) 1-8

[10] Rodzevich A P, Gazenaur E G, Belokurov G M 2014 J. Applied Mechanics and Materials 682 206-209.

[11] Mulyavko N M 2001 J. Proceedings of the Chelyabinsk Scientific Center 4 28-30.

[12] Iljyashchenko D P, Chinakhov D A, Gotovshchik Y M Influence of power supply energy characteristics upon the stability of MMA process 2015 J. Applied Mechanics and Materials 756 97-100

IOP Conf. Series: Materials Science and Engineering **125** (2016) 012002 doi:10.1088/1757-899X/125/1/012002

Hybrid Laser-Arc Welding Tanks Steels

G Turichin, I Tsibulskiy, M Kuznetsov, A Akhmetov, O Klimova-Korsmik

Institute of Laser and Welding Technologies SPbPU Polytechnicheskaya str. 29, Saint Petersburg, Russia

e-mail: kuznetsov_mich@ltc.ru

Abstract. The results investigate hybrid laser-arc welding of high strength steels using design responsible metallic construction and the highest strength body of vehicles. Welds from modern high strength steels grade Hardox 400, Hardox 450, Armox 600T and AB were created. High power fiber laser LS-15 with output 15 kW and arc rectifier VDU - 1500 DC were used in the experiment. Results of the metallographic research and mechanical tests are presented.

1. Introduction

At present high-strength steels are used at the creation of responsible engineering constructions widely because the steels have high strength and hardness with sufficient ductility and toughness properties. The modern steels increase life cycle of constructions, which working in difficult conditions, vehicle capacity at the maintaining mass of the construction and the dynamic characteristics of vehicles at reducing mass and maintaining strength characteristics of the construction [1].

More than half of gross revenue is created by welding and relation technologies in the heavy machinery. Up to two-thirds of the world's consumption of rolled steel is used for manufacturing welded constructions and facilities. However, most of the welding technologies used in Russia are not completely suitable for welding new material due to low productivity and high cost. Therefore, the critical issue of a technological development of the Russian Federation is the introduction of a new welding technology.

The most perspective solution of the problem is the technology of hybrid laser-arc welding (HLAW). The main advantages of the HLAW are high-performance, low welding deformation, the possibility of the welding with gap, the possibility of varying the chemical composition and mechanical properties of welds due to the melting of the electrode [2]. HLAW is more actually for using in the pipe building [3], shipbuilding [4] bridge building [5], rail carriage and automotive industry [6] thanks to its advantages.

Technical and economic calculation of the implementation of HLAW at the welding of a root pass at the large-diameter pipes creation, on the example Chelyabinsk pipe factory showed an increase welding productivity in 2-5 times, reducing energy consumption in 2 times, the flow of protective gas in 5 times, consumption of welding wire in 3 times at higher weld quality and overall performance of manufacturing large diameter pipes on 7%.

One of the important task at the HLAW of the high-strength steels is to get uniform strength of the weld metal and base metal, which depend from cooling rate of a weld joint, chemical composition and type of a welding wire [7]. Also arc type [8,9], width of a gap [10] and welding position of the HLAW in the space [11] effect on the wire alloying elements penetration.

This research is devoted to the experimental investigation technological possibilities of the HLAW of high-strength steel Hardox 400, Hardox 450, Armox 600T and AB with thicknesses from 7 mm to 20 mm in different welding positions in a single pass using different welding wires.

2. Experimental methods

The HLAW was carried out using a laser-arc technological complex based on a high power fiber laser LS-15 and an arc rectifier VDU - 1500 DC.

Swedish steel: Hardox 450 (thickness 12mm), Hardox 400 (15mm), Armox 600T (7 mm) and the Russian steel AB (48mm) were welded in the experiments. Mechanical properties of the steels are shown in the Table 1. Metal powder and solid welding wires Power Bridge 60M (Pb 60M), Sv-10H19N11M4F and Sv-08GSMT were used as welding electrode. The chemical compositions of the wires are given in the Table 2.

The HLAW of Swedish steel was carried out without mechanical preparation edges of the plates. The plates from AB steel had X-shaped edge preparation with dulling of 20 mm and edge angle 200. Plates from Hardox 450 and Armox 600T were welded in the PA welding position, Hardox 400 and AB – in the PG welding position. Photo of the experimental setup is shown on the Figure 1.

Table 1. Mechanical properties of the steels.

Grade of steel	Yield strength, MPa	Tensile strength, MPa	Hardness, HBW	KCV, J (-20^{0}C)
Hardox 400	1000	1250	370-430	45
Hardox 450	1100-1300	1400-1600	425-475	27
Armox 600T	1500	2000	570-640	12
AB	-	1000	-	-

Table 2. Chemical composition of the welding wires.

Grade of wire	C	Si	Mn	P	S	Cr	Ni	Cu	Mo	N_2	As	Ti	Al	V
Power Bridge 60M	0.04-0.08	0.4-0.8	1.3-1.6	<0.015	<0.015	-	0.4-0.6	-	-	-	-	-	-	-
Sv-08GSMT	0.09	0.47	1.07	0.013	0.016	0.22	0.15	0.17	0.23	0.01	0.008	0.06	0.043	0.005
Sv-10H19N11M4F	0.1	1.0-1.5	1.0-2.0	<0.025	<0.02	18.0-20.0	10.0-11.0	-	3.5-4.5	-	-	-	-	1.0-1.6

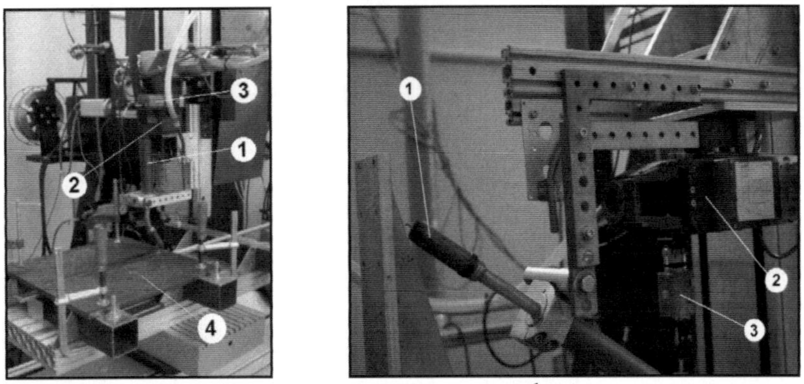

a b

Figure 1 – Location of the laser head and the torch at HLAW: a – PA welding position; b – PG welding position: 1 - torch, 2 - laser head, 3 - camera, 4 - weld samples.

IOP Conf. Series: Materials Science and Engineering **125** (2016) 012002 doi:10.1088/1757-899X/125/1/012002

Surface of the plates were cleaned up to metallic luster by angle grinders and degreased before welding by acetone. The plates were tacked before welding. Welding was carried out by straight butt joints. Mixture of shielding gas 80% Ar-20% CO2 and Ar of high purity with rate 25 l/min were used for protection of the welding bath from the atmosphere gases on the top and root sides of the seam. Metallographic studies were carried out on microscope DMI 5000 (Leica) with Tixomet software. Microhardness (HV1) of samples was determined using Microhardness tester Buehler Micromet 5103.

3. Results and discussion

3.1 HLAW of high-strength steel Hardox 450 and Hardox 400

Metal powder wire Power Bridge 60M, 1.6 mm in diameter was used at the HLAW of the Hardox 450; Sv-08GSMT with diameter 1.2 mm was used at the HLAW of the Hardox 400. The HLAW was carried out at the following values of the regime parameters: laser power (P) 14 kW, arc current (I) 396 A, arc voltage (U) 25.2 V, welding speed (V) 35 mm/sec, wire rate (Vw) 6.6 m/min (Hardox 450) and P=15 kW, I=276A, U=25.2V, V=35 mm/sec, Vw=9.12 m/min (Hardox 400). The appearance and cross macrosections of welds are shown on the Figures 2 and Figure 3, respectively.

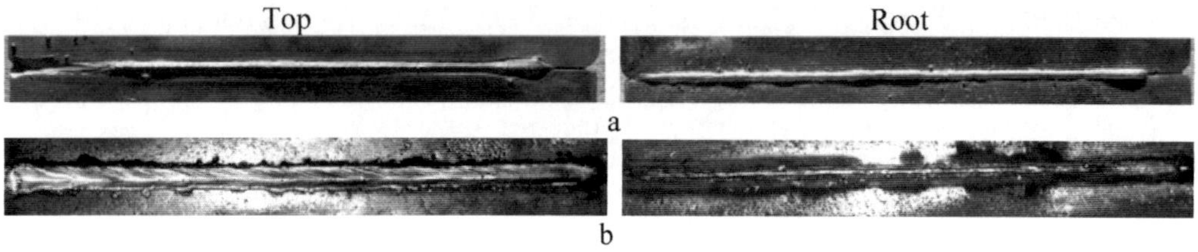

Figure 2 – The appearance of the welds: a - Hardox 450; b - Hardox 400.

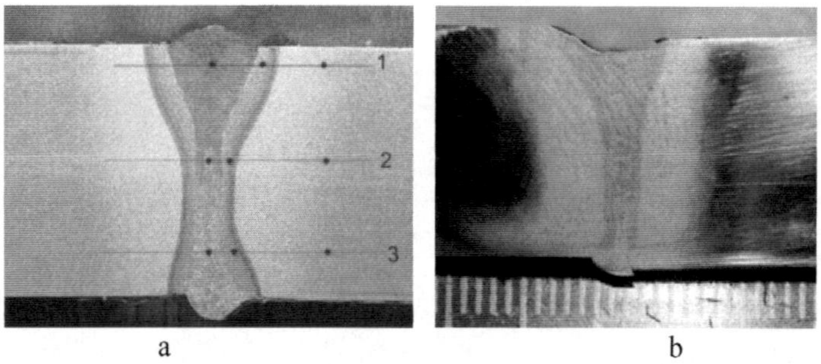

Figure 3 – Cross macrosection of the welds: a - Hardox 450; b - Hardox 400.

External and internal defects weren't detected at the visual inspection of the welds and of the cross macrosections. The HLAW of steel Hardox 400 has been implemented with offset by 1 mm of the welding edges, Figure 3b. The hardness test results for weld from steel Hardox 450 showed a discrepancy between real value of the base metal (BM) hardness and specification value and similar hardness of the weld metal (WM) and BM, Table 3.

Table 3. The results of measuring the hardness HV1 of the Hardox 450.

Measuring location	1	2	3	Av.v.
WM	354	362	378	364.7
HAZ	386	420	324	377
BM	367	359	374	366.7

The average hardness of the heat affected zone (HAZ) was higher approximately on a 10 HV1 than average hardness of the BM and WM. It should be noted, that the width of the HAZ in both cases was 1-1.5 mm, which incomparably smaller than the width of the HAZ of welded joints formed by arc welding processes.

3.2 HLAW of high-strength steels Armox 600T

The plates from steel Armox 600T with 7 mm thick was welded in a single pass in PA welding position using welding wires Pb 60M, Sv-10H19N11M4F and Sv-08GSMT.

The HLAW was carried out with parameters: P=7 kW, I=250 A, U=20 V, V=40 mm/sec. Appearance and cross macrosection of the weld formed at the HLAW with welding electrode Sv-10H19N11M4F is shown on the Figure 4.

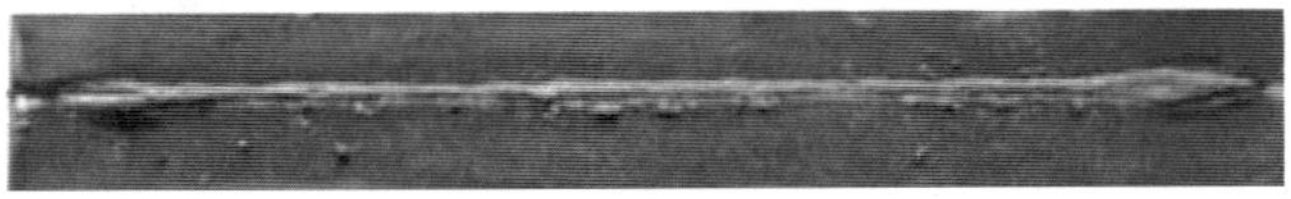

Top

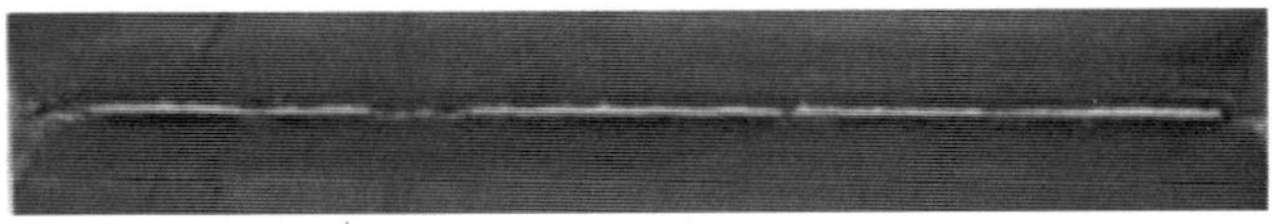

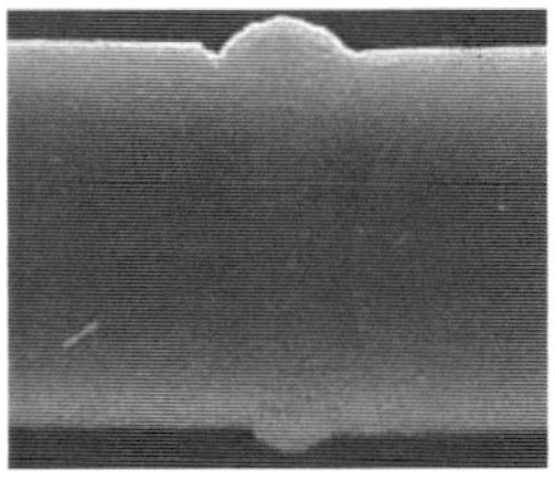

Root Cross macrosection

Figure 4 - The appearance of weld (Sv-10H19N11M4F)

Internal or external defects weren't detected by visual inspection and metallographic researches of the welds. The results of the microhardness tests for welds formed at the HLAW and laser welding are shown in the Table 4.

Table 4. The results of measuring the hardness HV1 of the Hardox 450.

Grade of the wire	WM	BM	HAZ
Sv-10H19N11M4F	571	558	617
Sv-08GSMT	563	584	645
Pb 60M	542	601	635
Laser welding	659	581	700

The maximum values of the microhardness found in the HAZ. The average microhardness of the WM, in the case of laser welding, is higher than the average microhardness of WM at the HLAW. Maximum values average microhardness of the WM and HAZ at laser welding can explain because of highest metal cooling rate and the formation of the martensitic structure.

The maximum value of the WM average microhardness was detected at the HLAW with electrode Sv-10H19N11M4F (570.5 HV1). It can be explained because of the nickel presence in an amount sufficient to completely or partially suppress the diffusion of a carbon at the $\gamma Fe \rightarrow \alpha Fe$ polymorphic transition with forming a martensitic or beanitic-martensitic structure of WM. The minimum value of microhardness (542 HV1) was detected at the HLAW with electrode Pb 60M. The width of the HAZ was less than 1mm.

From the weld formed during the HLAW with electrode Sv-10H19N11M4F were made samples for static tension test. Test samples were broke at the WM at values on the 15% lower than tensile strength of the BM. The results of the static tension test are shown in the Table 5.

Table 5. Results of static tensile test.

Test sample	Size a, mm	Size b, mm	Location of destruction	Maximal load, kN	Tensile strength, MPa
1–1	5.22	14.95	WM	139.26	1784.5
1–2	5.29	14.97	WM	137.02	1730.3

3.3 HLAW of the high strength steel AB

The HLAW of the steel AB was carried out in a single pass in the PG welding position. Welding wire Sv-08GSMT was used. The HLAW was carried out with the parameters: P=15 kW, V=18mm/sec, Vw=18m/min, I=410 A, U=40 V. Appearance and cross macrosection of the weld are shown on the Figure 5.

The appearance of the welded joint Cross macrosection

Figure 5– The welded joint (steel AB)

The weld had through penetration without visible external and internal defects. Results of the static tensile test of the weld and BM are shown in Table 6. WM and BM had similar tensile strength values.

IOP Conf. Series: Materials Science and Engineering **125** (2016) 012002 doi:10.1088/1757-899X/125/1/012002

Table 6. Results of sample measurements and average results of tests at static tensile .

Sample	Destruction	Sample thickness, mm	The width of the sample, mm	The maximum load, kN	Tensile strength, MPa
WM	WM	9.97	20.57	203.7	993.3
BM	BM	diameter – 5.95		28.72	1000

3.4 Metallographic research of welded joints
Metallographic research was carried out for Hardox 450 steel, Figure 6.

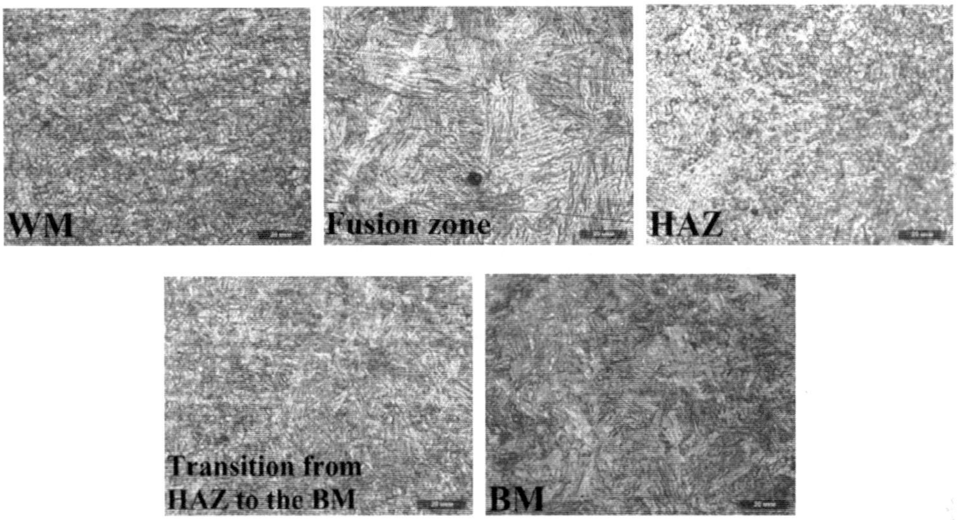

Figure 6 - The microstructure of the weld in different zones (of steel Hardox 450)

Martensite was found in the some areas of the weld structure. In the structure of WM was found martensite and carbides. Also fusion zone mostly consists of martensite. Less than 5 % of the martensite was found in the middle area of a HAZ. The main structural component of the HAZ middle part was the beanite with upper and granular morphology. Fine structure with predominantly beanite with granular morphology was defined in the transition zone from HAZ to the base metal. The microstructure of BM had beanite with lower and granular morphology and carbides.

Also it can be conclude, that thermal effect from HLAW slightly distorts of the base metal original structure in the weld joint metal, whereby the mechanical properties of the metal are stored.

4. Conclusions

The results of the research demonstrate not only the technological ability, but also the efficiency of the using hybrid laser-arc process for welding of the modern high-strength steel of large thicknesses.

At the HLAW of the high strength martensitic steel with thickness of 7 mm to 20 mm in different welding positions welds with mechanical properties on the level of base metal can be formed.

The effect of alloying elements of the welding electrode on the mechanical properties of welds was experimentally determined.

The results of metallographic research showed slight divergence of the weld structure with the structure of the base metal.

References

[1] Hillenbrand H G, Kalwa C 2002 High Strength Line Pipe for Project Cost Reduction World Pipelines 2
[2] Kah P 2012 Current trends in welding processes and materials: improve in effectiveness Advanced Materials Science 30 189-200
[3] Turichin G, Velichko O, Kuznetsov A, Pevzner J, Grinin O and Kuznetsov M 2014 Proc. Int. Conf. Laser Optics(Saint-Petersburg) p 1-4
[4] Cao X, Wanjara P, Huang J, Munro C and Nolting A 2011 Hybrid fiber laser – arc welding of thick section high strength low alloy steel Materials & Design 32 3399–3413
[5] Defalco J 2007 Practical Applications for Hybrid Laser Welding Welding J. 86 47 – 51
[6] Staufer H 2007 Laser Hybrid Welding in the Automotive Industry Welding J. 86 36 – 40
[7] Turichin G, Kuznetsov M, Sokolov M and Salminen A 2015 Hybrid Laser Arc Welding of X80 Steel: Influence of Welding Speed and Preheating on the Microstructure and Mechanical Properties Proc. Conf. on Nolamp (Lappeenranta)
[8] Gook S, Gumenyuk A and Rethmeier M 2014 Hybrid laser arc welding of X80 and X120 steel grade Science and technology of welding & joining 19 15-24
[9] Li G, Zhang C, Gao M and Zeng X 2014 Role of arc mode in laser-metal active gas arc hybrid welding of mild steel Materials and Design 61 239-250
[10] Turichin G, Tsibulskiy I, Kuznetsov M, Akhmetov A, Mildebrath M and Hassel T 2015 Influence of the Gap Width on the Geometry of the Welded Joint in Hybrid Laser-Arc Welding Proc. Conf. on Nolamp (Lappeenranta)
[11] Tsibulskiy I, Kuznetsov M and Akhmetov A 2014 Effect of Welding Position and Gap between Samples on Hybrid Laser-Arc Welding Efficiency Applied Mechanics and Materials 682 35-40

Forming Completely Penetrated Welded T-joints when Pulsed Arc Welding

N Yu Krampit[1a], PhD in Technical Sciences, M A Krampit[1b], A S Sapozhkov[1c]

[1]Yurga Institute of Technology, TPU Affiliate, Russia, Kemerovo Region, Yurga

e-mail: [a]nkrampit@mail.ru, [b]krampitm@tpu.ru, [c]sapogkov94@mail.ru

Abstract. The paper is focused on revealing the influence of welding parameters on weld formation when pulsed arc welding. As an experimental sample a T-joint over 10 mm was selected. Welding was carried out in flat position, which required no edge preparation but provided mono-directional guaranteed root penetration. The following parameters of welding were subjected to investigation: gap in the joint, wire feed rate and incline angles of the torch along and across the weld axis. Technological recommendations have been made with respect to pulsed arc welding; the cost price of product manufacturing can be reduced on their basis due to reduction of labor input required by machining, lowering consumption of welding materials and electric power.

1. Introduction

At present welding is one of key procedures to manufacture permanent joints in mechanical engineering and in other branches of industry. Thicknesses to be welded are diverse, varying from millimeter shares to dozens centimeters. It is not difficult to weld 2 to 4 mm thicknesses. Thinner metal is welded if heat input is reduced [1-5]. To achieve guaranteed root penetration constructions containing metal sheets over 4 mm are welded with edge preparation. As the consequence, manufacturing process gets more labor-intensive, and more welding materials and electric energy are consumed. Hence, the cost price of a product increases.

The key difficulty of arc welding steel plates and components without prepared edges is to form a correct back of a weld. Incomplete fusion of the section can be caused by insufficient heat supply to the place of welding. Excessive heat supply leads to complete penetration, molten metal flows out of the weld creating holes (burn-through) or sinters on the back. It is quite hard to have complete penetration of the total section thickness free of sinters or burn-through. The back can't be seen when welding, and a welder works in conditions leading to incomplete fusion. The gap being the result of incomplete fusion is the center of concentrating stresses; i.a. a weld destroys fast, especially under permanent or constant loads. Incomplete fusion can be eliminated if a joint is welded from the back for a long time. However, additional welding is difficult because of partial (overhead position) or complete inaccessibility (pipe welding in a butt joint etc.) of the back. Additional welding also causes growing labor intensity (by 30 – 40%) [6].

The paper outlines investigations into weld formation when T-joint welding in flat position without edge preparation, mono-directional guaranteed root penetration over 10 mm is provided.

At present there are a few recommendations available concerning welding this kind of joints in literature, moreover, the most of them are for advertising purposes and can hardly be used in manufacturing process because include no certain data. A great number of parameters affecting weld formation and its geometry [7] do not support the search for efficient conditions [8]. Assessment of arc pressure on the weld pool, gas-dynamic impact on the electrode metal bead [9], and thermal cycle of welding is required [10]. Such parameters as joint gap, incline angle of the torch, circuit inductance etc. were determined experimentally in this paper.

As a welding process we selected gas-shielded welding distinguished by better technological characteristics and lower expenditures on equipment and materials in comparison with submerged-arc, laser, electron-beam, and hybrid types of welding. When gas metal arc welding it is necessary to submerge the arc in the gap and support its stable burning between two edges. The authors of paper [11] dealt with various procedures of pulsed arc welding. We concentrate on short focused pulsed arc welding for research purposes. Complete penetration is possible provided that breadth of the gap, wire feed rate, arc voltage, as well as incline angles of the torch along the weld and towards the upright edge are selected correctly, therefore, results of experimental studying these parameters are given below.

2. Methods of experiments

We used samples made of 300x50 mm plates for carrying out experiments. The surface of plates was pre-polished so it has metallic luster. Pulsed arc welding was with short circuits. Recommended voltage for this thickness was 33.6 V (33 V). Welded samples were band-sawed and polished. Samples without visible defects were subjected to etching.

The experiment included several phases.

The gap in the joint was specified during the first phase. The plates were assembled so the gap changed 0 to 8 mm. The assembled sample was undergone pulsed arc welding with short circuits. We selected a zone with good penetration but without burn-through. Then, the gap was varied with small deviation in this zone.

The second phase was focused on influence assessment of incline angle of the torch along the weld axis. The gap was selected according to the data determined in the first experiment. Short beads (50 mm) were welded on the sample via pulsed arc welding with short circuits. Incline angle of the torch was varied within the range ±60° from the weld axis, with 10° step.

The third phase involved influence assessment of the incline angle of the torch across the weld axis. The gap was selected according to the first experiment. Short beads (50 mm) were welded on the sample via pulsed arc welding with short circuits. Incline angle of the torch was varied within the range 40-90° from the weld axis, with 10° step.

We specified the wire feed rate in the fourth phase. As voltage and wire feed rate are interdependent in up-to-date welding power sources, the change in wire feed rate was evaluated in view of the recommended value. Decreasing the electrode wire feed rate below the recommended value gave rise to growing voltage, whereas its increasing led to falling voltage and current strength. The wire feed rate was selected so the arc column could cover both edges to be welded.

3. Results of experiment

Specifying gap in the joint. When welding without a gap, even under high current strength guaranteed penetration was not provided (Figure 1, 2 a). Welding with big gap under low current strength did not support complete penetration too (Figure 1, 2 b). Complete penetration is possible in a very narrow area of the gap. Slight deviations from any parameters or uneven movement of the electrode caused incomplete fusion or burn-through.

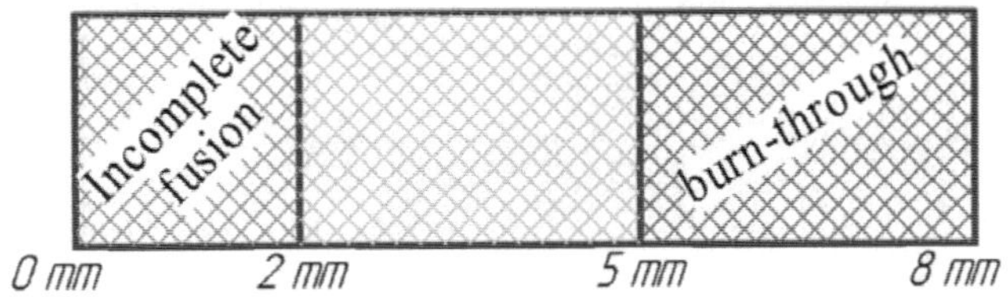

Figure 1 Influence of clearance on penetration

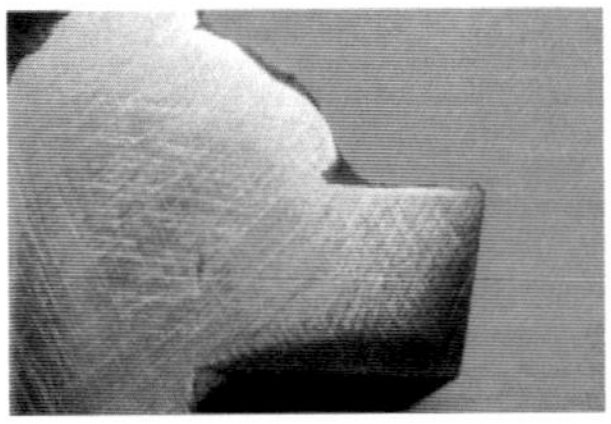

a

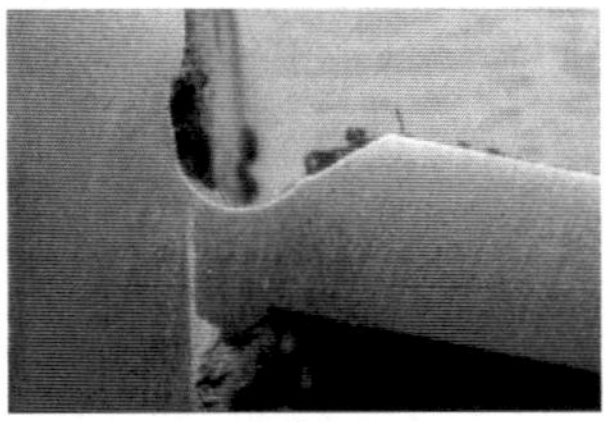

b

Figure 2 Incomplete fusion without the gap (a), with big gap but insufficient current strength (b)

Determination of the incline angle along the weld axis. Metal is on different levels in head and tail ends of the weld pool because of its displacement from under the arc root in the direction opposite welding wire movement [12]. Molten metal is displaced as long as the horizontal component of the arc column force Pa, directed towards displaced molten metal is balanced by the counterforce of hydrostatic pressure Ph of molten metal and weld dross (Figure 3). Welding process is stable if these two forces are counterbalanced. Displacement of equilibrium in any direction can cause either further arc deepening into base metal or molten metal dribbling under the arc root, as the consequence, the depth of penetration is incomplete due to worsening conditions of heat transfer through the liquid layer of metal [13].

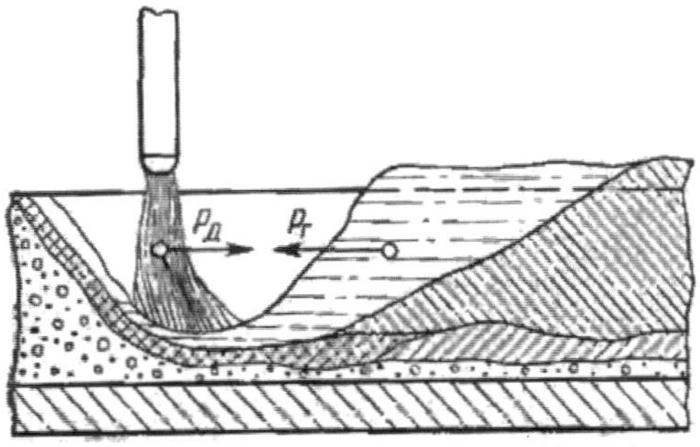

Figure 3 Forces affecting the weld pool

The experiment has determined the most efficient angle in conditions of welding with electrode inclined under obtuse angle is one of 5-10° (Figure 4).

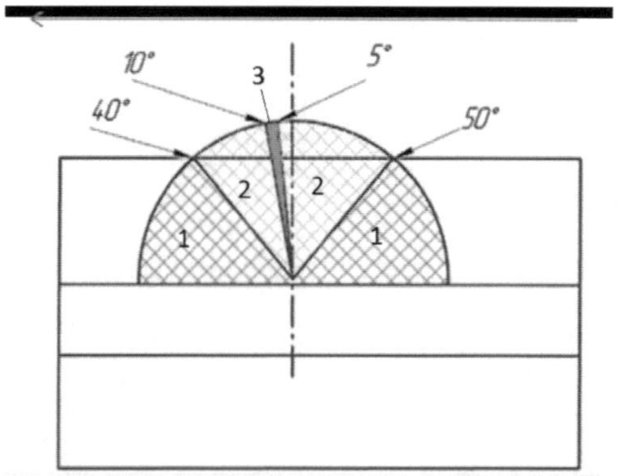

Figure 4 Defectless formation zones according to incline angle of the torch along the weld axis (zone 1 – zone of incline angle of the torch providing no complete penetration; zone 2 – zone of higher splattering; zone 3 – zone with recommended incline angle of the torch)

Specifying the incline angle across the weld axis. The experiment has determined the incline angle of the torch 80-90° from the horizontal plate is the most efficient one. As one can see in Figure 5 there is no edge undercut and incomplete fusion in zone 3.

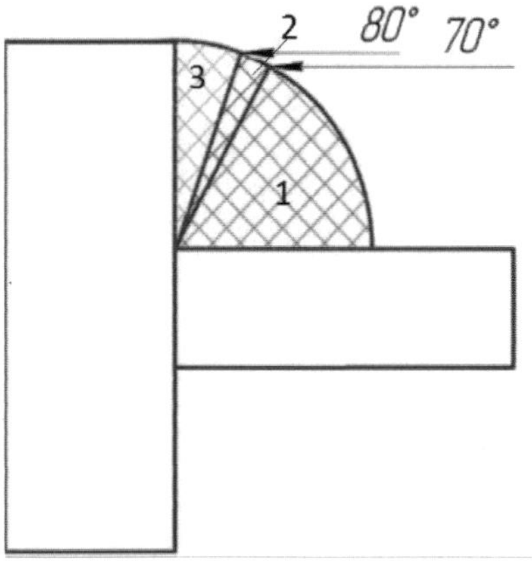

a

b

Figure 5 The importance of incline angle of the torch across the weld axis for penetration depth (a) (zone 1 – incomplete penetration; zone 2 – undercutting; zone 3 – efficient angle) and wall undercutting (b)

Specifying the rate of electrode wire feed. For slowed down rate of wire feed we selected such a long arc that was burning between two edges (Figure 6) when welding wire and arc were submerged in the gap. The efficient wire feed rate was in the range 9.5-10.5 m/min. The mean current strength was 325-340 A.

IOP Conf. Series: Materials Science and Engineering **125** (2016) 012003 doi:10.1088/1757-899X/125/1/012003

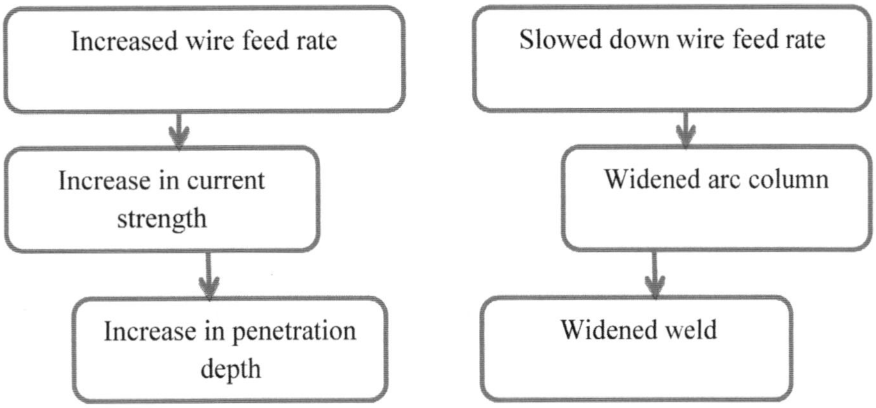

Figure 6 The dependence of weld parameters on wire feed rate

Welding was unstable even though all the afore-mentioned requirements and recommendations were met. That is why we decided to use the permanent backing of wire with diameter similar to the gap size. The permanent backing was mounted when assembling so the gad size was checked. A burn-through and molten metal outflow occurred in conditions of a short arc and wire support (Figure 7, a). One edge was cut off and there was an undercut on the other one with inclined torch and short arc (Figure 7, b). In conditions of increased voltage and slowed down wire feed rate there was arcing between two edges without penetration into the joint though (Figure 7, c). To provide simultaneous heat input into both edges and permanent backing we specified needed voltage and arc column diameter similar to the gap size (Figure 7, d). So weld root formation and fusion with edges were provided.

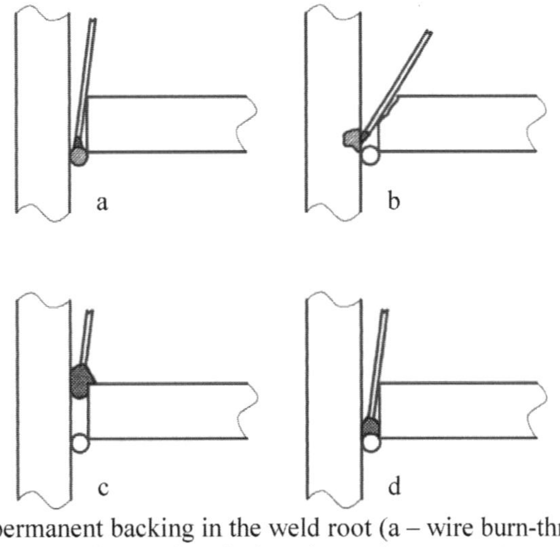

Figure 7 Welding with permanent backing in the weld root (a – wire burn-through, b – edge undercut, c – incomplete fusion, d – perfect process)

Figure 8 outlines etched metallographic section of a root weld. As one can see, a back weld bead was formed, and edges were fused along the whole gap thickness. These welding parameters facilitated complete filling the gap. If required a facing bead can be made.

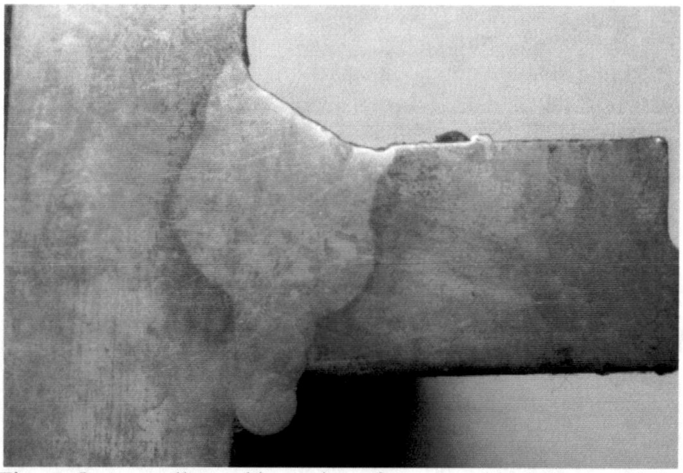

Figure 8. A metallographic section of a root weld of 12 mm T-joint

Had analyzed the experimental data the following conditions and operating recommendations were offered: voltage 33.6 V, wire feed rate 9.5-10.5 m/min, mean current strength 325-340A, gap in the joint 4±0.5 mm, incline angle of the torch along the weld axis 5-10° when welding with electrode inclined under obtuse angle, and incline angle across the weld axis 80-90°.

Therefore, on the basis of carried out investigations the interdependence of pulsed arc welding and formation of a T-joint over 10 mm has been revealed. Technological recommendations have been made with respect to pulsed arc welding with short circuits; the cost price of product manufacturing can be reduced on their basis due to reduction of labor input required by machining, lowering consumption of welding materials and electric power.

4. Conclusions

1. Complete penetration is possible provided that gap breadth is 4±0.5 mm, wire feed rate is 9.5-10.5 m/min, mean current strength 325-340A, and incline angle of the torch along the weld axis 5-10° and incline angle of the torch towards the vertical edge 80-90°.
2. The efficient welding conditions are those when welding arc melts both edges and backing simultaneously. These conditions are provided at arc voltage 33 V and gap breadth 4±0.5 mm.

References

[1] S F Goeke. Arc welding sensitive to heat materials with lowered energy output. URL: http://www.ewm-russia.ru/articles/coldarc.php (Accessed data 15.01.2015)
[2] Alyoshin N P, Gladkov E A, Brodyagin V N, Kuznetsov P S, Kopotneva E N, Sholokhov M A Impulse technologies of control over bead transfer when MIG/MAG welding *Welding and Diagnostics* – 2014 - No3
[3] Process Merkle ColdMIG. URL: http://merkle-russia.ru/files/merkle3.pdf (Accessed data 15.01.2015)
[4] Precision Pulse Weld Process Overview. URL: http://www.lincolnelectric.com/assets/us/en/literature/te12007.pdf____(Accessed data 15.01.2015)
[5] EWM-coldArc®. Cold Arc... impossible gets real! URL: http://www.svarkainfo.ru/rus/lib/blog/coldarc.php (Accessed data 23.05.2015).
[6] Welding Peculiarities of diverse materials in various conditions. Part 1. URL: http://penzaelektrod.ru/articles/art21.htm (Accessed data 20.06.2015).
[7] Krampit A G , Krampit N Y A method for the determination of the geometrical dimensions and area of the welded joint *Welding International*. - 2013 - Vol. 27 - №. 10. - pp 834-836

[8] Krampit A G , Krampit N Y, Krampit M A Effect of the parameters of pulsed arc welding on the formation of a root layer in a narrow gap *Welding International* - 2013 - №. 66(8). - pp. 3-5
[9] Chinakhov D A Calculation of Gas-dynamic Impact of the Active Shielding Gas on the Electrode Metal Drop in Gas Jet Shielded Welding *Applied Mechanics and Materials* Vol. 379 (2013). pp. 188-194
[10] Chinakhov D A Study of thermal cycle and cooling rate of steel 30ХГСА single-pass weld joints *Applied Mechanics and Materials* – Vols. 52-54. – 2011. – p. 442-447. URL: http://www.scientific.net/AMM.52-54.442
[11] Krampit M A Up to date modes of pulsed arc welding developed abroad *International journal of applied and fundamental research* – 2015 - №5 – pp.18-21
[12] Frolov B V Theoretical fundamentals of welding. M., «High School», 1970.
[13] Krampit A G, Dmitrieva A V Movement of liquid metal in a weld pool when welding with pulsed arc *Applied Mechanics and Materials* - 2014 - Vol. 682. - p. 386-391

Influence of Lateral Incision on Inhomogeneous Deformation of a Nickel [001] – Single Cristal at Axial Compression

D V Lychagin[1,2], E A Alfyorova[1,2] , M V Lychagin[3], Czyan Dilun[1]

[1] National Research Tomsk Polytechnic University 634050, Russia, Tomsk, pr. Lenina, 30
[2] National Research Tomsk State University 634050, Russia, Tomsk, pr. Lenina, 36
[3]Novosibirsk State University, 630090, Russia, Novosibirsk, Pirogova str., 2,

e-mail: dvl-tomsk@mail.ru , lychagin@nsu.ru, katerina525@mail.ru

Abstract. We used the scanning electron microscopy and the electron backscattered diffraction to investigate a deformation relief and a crystallographic disorientation of a nickel [001] single crystal with {100} faces with a lateral incision. We identified that the lateral incision can change the shear domains distribution pattern in the sample by creating additional deformation domains near the incision. The change in the patterns of the misorientation accumulation on the mutually perpendicular faces accompanies this deformation. We established that the orientation alteration occurs toward the increase of the Schmid factor for the slip systems in two of the four (previously equally loaded) slip planes. This method of shear deformation contributes to an optimal mutually consistent deformation in the adjacent areas of the single crystal.

1. Introduction

A fracture process ductile materials and possible failure play important role in material science and practice. The study of this process dealing with no only nucleation and propagation of dislocations, but also the initiation and extension of a crack. Fracture failure has been investigated by simulations and experiments at various length scales (see some examples in [1-5]). The literature gives us a set of different models proposed to describe dislocation nucleation from a crack tip [6-10].

From a practical point of view, an assessment of the product efficiency with stress concentrators is very important. The stress concentration areas intensify the processes of local deformation and destruction. They are a cause of reduction of generic plastic deformation before destruction. Numerous theoretical and experimental studies (e.g. [11, 12]) and handbooks (e.g. [13]) are devoted to investigations of the stress concentration. Mainly, the studies are dedicated to investigations of the incision shape, location and influence of an external load. The majority of studies on stress concentrators focus on the modeling, explore the processes of stress and deformation distribution in the incision areas, and do not examine the strain relief and the nature of the misorientation in this area. Meanwhile, the size, shape and orientation of the incision specify the deformation process characteristics.

The studies on single crystals represent the characteristics of the slip evolution under an inhomogeneous stress caused by the incision. The results, obtained by the authors earlier [14, 15], allowed to take into account the effects of the basic stress concentrators, and compression axis and lateral faces orientations on a plastic deformation in FCC single crystals. The most detailed studies were carried out on nickel single crystals [16, 17]. In these studies, we examine the inhomogeneity of the plastic deformation under the above-mentioned factors the lattice method. The electron-backscattered diffraction (EBSD) allows to explore the development of misorientations in relation to strain relief elements and formability of single crystals.

The aim of this paper is to investigate the deformation in nickel [001] single crystals with incisions on lateral faces, analyzing the deformation relief and crystallographic reorientation areas.

2. Materials and method

The object of the investigation is a nickel single crystal with compression axis orientation [001] and lateral faces {100}. The single crystals were prepared of a pure nickel containing less than 0.01% of impurities. The Bridgman method was used to grow the single crystals. The orientation was performed on the X-ray diffractometer. After the electroerosion cutting and polishing the deviation from the cubic orientation of the sample faces was ± 1 °. The study was carried out with samples of a size 3x3x3 mm and with a horizontal incision on a lateral face. The incision was made by the electroerosion cutting. It had a semicircular shape with radius at vertices R = 0,05 mm and depth t = 0,3 mm. The surface of samples was firstly mechanically polished and then electropolished in the saturated solution of the chromic anhydride and phosphoric acid at 20 V. The compressive deformation was done by the Instron ElectroPuls E10000 at rate 1,4 · 10-3 s-1 at room temperature. The graphite lubricant was used to reduce friction on lateral faces. The total sample strain was 0.41in relative units. The optical microscope Leica DM 2500P and scanning electron microscope Tescan Vega II LMU equipped with EBSD were used to investigate the deformation relief.

3. Experimental results

The presence of a horizontal incision on a lateral surface {100} of the [001] nickel single crystal changes the distribution of stresses and strains in a sample, when compared to the common deformation pattern where the sample is compressed in the form of a hexahedron and in the presence of an end friction. In the incision area, the stress conditions change. The incision area manifests as a stress concentration area. The magnitude of the stress concentration depends on: 1) the incision orientation with respect to the applied load and temperature; 2) the incision size compared to the size of the sample; 3) the shape of incision, determined by its geometric parameters. The geometric parameters of the stress concentrator are its linear dimensions (depth and radius) and an angle between the faces of the concentrator, in our case $\alpha = 0$.

When a strain increase, the incision edges come closer, and the radius of incision tip decreases. Under compression, the width of a sample a increases together with the depth of an incision t. The ratio of these parameters a / t = 10. We can assume that a >> t. In this case, the stress concentration coefficient $\alpha\sigma$ depends very little on a, and largely it depends on t. The greater the t, the higher the stress concentration factor is, i.e. $\approx \sqrt{t}$. Moreover, as the radius R decrease, the value of the stress concentration coefficient increases. When $\alpha = 0$, the value of the stress concentration coefficient depends on the geometrical parameters as following $\alpha\sigma \approx \sqrt{t}/R$. Therefore, during the deformation process, the maximum stress in the incision area is constantly greater than the average. It is directly proportional to the incision depth increase and inversely proportional to the radius of the incision tip. This creates conditions for a more active deformation in the incision area.

For a [001] nickel single crystal there are four equally loaded octahedral planes with two slip directions in each. The total number of slip systems is eight. The Schmid factor is equal to 0.41. Despite the fact that the slip systems of the given single crystal are equally loaded, the condition of an uniform volume alteration at compression of sample results in enabling of only some slip systems, not all of them. In single crystals without incisions first to activate are the slip systems near the primary stress concentrators, which are the near-end areas of the sample. The crystallographic diagram, the fragmentation pattern and the deformation relief are discussed elsewhere [14, 16]. Note, that the relief and the nature of the sample fragmentation, with concentrators and without concentrators, differ

IOP Conf. Series: Materials Science and Engineering **125** (2016) 012004 doi:10.1088/1757-899X/125/1/012004

substantially (Figure 1). The analysis of the deformation relief shows that the introduction of the incision creates an extra surface and changes the deformation around the incision. The increase of density of shear bands accompanies the facilitated slip in the extra surface.

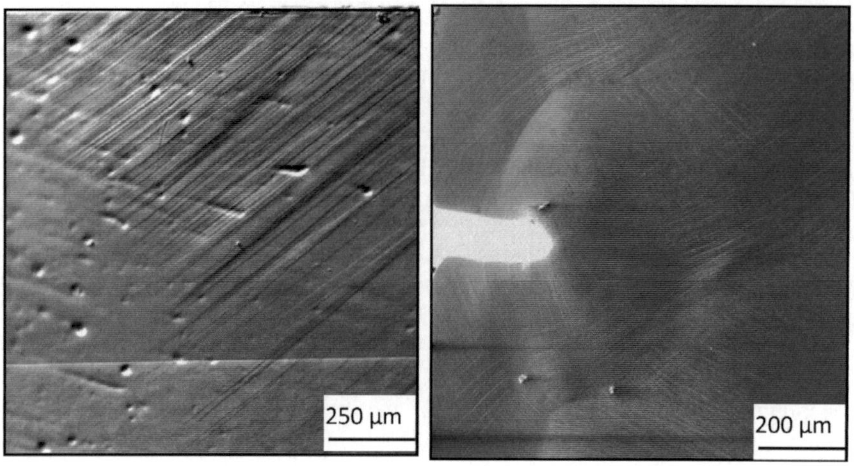

a) b)

Figure 1. The deformation relief of [001] single crystals without lateral incisions (a) and with the incision (b) on the lateral face (100).

We are able to observe the activation of two systems of shear bands according to two sides of the incision. The convergence of the incision edges leads to the curvature of the lateral edges, and as a result to the bending of the shear bands. The appearance of additional active slip systems is due to the higher degrees of stress at the incision tip. The activation of these local slip systems lead to the appearance of an additional strain domains inside the single crystal.

Note, that the atomistic simulation study for two FCC materials (Cu and Al) indicates the favorable factors dominating dislocation nucleation from a crack tip [18].

High and complex stress in the incision tip creates an environment for consecutive activation of possible slip systems, as well as active interactions between them. As a result, at the incision tip appear folds (Figure 2a). We can see a system of steps as the result of deformation along to conjugated slip planes (Figure 2b). The high density of the shear bands in the incision tip indicates the presence of intensive plastic deformation in this area.

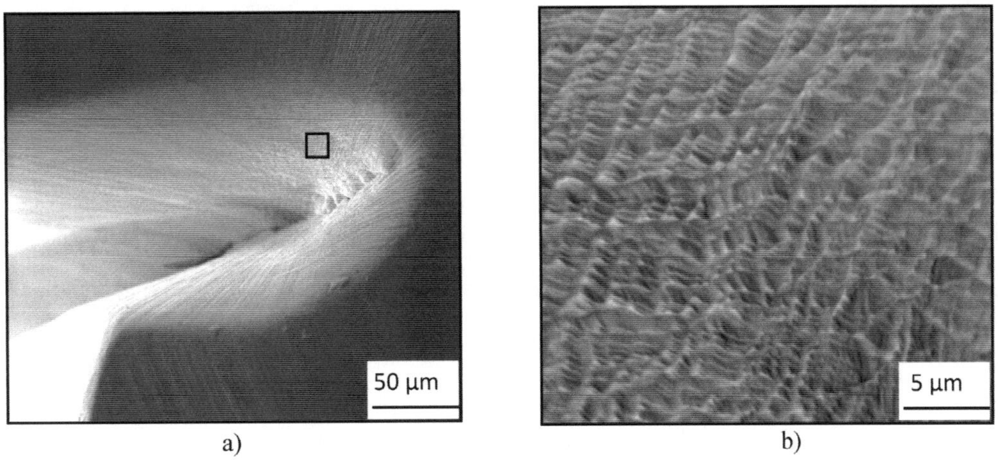

Figure 2. The deformation relief at the stress concentration area (a). Amplified view of the selected area (b).

The EBSD analysis was performed on the lateral face (010) perpendicular to the incision. It has been found that in this zone there is an alternation of areas with the orientation close to the initial [001] with respect to the compression axis (Y), and reoriented in the directions [105] and [117] (Figure 3) (Z-axis normal to the lateral face). The width of the areas with the orientation close to the initial is about 4 ... 7 microns, with the new orientations - 2 ... 4 microns. In this case, it is interesting to note that the value of the Schmid factor in reoriented areas increases to 0.43 ... 0.45. Consequently, the reorientation occurs in the direction of the most active slip.

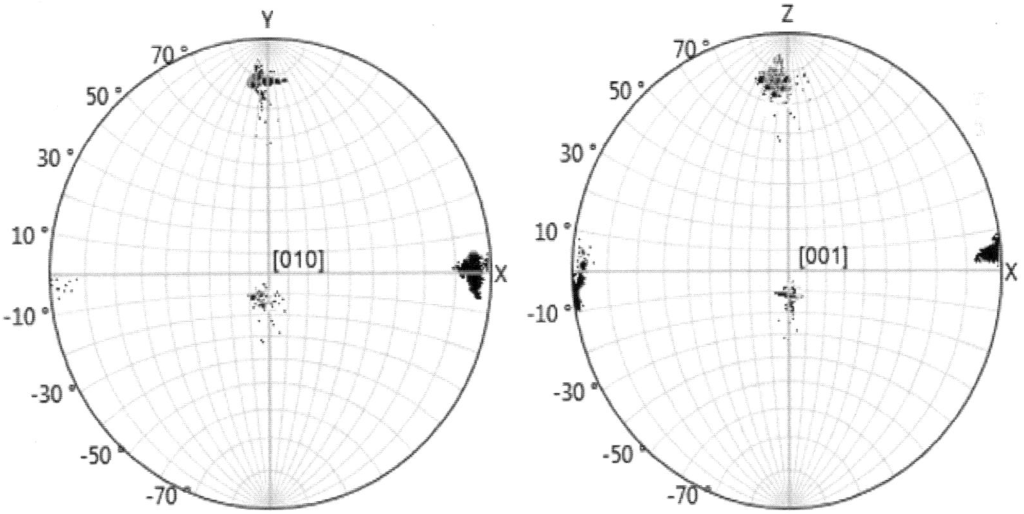

Figure 3. The direct pole figures obtained with the face (010) perpendicular to the incision.

Character of the folds formation at the incision tip is similar to the folds formation patterns occurring in a [111] nickel single crystals in areas of the macroscopic bending of face [17]. In fold structures the value of misorientation angles ranges from 2° to 5°. The maximum values of a misorientation angle are observed with a step of about 15 microns. The step in folds in a nickel [111] single crystal deformed by compression is about 5 microns. This fact agrees well with the misorientation value accumulating during a deformation in the cellular dislocation substructure.

The strain of the compression sample was equal to 0.16 (in relative units). The indicator for the sample with the concentrator was equal to 0.41. A cellular dislocation substructure occurs in nickel, when the strain is equal to 0.16. We are able to fix a microband substructure and a fragmented substructure, when the strain is equal to 0.41 [19]. In cellular substructure, it is possible to mark the accumulation of misorientations. This process goes through the excess density of dislocations in cell walls (about 10 cells are needed to accumulate this misorientation value). In the case of microband and fragmented structures, the misorientation emerges at the edges of microbands and fragments. Each edge may have a different misorientation value. Hatherly, Hutchinson, Karduck et. al. [20-22] and other researchers studied the accumulation of misorientation in microband and fragmented structure. They have shown that the accumulation of misorientation under condition of the transition from one microband to another contains three variants: 1) monotonous increase, 2) monotonous decrease, 3) periodically change of a misorientation. A comparison of our results with previously obtained shows that the greatest misorientation values are observed at the edges of microband deformation groups.

Misorientation gradient on the face lying in the incision plane is different from the one on the face perpendicular to the incision. It represents the spatial distribution of stresses near the incision. Therefore, the analysis of misorientation was also performed on the (100) face with the incision (Z-axis). Near the sample end, one can observe the slip systems discussed above (Figure 1). The EBSD analysis shows that the reorientation changes smoothly and in the same directions as in the mutually perpendicular face. However, if in the previous case, we observed an alternation of areas with different orientation, now we see the gradient change from one orientation (Figure 4, point 1) to another (Figure 4, point 2). The misorientation value of the adjacent areas is within 1°...2°, the maximum value does not exceed 5°. The Schmid factor value approaches the maximum possible value of – 0.5.

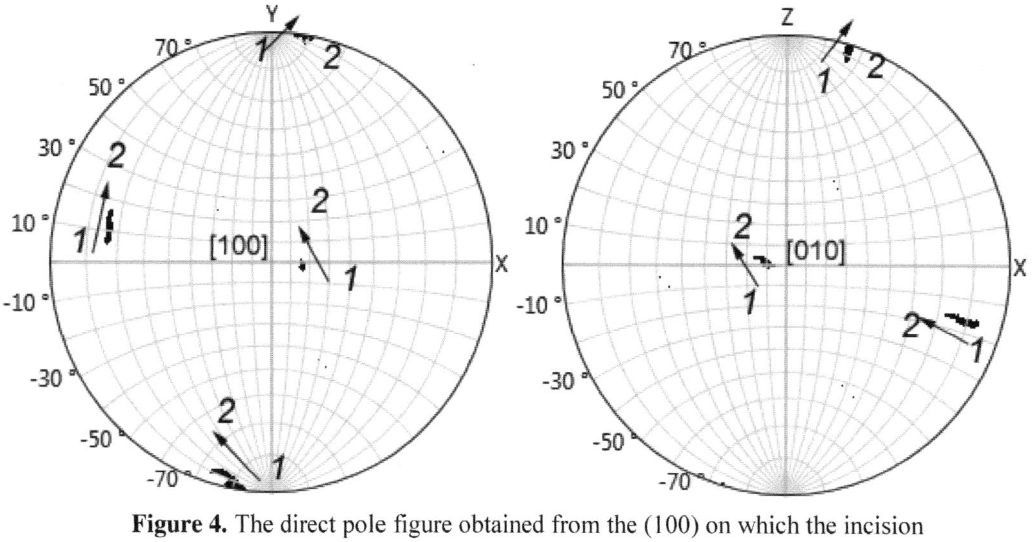

Figure 4. The direct pole figure obtained from the (100) on which the incision

We are able to observe the involvement of all equally loaded slip systems only in the situation when we consider a single crystal as a whole. The single crystal is broken on deformation domains (size 0.5 ... 2 mm). In each domain, only some of all possible slip systems are active. As a result, these macroscopic areas are the misorientation. This is due to accumulation of misorientation in meso-volumes of a material. The study presented in this article confirms this statement. On the meso-level there is a correlation between the change of orientation and active slip systems in them. When moving from the primary orientation [001] to the orientations [117] and [105], the Schmid factor changes for eight equally loaded slip systems. For four slip systems, it varies from 0.41 to 0.43, while for other four from 0.41 to 0.37. It is interesting to note that the increase of the Schmid factor is observed for systems of adjacent slip planes. According to authors, this slip feature contributes to the slip deformation of adjacent areas without disrupting the solidity of the material. This reorientation pattern and the Schmid factor alteration is typical for face with the incision and for the perpendicular to it face. The gradient increase of misorientation for the face with an incision can be explained by a smooth conversion of the curvature face due to the migration of the material to the incision area. On the face perpendicular to the incision, due to the nature of stresses, we observe a wavy surface alteration, which stimulates the formation of folded structures and periodic orientation change in slip systems at meso-level.

The deformation relief on the inner surface of the concentrator is a system of intersecting slip band that form a complex system of steps and folds. Substantial surface curvature in the tip of the concentrator, which occurs under the compressive deformation, contributes to folding. Involvement of a large number of slip systems becomes possible because of activation of the maximum value of all stress tensor components. As a result of the incision edges convergence and alteration of the surface orientation, inside the incision occurs the alteration of the crystallographic orientation in meso-areas toward favorable conditions for the slip.

4. Conclusions

Thus, the analysis of deformation relief and misorientation showed an interconnection between slipping and reorientation. The macroscopic forming in the incision area correlates with the nature of misorientation alterations. Its macroscopic alteration is accompanied by the formation of misoriented meso-areas where favorable to shear deformation in local domains slip system are activated.

Nature of change and accumulation of misorientations obtained by the EBSD method is in good agreement with the development and accumulation of misorientations in the dislocation structure observed in earlier works by authors and other researchers.

Significant surface curvature in the concentrator tip during the compressive deformation, as well as the maximum value of stress, leads to the formation of intense folding and to activation of the maximum number of slip systems. In addition, constant change of the surface orientation inside the incision creates favorable conditions for activation of new slip systems. Such deformation conditions should contribute to the intensive formation of the misoriented structure of micro- and nano-size.

Acknowledgements

The reported study was funded by RFBR, according to the research project No. 16-32-60007 mol_a_dk.

References

[1] Zhou S J, Beazley D M, Lomdahl P S and Holian B L 1997 *Phys. Rev. Lett.* 78 479
[2] Li H, Chandra N. 2003 *Int. J. Plasticity* 19 849
[3] Gourgiotis P A, Georgiadis H G 2008 *Int. J. Solids Struct.* 45 5521
[4] Song J, Curtin W A, Bhandakkar T K and Gao H J 2010 *Acta Mater.* 58 5933
[5] Michot G. 2011 *Acta Mater.* 59 3864.
[6] Armstrong R W 1966 *Mater. Sci. Eng.* 1 251.
[7] Anderson P M, Rice J R 1986 Scr. Metall. 20 1467.
[8] Rice J R 1992 *J. Mech. Phys. Solids* 40 239.
[9] Sun Y, Beltz G 1994 *J. Mech. Phys. Solids* 42 1905.
[10] Musazadeh M H, Dehghani K 2011 *Comput. Mater. Sci.* 50 3075.
[11] Mokryakov V V 2012 *Comput. Continuum Mech.* 3 168.
[12] Karpov E V 2002 J. *Appl. Mech. Tech. Phys*. 4, 43 630.
[13] Savin G N, Tul'chiy V I 1976 Vishcha shkola, Kiev [in Russian].
[14] Lychagin D V 2006 *Phys. Mesomech.* 3, 9 103.

[15] Alferova E A, Lychagin D V and Chernyakov A A 2014 *Appl. Mech. Mater.* 682 485.
[16] Alfyorova E A, Lychagin D V 2010 *Russ. Metall. Met.* 10 1.
[17] Alfyorova E A, Lychagin D V and Starenchenko V A 2011 *Phys. Mesomech.* 1-2, 14 66.
[18] Cheng X Y., Shi M X and Zhang Y W 2012 *Int. J. Solids Struct.* 49 3345.
[19] Starenchenko V A, Lychagin D V, Shaekhov R V and Kozlov E V 1999 *Russ. Phys. J.* 7 71.
[20] Duggan B J, Hatherly M, Hutchinson W B and Wakefield P T 1978 Met. Sci. 8, 12 343.
[21] Karduck P, Goux J M and Gottstein G 1979 Proceedings of the V International Conference: *Strength of Metals and Alloys*, Aachen, Germany [in Germany] 107.
[22] Ridha A A, Hutchinson W B, 1982 Acta Metall. 30 1929

Microstructure of Modified Layer Produced Using Aluminum Oxy-Hydroxide Nanostructured Powders

M A Kuznetsov [1a], G V Shlyakhova [2b] , PhD in Technical Sciences, V I Danilov [1,3c], Doctor of Technical Sciences, E A Zernin [1d] , PhD in Technical Sciences, S V Dementyev [1e]

[1]Yurga Institute of Technology, TPU Affiliate, Russia, Kemerovo Region, Yurga,
[2]Seversk Technological Institute MEPhi, Russia, Tomsk Region, Seversk,
[3]Institute of Strength Physics and Material Science of Siberian Branch of Russian Academy of Sciences, Russia, Tomsk

e-mail: [a]kyznechik_85@mail.ru, [b]shgv@ispms.tsc.ru, [c]dvi@ispms.tsc.ru, [d]yuti_sp@bk.ru, [e]hedg_h@mail.ru

Abstract. The paper provides the results of experimental research into the influence of aluminum oxy-hydroxide nano-structured powders on the microstructure of modified layers. It has been demonstrated aluminum oxy-hydroxide nano-structured powders AlO(OH) applied as modifiers bring about the decrease in dendrite dimensions, support equilibrium microstructure formation, and cause the growth of microhardness.

1. Introduction

Mechanical, technological and service properties of modified layers are strongly dependent on form and dimensions of grains arisen from crystallization [1, 2, 3, 4, 5, 6, 7, 8]. It is modifying that can have a significant effect on grain dimensions. The most-pronounced outcome of modifying is provided under input of fine-grained particles into the melt. The fusion temperature of these particles exceeds that of steel. At present input of nano-dispersed metallic and non-metallic powders into metal to improve steel structural strength in modified layers is considered a promising method of modifying [9, 10, 11, 12].

The paper provides consideration of surface layer modification by aluminum oxy-hydroxide nano-structured powders.

2. Methods of Research

Samples of modified layers to carry out experiments were produced from austenite grade steel (chemical composition of the wire: 0.12% carbon, 18% chromium, 9% nickel; 1-15% titanium) via welding with consumable solid electrode wire 1.2 mm in diameter in argon atmosphere.

The samples were made according to two technologies: 1 – solid wire in argon atmosphere; 2 – solid wire in argon atmosphere supplemented with nanopowder AlO(OH) with the appliance designed by the authors and specified in [13]. Rational concentration of nano-structured powder in the shielding gas was calculated in view of methods provided in paper [14].

Metallographic sections were made and etched in each sample in terms of the mode described in [13]. Investigation into the structure of surfaced metal was carried out by atomic-force microscopy (AFM) with microscope Solver PH47-PRO.

3. Results and Discussion

A modified layer has been found out to consist of two sub-layers with considerably various structures. Main microstructure of the first adjacent to the free surface sub-layer consists of relatively short, highly-branched dendrites without a preferred orientation (Fig. 1).

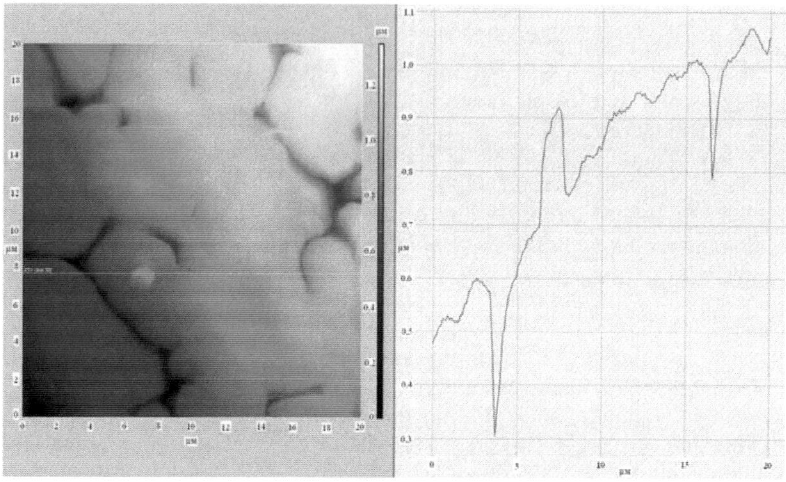

a

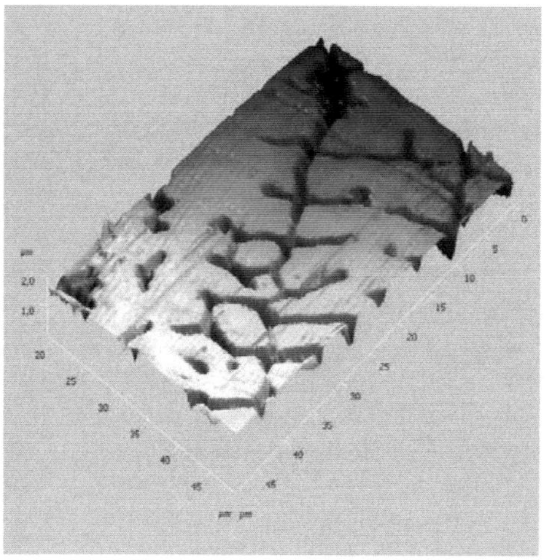

b

Figure 1 AFM 2D-image (a) and AFM 3D-image (b) of dendrites in the first sub-layer of Sample 1

The profilogram of 2D-image makes it possible to identify dendrite boundary width, which does not exceed 300 nm. 3D-image (Fig. 1 b) shows the overall view of short dendrites. They actually do not have a highly pronounced orientation towards the heat flow directions. Branches («branch pieces») of dendrites in this layer of Sample 1 without a modifier are not long (5…10 μm) as if compared with the axial length of a dendrite.

The layer of non-oriented dendrites is mostly pronounced in Sample 2 modified by nano-structured powder AlO(OH) (Fig. 2), the thickness of this sub-layer approximates to 1.5 mm. The dendrites build a practically gapless net in the sample without a modifier, whereas free surface spots can be found in the sample modified by nano-structured powder AlO(OH); the grain boundaries can't be identified there. The layer thickness of weakly-oriented dendrites in Sample 2 approximates to 1.2 mm.

Atomic-force microscopy has revealed the identity of dendrite boundary width in Sample 2 modified by nano-structured powder AlO(OH) to that in Sample 1 without a modifier (Fig. 2 a). It increases up to 1 μm in the contact points of adjacent dendrites.

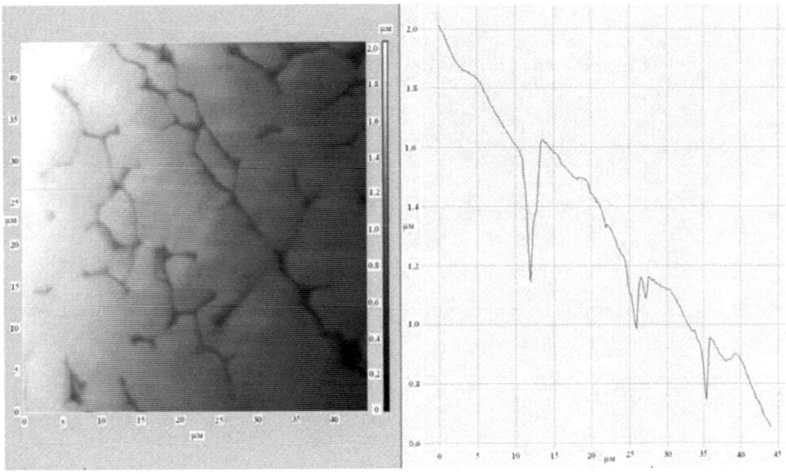

a

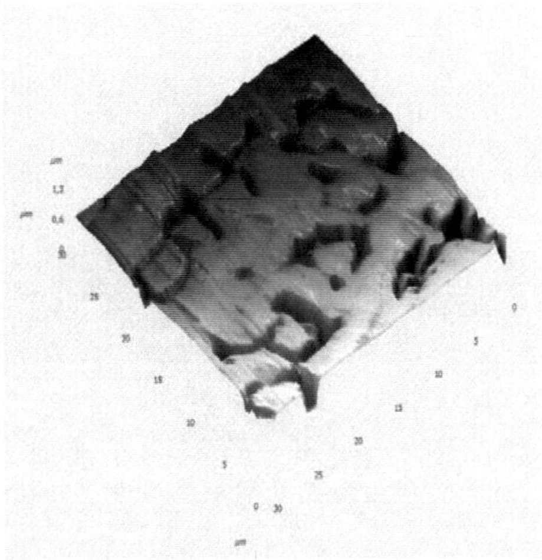

b

Figure 2 AFM 2D-image (a) and AFM 3D-image (b) of dendrites in the first sub-layer of Sample 2

IOP Conf. Series: Materials Science and Engineering **125** (2016) 012005 doi:10.1088/1757-899X/125/1/012005

The overall view and orientation of short dendrites in this layer of Sample 2 modified by nano-structured powder AlO(OH) (Fig. 2 b) is similar to the overall view of dendrites in Sample 1 without a modifier.

The layer of relatively short, highly-branched dendrites without a preferred orientation turns gradually into the next layer of oriented dendrites.

The orientation of dendrite long axes in the layer under consideration gets normal towards the fusion boundary – along the direction of heat flow into the base metal.

They form stripes separated by free surface spaces with unidentifiable grain boundaries close to the fusion boundary in Sample 1 without a modifier (Fig. 3). Long axes of dendrites are nearly parallel in each stripe, and there is no branching there.

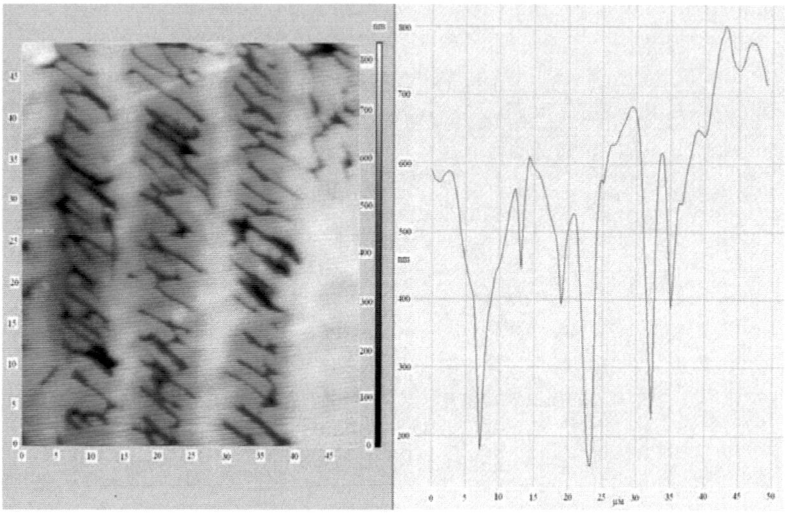

a

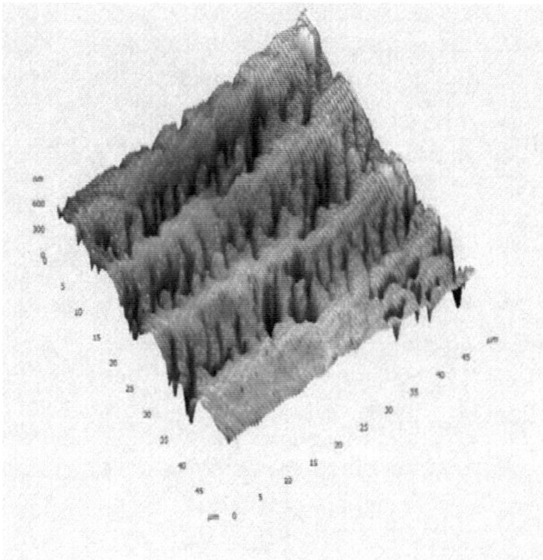

b

Figure 3 AFM 2D-image (a) and AFM 3D-image (b) of dendrites in the second layer of Sample 1

There are no dendrite stripes in Sample 2 modified by nano-structured powder AlO(OH), although parallelism of dendrite long axes is distinct there (Fig. 4).

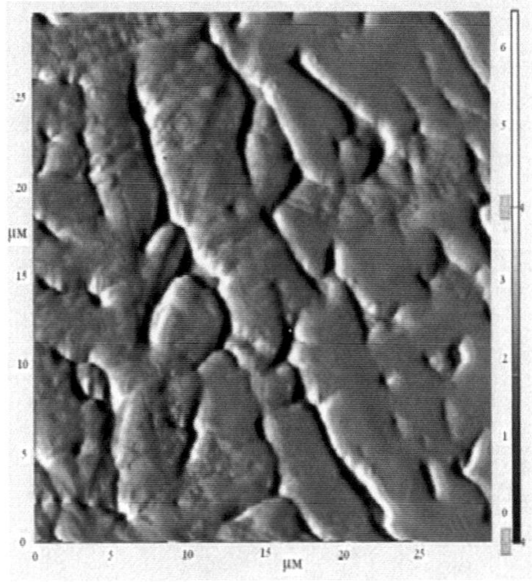

a

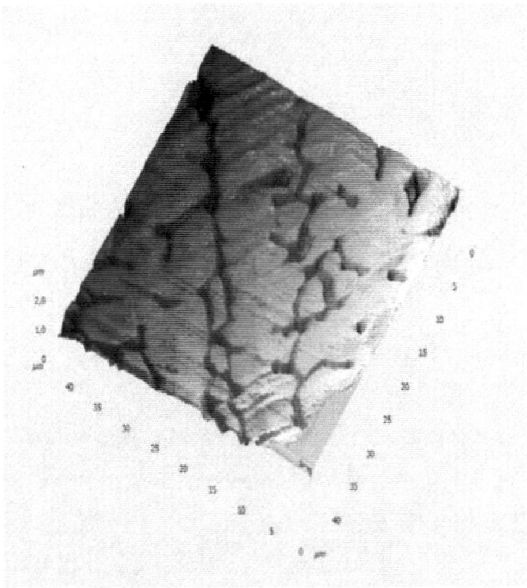

b

Figure 4 AFM 2D-image (a) and AFM 3D-image (b) of dendrites in the second layer of Sample 2

A more equilibrium structure of deposit according to dendrite dimensions is formed in the sample modified by nano-structured powder AlO(OH). That is, the average dendrite of the non-modified sample is 20-25 µm wide, and 1.2-1.6 µm thick, whereas the width and the thickness of the average dendrite in the sample modified by nano-structured powder AlO(OH) are 5-7 µm and 09-1 µm, respectively.

Application of nano-structured powders AlO(OH) affords ground for the influence on the composition and structure of the surface layer. It has been found out the surface layer has a layered structure due to the changing conditions of heat removal alongside with the depth of the liquid pool. Oriented dendrites occupy approximately a half of the surface layer volume. There is also a layer of non-oriented dendrites.

In view of generally accepted notions the more distinct the dendrite structure of the surface layer and the coarser structure of a dendrite are, the worse service properties products have.

Viewed in this way Sample 1 without modifiers yields to Sample 2 modified by nano-structured powder AlO(OH). That is, microstructure of a surface layer modified by nano-structured powder AlO(OH) is finer-grained, and equilibrium.

Atomic-force microscopy made it possible to discover secondary phase particles in the body of dendrites, especially near their boundaries. These particles are found both in Sample 1 without a modifier and (Fig. 5) and Sample 2 modified by nano-structured powder AlO(OH) (Fig. 6).

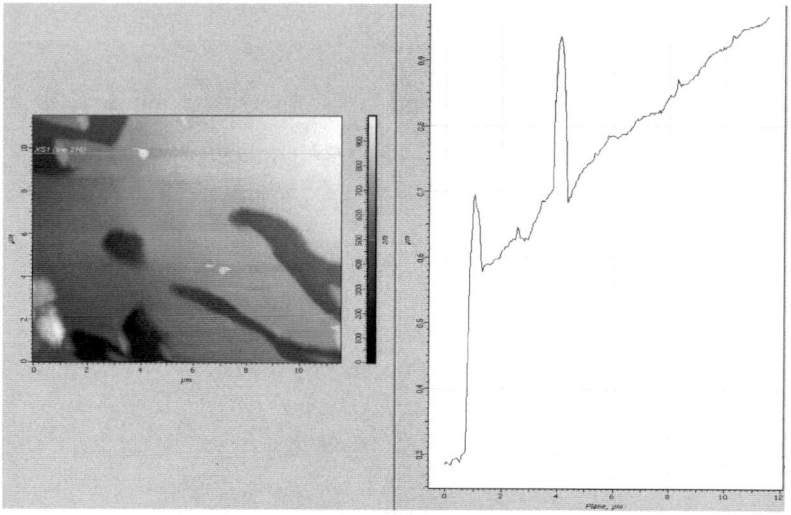

a

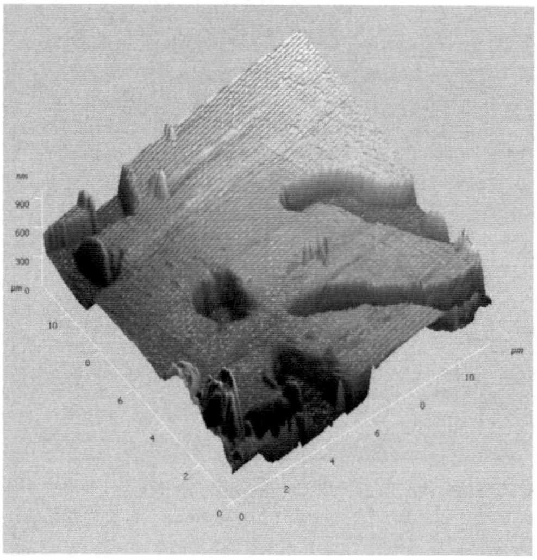

b

Figure 5 Secondary phase impurities in Sample 1 AFM 2D-image (a), and AFM 3D-image (b)

The mean diameter of these particles is 0.5…2 μm; and the height over the surface is 150…300 μm. Although we could not identify the composition of these particles, an assumption has been made they are conditional on the process of modifying.

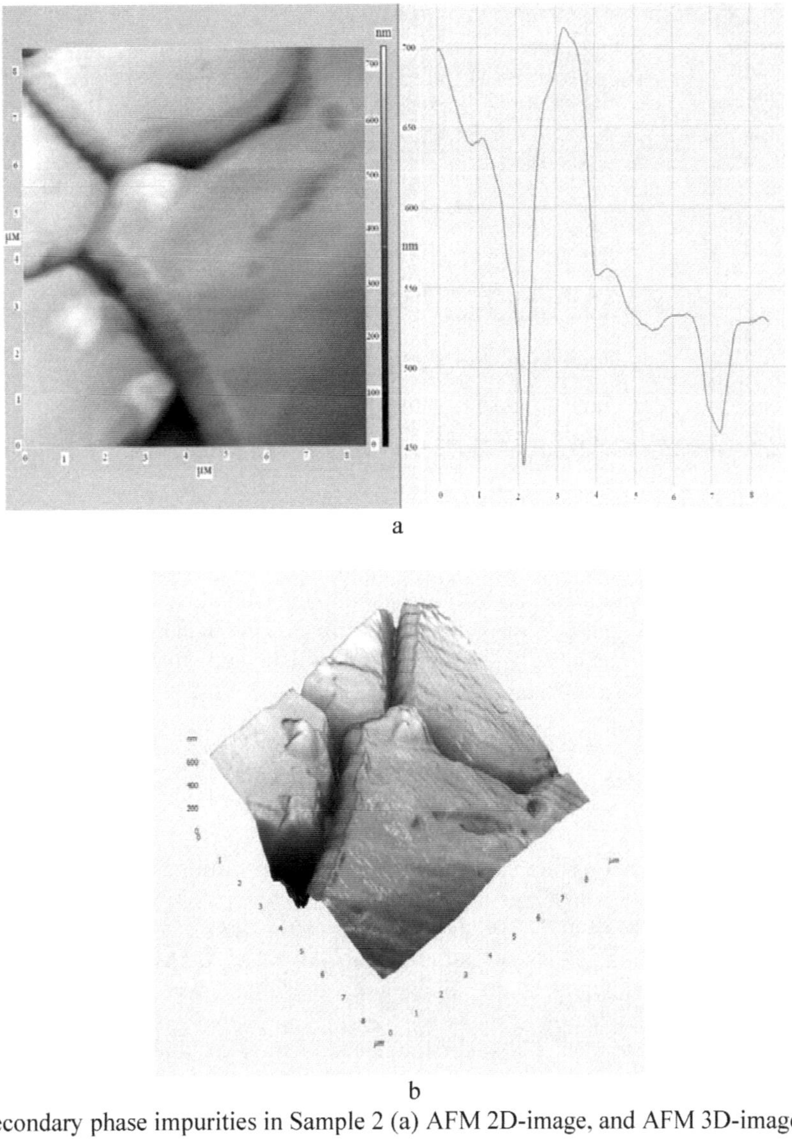

a

b

Figure 6 Secondary phase impurities in Sample 2 (a) AFM 2D-image, and AFM 3D-image (b)

The change in microhardness (MPa) is depicted in Figure 7.

The microhardness of sub-layers comes down near the base metal. In sub-layers with a modifier microhardness is higher, especially in the second sub-layer.

In conditions of using the modifier AlO(OH) obtained from interaction of electro-explosive aluminum nanopowder with water microhardness is 14-16% higher than without a modifier.

Therefore, microhardness confirms a layered structure of the surface layer and emphasizes the influence of modifiers making the structure finer.

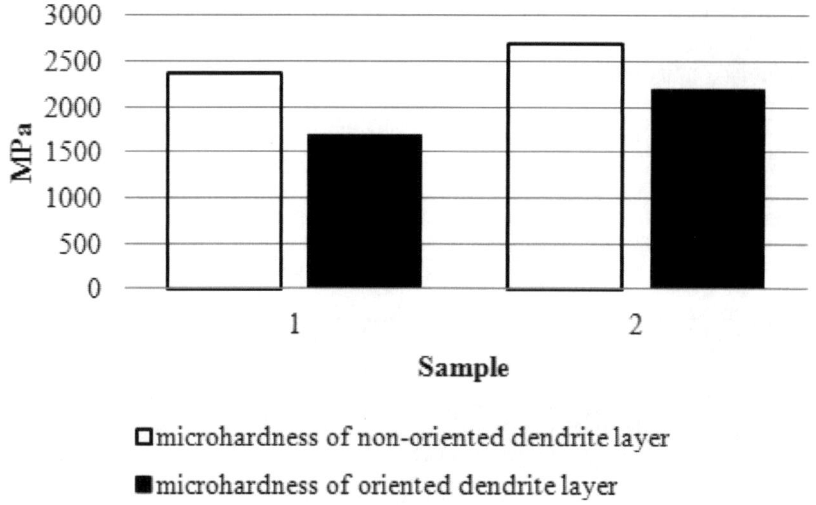

Figure 7 Influence of modifiers on the microhardness:

1 – Sample 1 without a modifier, 2 – Sample 2 modified by AlO(OH) powder

4. Conclusions

1. It has been found out nanostructured aluminum oxy-hydroxide powder supports surface layer modification.

2. We have determined the input of nanostructured AlO(OH) powders into the surface layer brings about the maximum effect of system Fe-C-Cr-Ni-Ti modification and improves service properties of the surface layer. It is caused by additional centers of crystallization forming in the melt. These centers are inoculators in crystallizing metal. The average dendrite is 1.5-2 times thinner (30-40%), and 2-3 times narrower (45-55%).

3. Nanostructured AlO(OH) powder applied as a modifier supports production of a more equilibrium structure of deposit according to dendrite dimensions.

References

[1] Material Science: A textbook for higher educational institutions / B N Arzamasov, V I Makarova, G G Mukhin et al; edited by B N Arzamasov, G G Mukhina. – 3d edition, stereotyped. – M.: MSTU by N.E. Bauman, 2002. – 648 p.: ill.

[2] Lakhtin Yu M Metal Science and Thermal Treatment of Metals: A textbook for higher educational institutions. 2nd edition revised and supplemented. – M.: Metallurgy, 1977. – 408 p.

[3] Lakhtin Yu M, Leontyeva V P Material Science: A textbook for higher educational institutions. – 3d edition revised and supplemented. – M: Machinebuilding, 1990. – 528 p. ill.

[4] Livshits L S Metal Science for Welders (welding of steels). – M.: Machinebuilding, 1979. – 253 p.

[5] Il'yaschenko D P, Chinakhov D A, Danilov V I, Schlyakhova G V and Gotovshchik Yu M 2015 IOP Conference Series: Materials Science and Engineering Physical Nature of the Processes in Forming Structures, Phase and Chemical Compositions of Medium-Carbon Steel Welds 91 pp http://iopscience.iop.org/article/10.1088/1757-899X/91/1/012006/pdf.

[6] Il'yaschenko D P, Chinakhov D A, Danilov V I, Schlyakhova G V and Gotovshchik Yu M 2015 IOP Conference Series: Materials Science and Engineering Increasing Strength and Operational Reliability of Fixed Joints of Tubes by MMA Welding 91 pp http://iopscience.iop.org/article/10.1088/1757-899X/91/1/012007/pdf.

[7] Pavlov N V, Kryukov A V, Zernin E A and Gritsenko V V 2015 IOP Conference Series: Materials Science and Engineering Systems to Control Molten Metal Transfer in Arc Welding 91 pp http://iopscience.iop.org/article/10.1088/1757-899X/91/1/012021/pdf.

[8] Kuznetsov M A, Kartsev D S, Zhuravkov S P, Barannikova S A 2015 Technologies and Materials Structure Formation and Corrosion Resistance of Weld Joints from Austenite Grade 1 Steels 1 pp. 18-24.

[9] Sokolov G N, Troshkov A S, Lysak I V, Samokhin A V, Blagoveshchensky Yu V, Alekseev A N, Tsvetkov Yu V 2011 Welding and Diagnostics Influence of Nano-dispersed Carbides WC and Nickel on the Structure and Properties of the Deposit 3 pp. 36-38.

[10] Sokolov G N, Lysak I V, Troshkov A S, Zorin I V, Goremykina S S, Samokhin A V, Alekseev A N, Tsvetkov Yu V 2009 Physics and Chemistry of Material Processing Modifying Structure of a Deposit by Nano-dispersed Tungsten Carbides 6 pp. 41 – 47.

[11] Eremin E N. 2009 Omsk Scientific Bulletin Application of Nano-particles of High-melting Compounds to Improve the Quality of Weld *Joints from Heat-resistant Alloys* 3 pp. 63-67.

[12] Kobernik N V, Mikheev R S, Pankratov A S, Linnik A A 2013 *Engineering Bulletin Modifying* the Deposit by Nano-sized Particles of Tungsten Carbide to Improve Service Properties of Weld Joints 4 pp. 9-12.

[13] Kuznetsov M A, Zernin E A, Kolmogorov D E, Schlyakhova G V, Danilov V I 2012 *Welding and Diagnostics Structure*, Morphology and Dispersion of Metal Welded by Consumable Electrode in Argon Supplemented with Nanostructured Modifiers 6. pp. 8-10.

[14] Kuznetsov M A, Barannikova S A, Zernin E A, Filonov A V, Kartsev D S 2015 *Applied Mechanics and Materials* Methods for defining the concentration of nanostructured powders in protective gas and its effect on the microstructure of deposit metal 770 pp. 28-33.

Technology of Manufacture of the Negative Matrices for Linear Fresnel Lenses[1]

V V Lapshin[1], E M Zakharevich[1], S V Grubyy[2]

[1]OJSC VNIIINSTRUMENT (All-Russian Scientific Research Tool Institute), City of Moscow
[2]Bauman MSTU (Moscow State Technical University), City of Moscow

e-mail: vasylap@mail.ru

Abstract. This article describes the main structural features of negative matrices which are intended for the production of positive copies of linear Fresnel lenses. Linear lenses are used in a space solar energy industry as solar concentrators in the photovoltaic modules. The article covers the essential requirements which are placed on the equipment and technology for the production of such matrices.

1. Introduction

The onboard power system of a spacecraft as a part of the solar battery has to to provide high specific capacity at low weight. To increase the electric capacity, radiation concentrators in the form of a linear Fresnel lens are included in the solar battery photovoltaic modules. Fresnel lenses are mainly used both as magnifying, collimating lenses and as solar energy concentrators [1 - 3]. The Fresnel lenses are available in two basic configurations: linear and circular. The linear lenses feature parallel grooves and sun light is focused in a line [4]. Such lenses are best suited to the production of composite lens panels of solar batteries for spacecraft. The use of optical concentrators allows to reduce the area of solar cells and to extend the battery life since concentrators protect the solar cells against exposure to space radiation.

Specialists at the Ioffe Physical-Technical Institute of the Russian Academy of Sciences developed a method of the production of Fresnel lenses for the concentrator photovoltaic module (RU 2456645) [5]. OJSC VNIIINSTRUMENT together with N. E. Bauman MSTU developed and implemented a technology of the production of a negative matrix for subsequent production of positive copies of a linear Fresnel lens. As is apparent from the proposed method, the matrix with a negative Fresnel profile will be fixed on a flat ground in a special appliance once it is manufactured using a diamond turning technology. A glass plate is installed in front of the processed profile, whereby from the side facing the matrix profile, a transparent adhesive coating is applied on the plate, and from the opposite side - an antireflection coating is applied. Subsequently, the gap between the plate and matrix is filled out with a liquid silicone elastomer and vulcanized. Afterwards, the ready-made positive linear Fresnel

[1] This study was performed within the Agreement No. 14.579.21.0042 of 25.08.2014 (unique identifier RFMEF157914X0042) signed between OJSC VNIIINSTRUMENT and the Ministry of Education and Science of the Russian Federation on the topic "Development of technology and equipment for nanomachining of optical materials with diamond monocrystal and abrasive tools in the quasiplastic cutting mode".

lens is separated from the plate. The developed technology for the production of linear lenses improves their optical quality and reduces the weight.

The negative linear Fresnel matrix (Fig.1) is a strip with the thickness of 2 mm and a width of 69.6 mm, with a cut profile. Polycarbonate, optical PMMA, oxygen-free copper or AlMg6 aluminium alloy may serve as suitable materials for the production of above mentioned parts. The profile of the Fresnel matrix consists of two equal parts, each with a width of 25 mm, located parallel to each other, each consisting of two segments symmetrical to each other. Each segment has 50 grooves with various depth and profile tilt angle. Thus, the profile of the Fresnel matrix consists totally of 4 segments and the total number of grooves is 200 pieces. Segment 2 is a mirror image of segment 1, and segment 4 – a mirror image of segment 3 (Fig. 1). The depth of profile grooves varies from 2.3 μm to 185 μm, whereby the tilt angle varies from 0.53° to 36.4°.

Matrices must meet demanding requirements for surface quality and accuracy of manufacturing. The roughness of structured surfaces should not exceed Ra 0.01 μm, tolerance on main linear dimensions *h* and *a* is less than 1 μm, tolerance on the angular size δ is 0.1°.

A diamond turning technology with an ultra-precision CNC machine is applied for the production of matrices with specified accuracy and quality requirements.

2. Requirements which are demanded from an ultra-precision machine for the production of matrices

Matrices are processed using an ultra-precision "Asferica" CNC machine. The machine is designed for diamond turning and milling of ultra-precision surfaces of parts made of various materials. The machine can process axisymmetric complex-shape surfaces with optical quality, including aspherical, axicons and Fresnel optics.

The "Asferica" machine has the following characteristics:

- aerostatic bearings in main shape-generating nodes;
- machine drives: synchronous, linear, insignificantly vibroactive drives;
- the machine frame is mounted on anti-vibration supports;
- the machine CNC system ensures a positioning error of less than 0.1 μm.

The machine features metrological equipment for setting up the tool in the rotational centre of the turntable. When processing optical elements, the machine ensures a surface roughness of less than Ra 0.01 μm and form error of less than 1 μm.

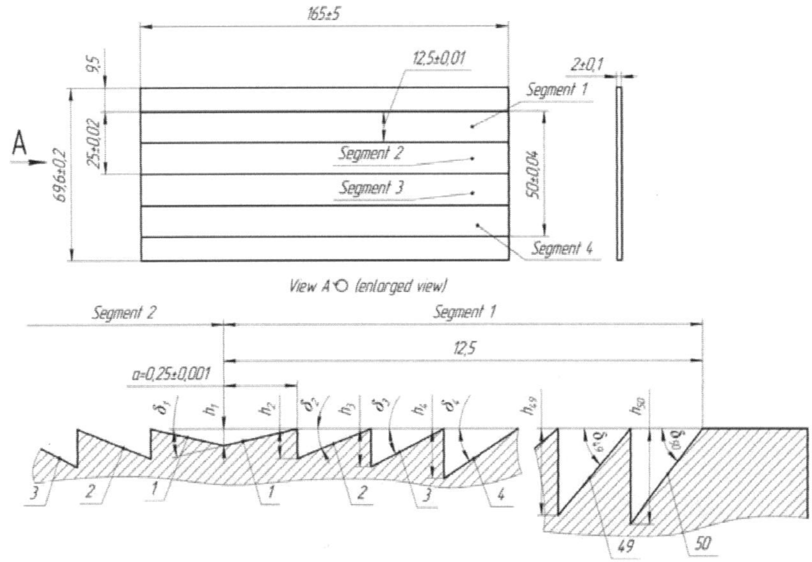

Figure 1. Linear negative Fresnel matrix

IOP Conf. Series: Materials Science and Engineering **125** (2016) 012006 doi:10.1088/1757-899X/125/1/012006

3. Cutting tool

Diamond cutters with a special geometry of the cutting point are used as cutting tool. The roughness of the working surfaces of such cutters must be less than Ra 0.01 μm, and the rounding radius of the cutting edge ρ should be 30-50 nm. Only cutters made of natural monocrystal diamond satisfy such conditions. Russian-made diamond cutters sharpened to a sharp edge (Fig. 2) and diamond cutters with a tip radius of 20 μm made by the Dutch company Contour (Fig. 3) are used for processing of matrices. Manufacturing error of the tip radius of the Dutch cutter is less than 1 μm. The use of tools with a set tip radius allows to adapt necessary radius correctors in the control software and increase the machining accuracy of the profile.

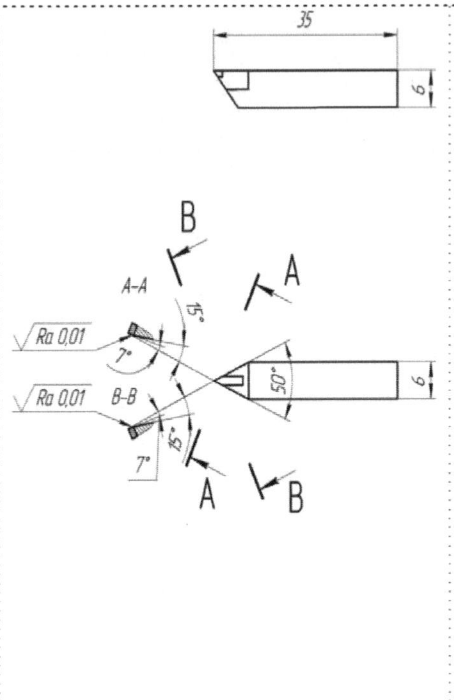

Figure 2. Diamond cutter sharpened to a sharp edge for processing of matrices

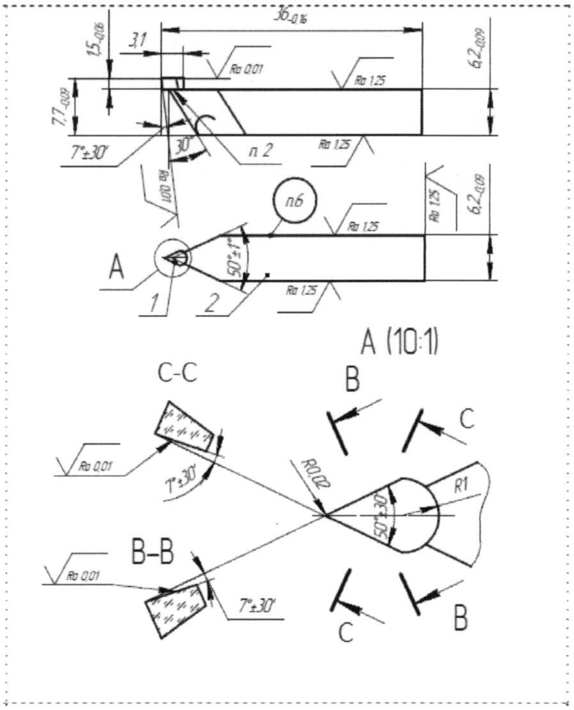

Figure 3. Diamond cutter with the radius of 0.02 mm

4. Manufacturing techniques and processing modes

Strip materials as follows are used for the production of matrices: copper, PMMA, aluminium. The strip with a width of 70 mm and a length of 1055 mm is mounted on a cylinder with a diameter of 350 mm and tensioned using a special mechanism so as to ensure a snug fit. Following the treatment, the strip is cut into 5 pieces, each with the length of 165 mm, and each of them may be used as a matrix. The diamond cutter is mounted in the cutter support which is placed on a turntable. The cutter support used makes it possible to adjust the tool height in order to precisely set up the cutter based on the height of spindle centre lines.

Prior to treatment of matrices, necessary preparatory work should be carried out. Once the strip is installed on the machine, it must be turned to the required size both in thickness and in width. Turning on the outer cylinder is performed by a diamond cutter with a tip radius of 2 mm, and end surfaces are trimmed using a rough cut tool sharpened to a sharp edge. Parts are processed at high speed modes since high precision and surface quality is not required for a pre-treatment. Subsequently, the diamond

IOP Conf. Series: Materials Science and Engineering **125** (2016) 012006 doi:10.1088/1757-899X/125/1/012006

cutter is installed in the cutter support for processing of matrices. The tool tip is aligned with the rotational centre of the turntable using a microscope whose reticule was aligned with the centre beforehand.

On the next stage, setting up is performed to determine the tool position in the coordinate system relative to the workpiece. For this purpose, a magnifying camera is placed on the machine und with the help of this camera contact of the tools with the workpiece is determined. Besides, angular coordinates are determined at which the right and left cutting edges of the cutter become parallel to the processed strip generatrix. Once the preparatory work is finished, correctors will be put in the machine CNC system which set the required coordinate systems. Due to peculiarities of setting up, the coordinate systems for left and right cutting edges of the tool must set individually. Machining pattern are shown in Figure 4.

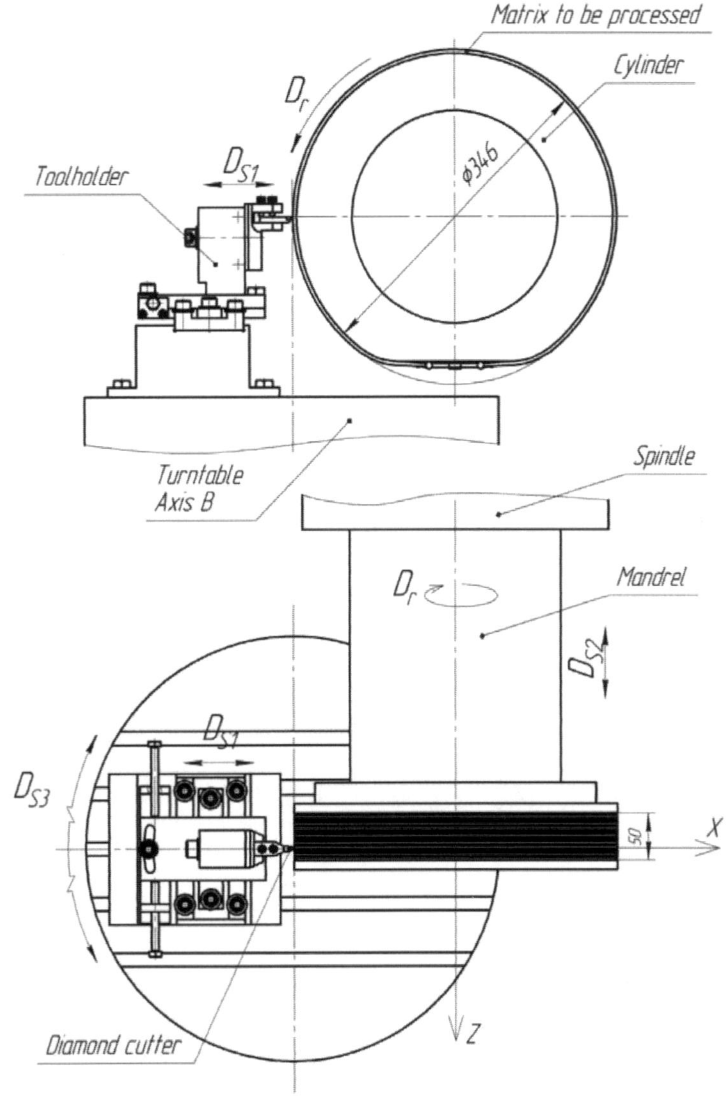

Figure 4. Kinematic diagram of the processing of matrices for Fresnel lenses

Once all preparatory setup works are finished and the program text is finalised, the profile of the matrix will be processed. The sequence of profile processing consists of the following main stages:

1. Advancing the cutter to the starting position and setting it to the desired angle.
2. Cutting of the first groove using a cut-in technology, leaving an allowance t for final cut.
3. Advancing the cutter to the starting point for machining of the groove using the profile method.
4. Processing of the groove using the profile method.
5. Final cutting of the groove end surface.
6. Steps 1 – 5 have to be repeated for the required number of grooves n, whereby each groove has different parameters δ, h.
7. Processing of the second symmetrical section with grooves (Figure 5).

Rational modes of processing are fixed: spindle rotation speed n = 270 rpm, advancing S = 2 μm/rev.

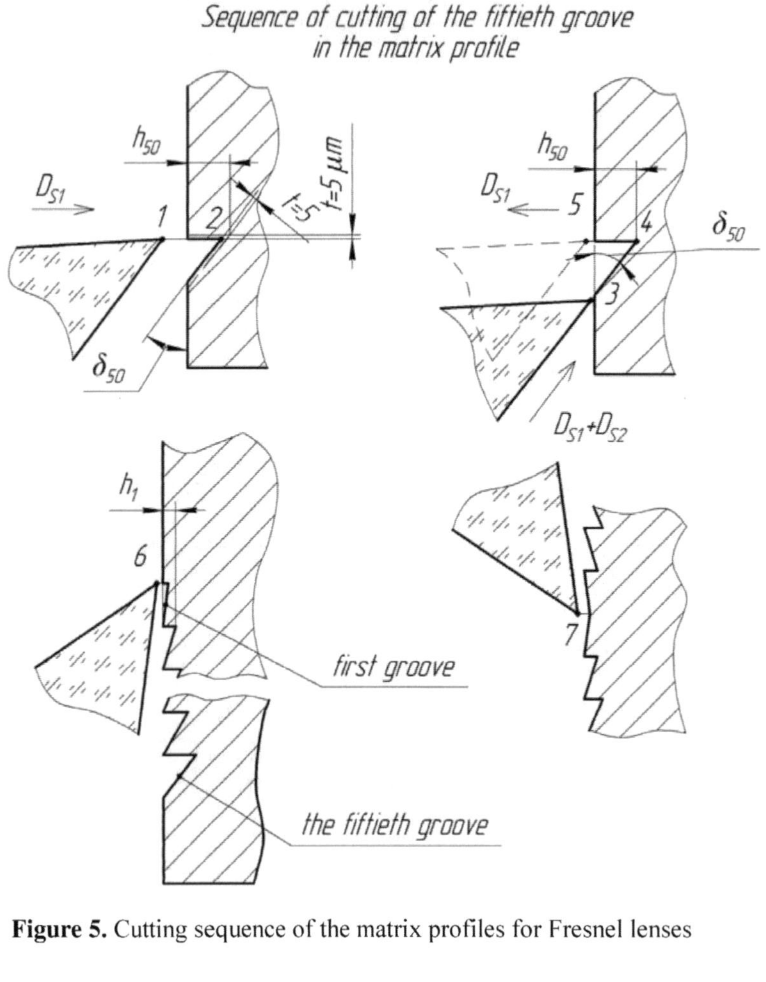

Figure 5. Cutting sequence of the matrix profiles for Fresnel lenses

5. Results

By order of Ioffe Physical-Technical Institute of the Russian Academy of Sciences, matrices made of polycarbonate, PMMA and copper were processed according to the developed technology using round-nosed cutters and cutters sharpened to a sharp edge. Following processing of each matrix, necessary metrological control of roughness and machining precision was carried out. The NanoFocus® μSurf system was used as measuring equipment and in this system a principle of confocal microscopy was implemented. Due to the fact that this device cannot scan surfaces with large tilts, roughness and machining precision was measured on the grooves with a tilt angle of not more than 10°.

According to roughness control outcomes it was found that the best results were achieved during processing of matrices made of PMMA using a cutter sharpened to a sharp edge:

- roughness Ra – 0.015 μm;
- spacing error a - less than 1 μm;
- angle error δ - less than 0.1°.

6. Conclusions

This article describes the structural features of matrices with structured surfaces and linear Fresnel lenses which are made using such matrices. The article also provides requirements which are placed on extremely-precise equipment and diamond monocrystal cutters used for machining of structured surfaces. The article outlines the technology for matrix processing, describes the modes and conditions of diamond turning. The article contains the results of metrological control of processed matrices in terms of profile accuracy and roughness of the machined surfaces.

References

[1] Davis A, Kuhnlenz F 2007 Optical design using Fresnel lenses. Basic principles and some practical experiment *Optic and Photonic* 4 52-55
[2] Gaurav A, Madhugiri, Karale S 2012 High solar energy concentration with a Fresnel lens: a review International *Journal of Modern Engineering Research* 2 1381-85
[3] Sierra C, Alfonso J Va'Zquez 2005 High solar energy concentration with a Fresnel Lens *Journal of Materials Science* 40 1339-43
[4] Khamooshi M, Salati H, Egelioglu F 2014 A rewiew of Solar Photovoltaic Concentrators *International Journal of Photoenergy* 14 1-17
[5] Patent RF No. 2456645, 20.07.2012
[6] Petrushin S I 2013 Calculation of residual stresses in multilayer composite materials *Applied Mechanics and Materials* 379 95-100

Ultrasound Velocity Measurements in High-Chromium Steel Under Plastic Deformation

Aleksey Lunev [1,2,a], **Anna Bochkareva** [1,2,b], **Svetlana Barannikova** [1,3,4,c], **Lev Zuev** [1,3,d]

[1]Institute of Strength Physics & Materials Science, SB RAS, 2/4 Akademicheskii Ave., Tomsk, 634055, Russia
[2]National Research Tomsk Polytechnic University, 30 Lenina ave., Tomsk, 634050, Russia
[3]National Research Tomsk State University, 36 Lenin Ave., Tomsk, 634050, Russia
[4]Tomsk State University of Architecture and Building, 2 Solyanaya Ave., 634003, Tomsk, Russia

e-mail: [a]agl@ispms.tsc.ru, [b]avb@ispms.tsc.ru, [c]bsa@ispms.tsc.ru (corresponding author), [d]lbz@ispms.tsc.ru

Abstract. In the present study, the variation of the propagation velocity of ultrasound in the plastic deformation of corrosion-resistant high-chromium steel 40X13 with ferrite-carbide (delivery status), martensitic (quenched) and sorbitol (after high-temperature tempering) structures have beem studied/ It is found that each state shows its view of the loading curve. In the delivery state diagram loading is substantially parabolic throughout, while in the martensitic state contains only linear strain hardening step and in the sorbitol state the plastic flow curve is three-step. The velocity of ultrasonic surface waves (Rayleigh waves) was measured simultaneously with the registration of the loading curve in the investigated steel in tension. It is shown that the dependence of the velocity of ultrasound in active loading is determined by the law of plastic flow, that is, the staging of the corresponding diagram of loading. Structural state of the investigated steel is not only changing the type of the deformation curve under uniaxial tension, but also changes the nature of ultrasound speed of deformation.

Introduction

The commonly accepted concepts of the process of plastic deformation are based on the data testifying that it occurs in stages [1], and the adequacy of the models developed for describing this process implies a strict correspondence between the stages and the governing microscopic mechanisms. The determination of the boundaries of these stages and the corresponding mechanisms presents a complicated problem, especially for polycrystals, because, in most cases, reliable and informative external manifestations of the changes in the deformation mechanisms are absent. For such purposes, the data on the integral characteristics of the material, e.g., magnetization and electric resistance, can be useful. They are especially valuable, because, in contrast to microscopic studies, they can be obtained immediately in the course of mechanical tests without any additional operations related to the preparation of special samples for the analysis.

From this point of view, acoustical methods of studying the properties of solids are quite promising [2-4]. In addition to the commonly used effects such as the acoustic emission from mechanically loaded samples or the amplitude dependence of internal friction, which can be interpreted on the basis of specially developed theoretical models, it is possible to use a more easily measured characteristic, namely, the velocity of ultrasound propagation V. Rayleigh waves, one kind of surface acoustic waves, are ideally suited to surface inspection problems [5]. The penetration depth depends on the frequency of Rayleigh waves. Approximately, the penetration depth is in the order of the wavelength [6]. Rayleigh waves with higher frequencies are more influenced by the surface layer, because they have a lower penetration depth. Damage usually tends to initiate and to concentrate at the surface. For the study of surface imperfections and damage, Rayleigh waves that form and propagate along boundaries, are better suited than body waves to study surface process [7].

It was shown [8] that in the course of plastic flow the propagation velocity of ultrasonic surface waves (Rayleigh waves) in the material changes, the dependence $V(\varepsilon)$ having a complicated N shape. In the present paper we present data on the dependence of V on the magnitude of the stress σ acting during deformation of high-chromium steel. The material possesses an appealing combination of strength and plastic properties, corrosion resistance in atmospheric, water-vapor, aquatic, and (on numerous occasions) acid environments, and is compositionally simple. This has led to its widespread use for production of critical machine parts and structural members. Certification of this kind of materials makes it essential to determine the following characteristics: ductility margin, life time, and fracture mode, all being dependent in many respects on strain-localization patterns [9, 10]. The deformation behavior of high-chromium steel was investigated in [11]. The variation of the propagation velocity of ultrasound in the plastic deformation of corrosion-resistant high-chromium steel with ferrite-carbide (delivery status), martensitic (quenched) and sorbitol (after high-temperature tempering) structures have been studied.

Materials and experimental methods

The test material was high-chromium stainless steel (0.4%C–0.6%Si–0.55%Mn–12.5%Cr). The specimens were configured as dog bones with gage section measuring 50×10×2 mm. The test pieces were cut out from sheet steel along the rolling direction. The material was examined with ferrite-carbide (delivery status), martensitic (quenched) and sorbitol (after high-temperature tempering) structures. The specimens were homogenized at T = 1320 K for 3 hrs and subjected to fast air cooling to provide martensitic structure with a small amount of chromium carbides. On high-temperature tempering from T = 870 K for 3 hrs and furnace cooling, sorbitic structure (ferrite and carbides) was generated. The structure of steel in various types of heat treatment was studied by using optical and atomic force microscopy. The specimens thus prepared were subjected to uniaxial tension at room temperature at the rate $6.67\cdot10^{-5}$ s^{-1} (0.2 mm/min).

The velocity of ultrasonic surface waves (Rayleigh waves) was measured simultaneously with the registration of the loading curve in the investigated steel in tension in a LFM-125 testing machine using an OWON RDS6062S device. The principle for measuring the velocity of propagation of Rayleigh waves is based on the pulse autocirculation method [12]. The error of the measurement amounts to $2\cdot10^{-4}$, and operation with the instrument does not require any special skills on the part of the operator. The basic principle of the autocirculation method consists of setting up a closed circuit to transmit the pulses. A radiating piezoelectric transducer, when acted upon by a short electric pulse, generates an acoustic wave in the sample. The wave traveling from the transmitting piezoelectric transducer to the receiving piezoelectric transducer is converted back into an electrical signal and again arrives at the radiating transducer. Hence, keeping the distance between the transducers fixed, the frequency at which a pulse appears at a certain point of the circuit will depend on the time taken for the acoustic signal to travel through the sample and the delay in the circuit. Since the delay in the circuit is negligibly small compared with the propagation time of the acoustic wave in the sample, the autocirculation frequency will characterize the velocity of propagation of the ultrasonic wave in the sample. In this case the Rayleigh surface waves have a frequency of 5 MHz.

The ultrasonic sensor, placed on the object being investigated, has two inclined piezoelectric transducers, situated at a fixed distance from one another, called the base. The inclination of the piezoelectric transducers is chosen in such a way that a surface Rayleigh wave is generated in the object. For reliable measurement of the velocity, it is necessary to ensure that the contact with the metal of the article being monitored is clean, the surface must be smooth, and the sensor must be pressed tightly in position. Acoustic contact with the piezoelectric transducer is provided by a no aggressive liquid lubrication, for example, transformer oil. One must bear in mind that the space between the piezoelectric transducers must remain dry and clean.

Experimental results

The stress-strain curves for the material are shown in Fig. 1. The work-hardening modes for the delivery status (1), quenched steel (2) and that subjected to quenching and high-temperature tempering (3) are seen to be essentially different. In the deforming sample of the original alloy (1), the onset of necking takes place, which is a forerunner of viscous fracture. The quenched steel (2) exhibits pseudo brittle fracture without distinct necking. In the tempered steel (3), a well-pronounced neck is formed, and the descending branch of the curve indicates that up to 2% of the total strain is built up for the fracture percent elongation δ = 9.6%.

The stress-strain curve for a polycrystalline material can be described by the Lüdwick relation [1]

$$\sigma = \sigma_0 + K\varepsilon^n \qquad (1)$$

where K is the coefficient of deformation hardening and n is the exponent of deformation hardening. The strain exponent n has different values in different portions of the stress-strain curve and changes in a stepwise manner depending on the degree of strain. To each of the deformation stages correspond particular constant values of n and K.

Using the method reported in [11], the loading curve can be represented in the system of functional logarithmic coordinates as $\ln(s - s_0) = f(\ln e)$ (here s is the true stress which takes no account of reduction in the work cross-section and e is the true deformation). This enables individual segments to be singled out on the curve for a constant value of the exponent n, which varies discretely on going from one segment to the next. All the loading diagrams obtained in this study were analyzed by the above method.

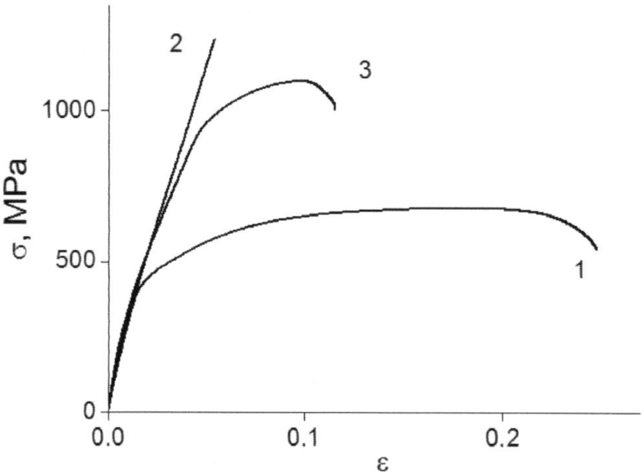

Fig. 1. Stress-strain curves for high-chromium steel: 1 – delivery status; 2 – quenched state; 3 – tempered state

The quantitative data on mechanical characteristics and deformation stages of alloys are listed in the Table 1 and Table 2. The deformation curves plotted for the original alloy (1) were represented in functional logarithmic coordinates. One rectilinear segment is distinguished on the flow curve for strains ε in the range 1.4-6.8% for a constant value $n \approx ½$. This segment corresponds to the parabolic work hardening stage.

Table 1. Mechanical characteristics of high-chromium stainless steel (1 – delivery status; 2 – quenched state; 3 – tempered state)

	$\sigma_{0.2}$ [MPa]	σ_B [MPa]	δ [%]	HV [MPa]	V_0 [m/s]
1	420.0	676.6	22.7	2060	3067
2	348.5	1233.4	1.97	3880	2958
3	278.6	1098.3	9.6	3498	3034

Footnotes: $\sigma_{0.2}$ – proof stress; σ_B – ultimate stress; δ – relative elongation to rupture; HV - microhardness values; V_0 - velocity of Rayleigh waves in the unstressed sample.

Thus the stress-strain curve for the quenched steel has been found to include a linear-hardening stage for strains $0.9 \leq \varepsilon \leq 4.5\%$ followed by a stage with constantly changing values of n and K (Fig. 1). The stress-strain curve for the tempered steel exhibits three deformation stages (Fig. 1). These are a linear-hardening stage ($1.9 \leq \varepsilon \leq 4.1\%$), a Taylor parabolic work-hardening stage ($5.3 \leq \varepsilon \leq 7.1\%$) with $n \approx ½$ and a prefracture stage ($7.7 \leq \varepsilon \leq 9.5\%$) with $n = 0.3$. The data agree with the results of previous studies [11].

It has been established [8] that the velocity of propagation of an ultrasonic wave in a sample deformed by stretching depends on the overall strain, the stress time and the structure of the material. The analysis of the dependences of the velocity V on the total strain ε and the actual stress σ for deformations up to fracture revealed a number of interesting effects (Fig. 2, Fig. 3). It was found that the ultrasonic velocity considerably varies with tension, and the dependences $V(\varepsilon)$ and $V(\sigma)$ are fairly complicated. For example, the curve $V(\varepsilon)$ obtained for tempered state and shown in Fig. 2 c is N-shaped with three well-defined stages [8]. Such a shape testifies to the difference between the mechanisms governing the relation between V and ε in the corresponding intervals of plastic deformation.

Table 2. Deformation stages of high-chromium stainless steel (1 – delivery status; 2 – quenched state; 3 – tempered state)

Material/State	Linear-hardening stage $\sigma \sim \varepsilon$		Taylor parabolic work-hardening stage $\sigma \sim \varepsilon^{1/2}$		Prefracture stage $\sigma \sim \varepsilon^{1/3}$	
	ε_{init} [%]	ε_{fin} [%]	ε_{init} [%]	ε_{fin} [%]	ε_{init} [%]	ε_{fin} [%]
1	-	-	1.4	6.8	-	-
2	0.9	4.5	-	-	-	-
3	1.9	4.1	5.3	7.1	7.7	9.5

In these investigations, attention was paid to the way in which the velocity of ultrasonic waves depends on the stress. Using dimensionless values of the velocity and the stress and approximating the separate stages by linear functions, we obtain the generalized relation [8]

$$\frac{V}{V_0} = \alpha_i \frac{\sigma}{\sigma_B} + \beta_i, \tag{2}$$

where V_0 is the velocity of Rayleigh waves in the unstressed sample, m/sec; α_i and β_i are empirical quantities, independent of the material; $i = 1, 2$ is the number of the linear part in Fig. 3; and σ_B is the tensile strength of the material being investigated, MPa. The calculated values of α_i and β_i for sections 1, 2 and 3 are listed in the Table 3.

The typical dependences V/V_0 and σ/σ_B shown in Fig. 3 for delivery status (1) and tempered state (3) have three stages, and for the quenched steel (2) - one stage. The linear character of the relation between V and σ at each stage is interesting. Similar three-stage dependences $V(\sigma)$ were obtained for the other materials studied in the course of the experiments described [8].

Table 3. Empirical quantities α_i and β_i for Eq. (2) for delivery status (1), quenched state (2), tempered state (3) of high-chromium stainless steel

Segment	Delivery status (1)		Quenched state (2)		Tempered state (3)	
	α_i	β_i	α_i	β_i	α_i	β_i
1	$-9.4\cdot10^{-4}$	1	$-4.37\cdot10^{-4}$	1	$5.97\cdot10^{-4}$	1
2	$-1.29\cdot10^{-2}$	1.008			$-3.42\cdot10^{-3}$	1.002
3	$-5.46\cdot10^{-2}$	1.048			$-4.07\cdot10^{-2}$	1.034

It follows from (2) that

$$\sigma_Б = \frac{\alpha_i \sigma}{(V/V_0 - \beta_i)} \tag{3}$$

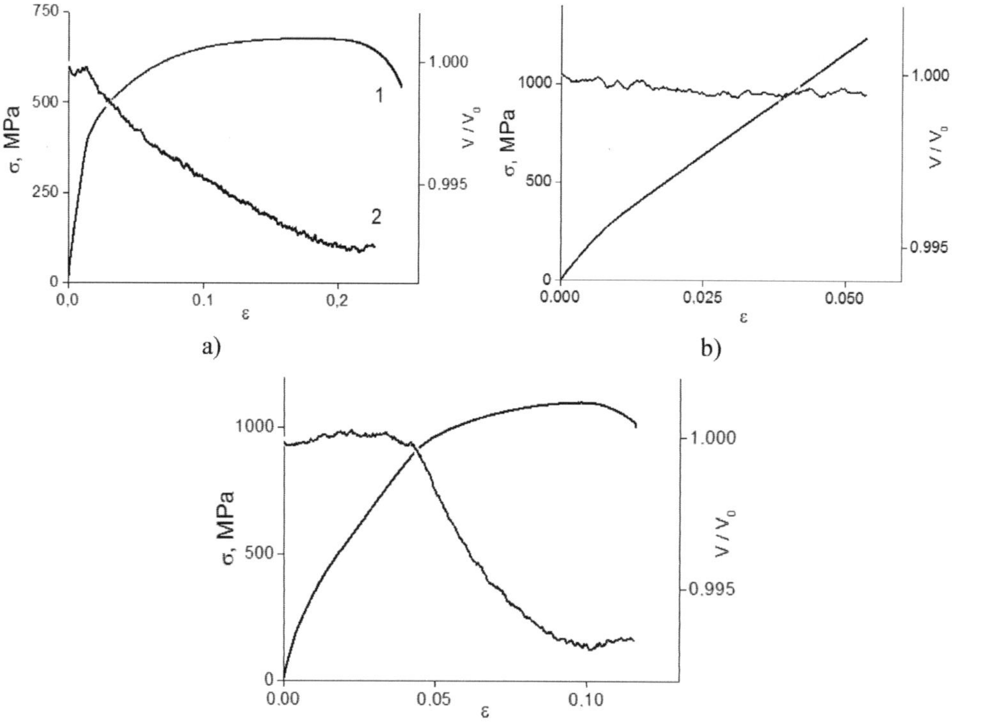

Figure 2. Stress–strain curve of the plastic flow (1) for delivery status (a), quenched state (b), tempered state (c) of high-chromium stainless steel and the dependence of the velocity of ultrasound propagation against the total strain (2)

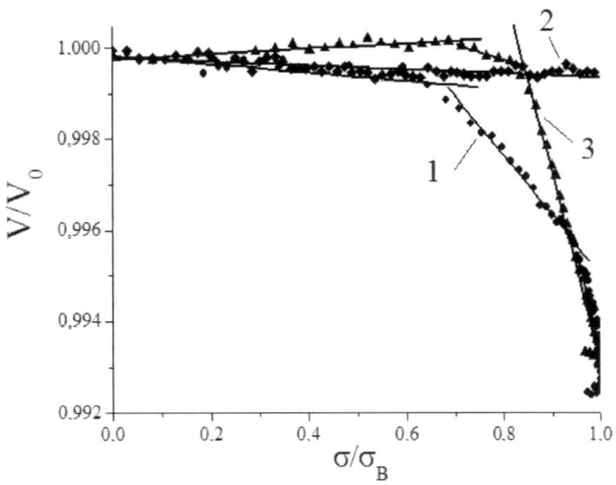

Figure 3. The dependence of the velocity of ultrasound propagation against the acting stresses for delivery status (1), quenched state (2), tempered state (3) of high-chromium stainless steel

Equation (3) can be used to estimate the tensile stress for small plastic deformations long before the sample fractures. Hence, in order to determine σ_B, it is sufficient to measure the velocity of ultrasonic waves for stresses in the sample in the range $\sigma_{0.2} < \sigma < 0.6\sigma_B$ (where $\sigma_{0.2}$ is the tensile strength), i.e., in the section of small plastic deformations.

Summary
The velocity of ultrasonic surface waves (Rayleigh waves) was measured simultaneously with the registration of the loading curve of high-chromium steel subjected with ferrite-carbide (delivery status), martensitic (quenched) and sorbitol (after high-temperature tempering) structures. In the delivery state diagram loading is substantially parabolic throughout, while in the martensitic state contains only linear strain hardening step and in the sorbitol state the plastic flow curve is three-step. It is shown that the dependence of the velocity of ultrasound in active loading is determined by the law of plastic flow, that is, the staging of the corresponding diagram of loading. Structural state of the investigated steel is not only changing the type of the deformation curve under uniaxial tension, but also changes the nature of ultrasound speed of deformation. The data obtained from this investigation are of practical use for developing theoretical foundations of plastic working of high-chromium steel. This can provide, among other things, a reliable choice of limiting degrees of strain for cold rolling of sheet slabs and tubular billets. The proposed method can be used to estimate the tensile strength of high-chromium steels before they fracture.

Acknowledgment
The work was performed in the frame of the Tomsk State University Academic D.I. Mendeleev Fund Program and the Program of Fundamental Research of State Academies of Sciences for the period 2014-2020 yrs.

References
[1] J Pelleg, Mechanical properties of materials, Springer, Dordrecht, Heidelberg, New York, London, 2013.
[2] M Kobayashi, S. Tang, S. Miura, K. Iwabuchi, S. Oomori, H. Fujiki, Ultrasonic nondestructive material evaluation method and study of texture and cross slip effects under simple and pure shear states, Int. J. Plast. 19 (2003) 771–804.
[3] V V Mishakin, S Dixon, M D G Potter, The use of wide band ultrasonic signals to estimate the stress condition of materials, J. Phys. D: Appl. Phys. 39 (2006) 4681-4687.

IOP Conf. Series: Materials Science and Engineering **125** (2016) 012007 doi:10.1088/1757-899X/125/1/012007

[4] V V Murav'ev, L V Volkova, E N Balobanov, Estimation of residual stresses in locomotive wheel treads using the acoustoelasticity method, Rus. J. Nondestr. Test. 49 (2013) 382-386.
[5] J D N Cheeke, Fundamentals and applications of ultrasonic waves, CRC Press, Boca Raton, 2002
[6] D Schneider, B Brenner, T Schwarz, Characterization of laser hardened steels by laser induced ultrasonic surface waves, J. Nondestr. Eval. 14 (1995) 21–9.
[7] A Zerwer, M A Polak, J C Santamarina, Wave propagation in thin plexiglas plates: implications for Rayleigh waves, NDTE Int. 33 (2000) 33–41.
[8] L B Zuev, B S Semukhin, A G Lunev, The use of measurements of the velocity of ultrasound to determine the stress-strain state of metal articles, Meas. Techn. 53 (2010) 439-443.
[9] L B Zuev, V I Danilov, S A Barannikova, V V Gorbatenko, Autowave model of localized plastic flow of solids, Phys. Wav. Phen. 17 (2009) 1–10.
[10] L B Zuev, S A Barannikova, Experimental study of plastic flow macro-scale localization process: pattern, propagation rate, dispersion, Int. J. Mech. Sci. 88 (2014) 1-8.
[11] V I Danilov, D V Orlova, L B Zuev, G V Shlyakhova, Special features of the localized plastic deformation and fracture of high-chromium steel of the martensitic class, Rus. Phys. J. 52 (2009) 525-531.
[12] R Truell, C Elbaum, B Chick, Ultrasonic Methods in Solid State Physics, Academic Press, New York, 1969.

IOP Conf. Series: Materials Science and Engineering **125** (2016) 012008 doi:10.1088/1757-899X/125/1/012008

Nanostructuring and Physical Properties of Metal-Ceramic Composites With a Different Content the Ceramic Components

V E Ovcharenko[1,2], Yu F Ivanov[2,3], A A Mokhovikov[2], K V Ivanov[1], V Yu Ul'yanitsky[4]

[1]Institute of Strength Physics and Materials Science SB RAS, 2/4, Pr. Academicheskii, Tomsk, 634021, Russia
[2]National Research Tomsk Polytechnic University, 30, Pr. Lenin, Tomsk, 634050, Russia
[3]Institute of High Current Electronics SB RAS, 4, Pr. Academicheckii, Tomsk, 634021, Russia
[4]Lavrent'ev Institute of Hydrodynamics SB RAS, 15, Prosp. Ak. Lavrent'eva, Novosibirsk, 630090

e-mail: ove45@mail.ru

Abstract. A comparative analysis of the microstructure and durability of hard-alloy plates treated with high-energy pulsed electron beam generated from the plasma of argon or xenon in steel cutting was conducted. It is shown that the choice of the plasma gas to generate electron from the cathode plasma-filled setting for pulsed electron-beam irradiation has a significant influence on the formation of structural-phase state in the surface layer and the durability of hard-alloy plates in steel cutting.

1. Introduction

A generally accepted strategy in developing new methods for increasing the mechanical strength of metal-cutting hard-alloy plates is the improvement of the metal-ceramic structure of the plates to be achieved through a proper doping of metal binder and a proper choice of both the ceramic component and powder-mixture sintering conditions. The latest achievement in this field is the sintering of hard alloys from nano-powders forming the starting metal-ceramic composition. 3D-nanostructured hard alloys feature an enhanced mechanical strength, hardness, crack growth resistance, and wear resistance [1-3]. The main drawback of such alloys is their high cost, defined by the high cost of the nano-powder components of the metal-ceramic composition and by some specific features of the sintering technology; this drawback seriously hampers the use of the above strategy in national industry.

An alternative solution of the problem on improving the mechanical strength of metal-cutting hard-alloy plates produced in Russia is the hardening of the surface layer of the plates based on the concept of formation of a multi-level phase-structural state of the material in

micro-, submicro-, and nano-size regions. As it was shown in [3], the physics behind the formation of the multi-level alloy structure consists in realization of conditions permitting localization of plastic deformation in the material volume at the lowest possible scale of the internal structure. The latter in

IOP Conf. Series: Materials Science and Engineering **125** (2016) 012008 doi:10.1088/1757-899X/125/1/012008

turn provides for a more uniform distribution of elastic stress in the material and increases the energy required for nucleation of critical stress concentrators. The above-said is of special significance for the discussion of the mechanisms of surface hardening of brittle materials such as hard alloys intended for applications in tools, in which the outer surface comes as a nucleation source of rupture microcracks.

A metal-ceramic alloy in the initial, as-sintered state involves three phase-structural-state levels: mesoscopic-scale particles of ceramic component with 1-10 μm sizes (carbides, carbonitrides, borides, etc.), interparticle metal-binder layers about 4 μm thick, and particle-binder transition zones of width ~1 μm (Fig. 1).

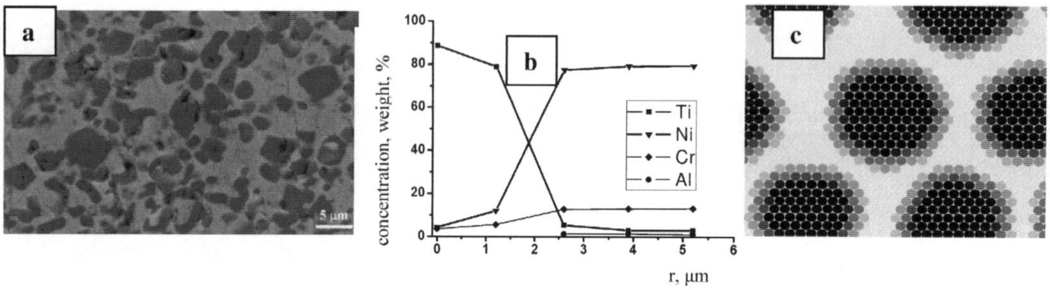

Fig. 1. The microstructure of the surface layer of 50 vol.%TiC – 50 vol.% (Ni-Cr) alloy (a); the distribution of the elemental composition of the metal-ceramic alloy at the interface between the carbide particles and the metal binder (b); and a computer model of the structure of the hard alloy with carbide particle-to-metal binder transition zones (c).

A numerical modeling of the effects due to individual phase-structural state levels on the strength of hard alloys under dynamic loading at a given volumetric proportion between the components of metal-ceramic composition has showed that, within the three-level structure, the mechanical strength of hard alloys is primarily defined by the mechanical strength and geometric sizes of the zone of transition from carbide particles to the metal binder: on increasing the strength and width of the transition regions within the actual range of parameters, a three-fold increase of the mechanical strength of hard alloys can only be achieved. In other words, the presently available possibilities in improving the material characteristics (physical, strength, and wear properties) of hard alloys with a metal-ceramic structure involving three levels of the phase-structural state by means of powder metallurgy (sintering of metal-ceramic powder compositions) are rather limited.

A cardinal improvement of the mechanical strength of traditional hard alloys is only possible via the formation of additional phase-structural-state levels in the metal binder of metal-ceramic compositions in submicro- and nano-size regions and via the dispersion of the particles of ceramic component down to nano-size state. An efficient method for such modification of the phase-structural state of the surface layer of hard alloys is the irradiation of the hard-alloy surface with a high-intensity, low energy electron beam with submillisecond duration of its action in plasma-forming gases with different values of ionization energy, which allows a cardinal modification of the structure of the surface layer several ten micrometer thick achieved through transferring the layer in a multimodal phase-structural state without any substantial modification of the phase-structural state of the material volume [4].

The purpose of this work was to study the laws of the formation of structural levels in the surface layers of the metal-ceramic composites with different ratios of the ceramic and metal components under pulsed electron-ion-plasma irradiated surface and influence the formation of additional structural levels of the physical properties of the surface layer of the composites.

2. Materials and methods of research

Studies were conducted on the sintered composite with a low content of ceramic component - 50% TiC / 50% (Ni + Cr + Al) and a high content of ceramic component - (75% WC + 14% (Ti, Ta, Nb) C) / 11 % Co. Pulsed electron-ion-plasma irradiation composite sample was performed at an energy density of the electron beam 20 ... 100 J / cm2 with a duration of irradiation pulses 50 ... 200 μs with a pulse repetition rate up to 10-1s. Irradiation was carried out in inert gas discharge plasma (argon, xenon) and reactive (nitrogen) gas.

The microstructure of the surface of a solid alloy after electron beam treatment was investigated in the scanning electron microscope SEM-515 «Philips» with energy-top box for the local elemental analysis EDAX ECON IV. Investigations of the phase composition and defect substructure was carried out with an electron microscope «Jeol - 200", phase composition was analyzed by indexing mikroelektronogramm using standard techniques of electronic diffraction microscopy.

Results and Discussion

It was found that, during pulsed electron-beam irradiation of hard alloys with a high content of metal binder in the metal-ceramic composition (50.0 vol.%TiC-50.0 vol.% (Ni-Cr)-alloy) in Ar-containing gas-discharge plasma (ionization energy of Ar 1519.6 kJ/mole), in the metal binder of the hard-alloy surface layer, as a result of the decay of the titanium-carbon solid solution, in the binder solution secondary titanium carbide particles with sizes down to 200 nm, which form the fourth phase-structural-state level, appear. Similarly, during pulsed electron-beam irradiation of the alloy in (Ar+N)-containing gas-discharge plasma (ionization energy of N 1401.5 kJ/mole) in the metal-binder interlayers 50-nm nanoparticles of aluminum nitride AlN appear; such particles form the fifth phase-structural-state level of the hard-alloy surface layer (Fig. 2).

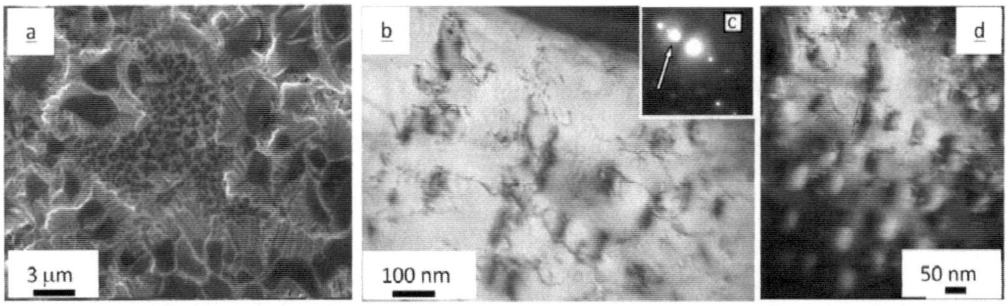

Fig. 2. The microstructure of the surface layer of TiC-(Ni-Cr) hard alloy subjected to electron-beam irradiation in Ar-containing gas-discharge plasma (electron-beam energy flux density 40 J/cm2, 15 pulses, pulse repetition frequency 1 s-1, width of irradiation pulses 200 μs) (a) and in (Ar+N)-containing gas-discharge plasma (40 J/cm2, 15 pulses, 1 s-1, 200 μs) (b-d). Scanning (a) and transmission (b - d) electron microscopy data.

Pulsed electron-beam irradiation of TiC-(Ni-Cr) hard alloy in plasma-forming gases with a smaller value of ionization energy (1350.0 kJ/mole for Kr and 1170.0 kJ/mole for Xe) exerts a more cardinal influence on the phase-structural state of the material in the surface layer. Like in the case of the irradiation of the material in Ar- and (Ar+N)-containing gas-discharge plasmas, at fixed regime of pulsed electron-beam irradiation nanostructuring of carbide-component particles occurs, - the carbide particles of the metal-ceramic composition get disintegrated in a multitude of nano-sized particles, thus forming a polycrystalline structure with a dendrite-like carbide skeleton in each crystal irradiated with electrons in Kr-containing plasma and with nano-dispersed distribution of the carbide component in crystals irradiated with electrons in Xe-containing gas-discharge plasma (Fig. 3).

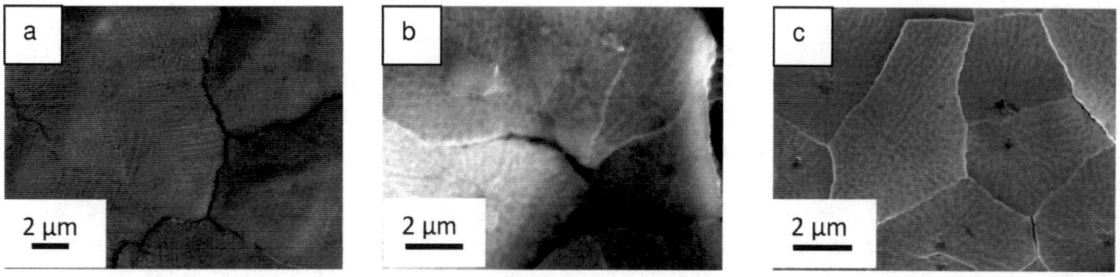

Fig. 3. The microstructure of the surface layer of TiC-(Ni-Cr) hard alloy subjected to pulsed electron-beam irradiation of the alloy in Kr-containing (a), Xe-containing (b), and (Xe+N)-containing (c) gas-discharge plasmas: a, b – 50 J/см^2 , c – 40 J/см^2; width of irradiation pulses - 150 μs, total number of pulses - 15.

By means of transmission electron microscopy, it was found that, right on the irradiated surface of hard-alloy samples, a two-phase mixture involving titanium carbide (TiC) nanocrystals and a (Ni-Cr) binder respectively sized 15...25 nm and 3...5 nm, forms (the circumference in Fig. 4, a singles out the foil area from which the electron-diffraction micropattern was taken).

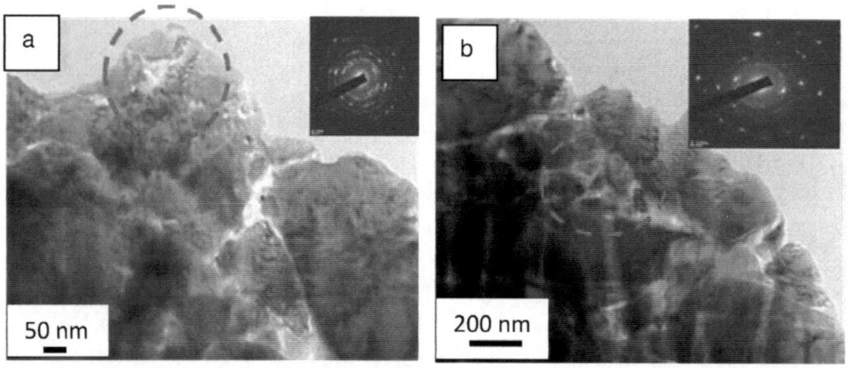

Fig. 4. The microstructures in the surface layer of TiC-(Ni-Cr) metal-ceramic alloy at depths 5-10 μm (a) and 25-30 μm (b) after processing the material in Xe-containing plasma.

With increasing distance from the extreme point of the outer surface (to 25-30 μm), fractured titanium carbide particles typical of the initial state of the material are observed (in the images, one such particle is singled out with the circumference), with regions involving carbide-phase particles sized 40...100 nm that occur at the junctions between the titanium carbide particles. In the electron diffraction micropatterns, reflections due to the carbide phase are only observed (Fig. 4, b). Further deepening to a depth of 50-60 μm has showed that the layer located at this depth contains comparatively large TiC carbide particles sized 0.5-1.0 μm and a (Ni-Cr) binder reinforced with secondary titanium carbide nanoparticles (Fig. 5, a). The deepening to depths in excess of 100 μm reveals coarse single-crystal titanium carbide particles and the (Ni-Cr)-binder in the initial state, i.e. without secondary titanium carbide nanoparticles (Fig. 5, b).

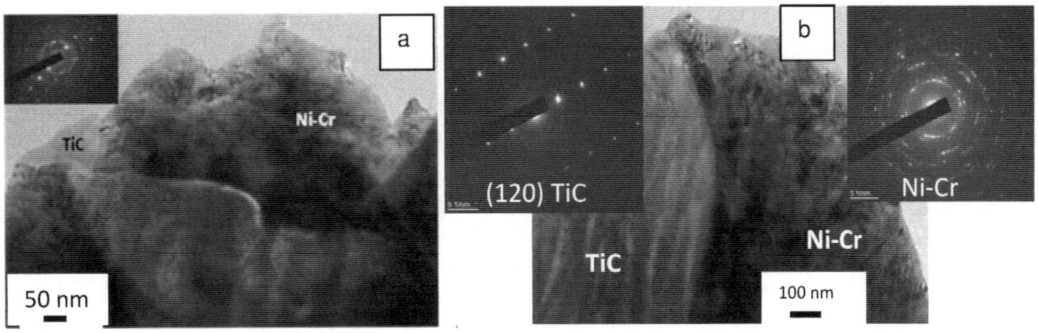

Fig. 5. The microstructure of the surface layer of TiC-(Ni-Cr) metal-ceramic alloy at depths 50-60 μm (a) and ~100 μm (b) after irradiation of the sample with electrons in Xe-containing gas-discharge plasma

The described formation pattern of the additional phase-structural-state levels of the surface layer of TiC-(Ni-Cr) hard alloys at pulsed electron-beam irradiation of samples in plasma-forming gases with different values of ionization energy has a direct relationship with the control of strength-defining properties of the surface layer of hard alloys used as tool materials. A kinematic diagram of a machine for performing abrasion tests according to ASTMG65, and also the rate of the abrasive wear of the surface layer of TiC-(Ni-Cr) hard-alloy samples prior to and after pulsed electron-beam irradiation of the samples in plasma-forming gases with different values of ionization energy versus the number of abrasive-disk rotations, are shown in Fig. 6. The planar sample under testing was pressed, with a 36-N force, to a 218-mm diameter rubbered disk rotating at a rotary speed of 200 rpm. To the friction surface, 13A-grade fused-alumina abrasive powder was supplied. The graininess of the powder was 20P according to GOST (Federal Standard) 28818-90 (particle size 200-250 μm). The abrasive consumption was 270-280 g/min.

At the initial time of the abrasion tests, the hard-alloy samples irradiated with the electron beam demonstrate a mechanical strength several times greater than the mechanical state of the initial (non-irradiated) alloy, this strength being the greater the larger is the number of the phase-structural-state levels available in the surface layer of the samples. With increasing the number of abrasive-disk rotations, the surface layer with increased mechanical strength becomes thinner, and the phase structural state of the material on the sample surface undergoes a variation with decreasing the number of phase-structural-state levels on the surface. As a result, the mechanical strength of the surface layer in irradiated samples approaches the mechanical strength of the hard alloy in the initial state of the material.

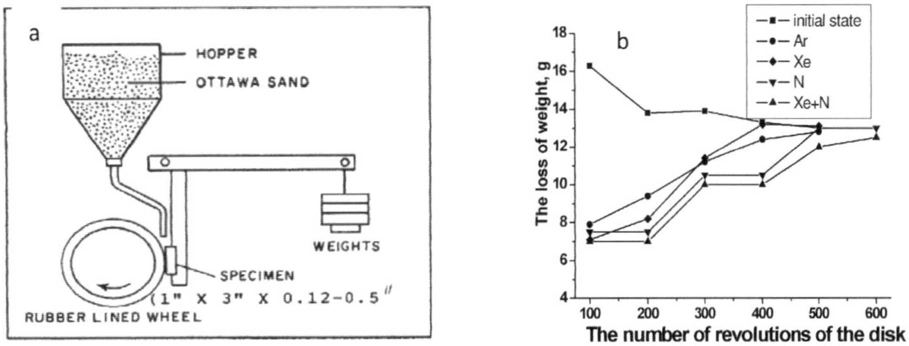

Fig. 6. A kinematic diagram of a machine used to perform abrasion tests according to ASTMG65 (a), and the rate of abrasion wear of TiC-(Ni-Cr) hard-alloy samples before and after pulsed electron-beam irradiation of the samples in plasma-forming gases with different values of ionization energy versus the number of abrasive-disk revolutions (b).

The above-said is in full measure confirmed by results of a study of the resistance of the surface layer of hard alloys during cutting of the samples with a diamond counter-body, and also by results of a study of the durability of the surface layer in cutting the samples with a metal hard-alloy plate on a turning machine. The temperature dependences of the resistance of the samples to cutting them with the diamond counter-body (as estimated from the depth of grooves cut on the sample surface) measured before and after pulsed electron-beam irradiation of the samples in Ar-, N-, and (Xe+N)-containing gas-discharge plasmas are shown in Fig. 7 (a). The same figure shows the durability of hard-alloy samples in metal cutting versus the energy flux density of the electron beam used for pulsed irradiation of the samples in gas-discharge plasmas based on Ar, N, Xe, and (Xe+N) gas mixtures (Fig. 7, b).

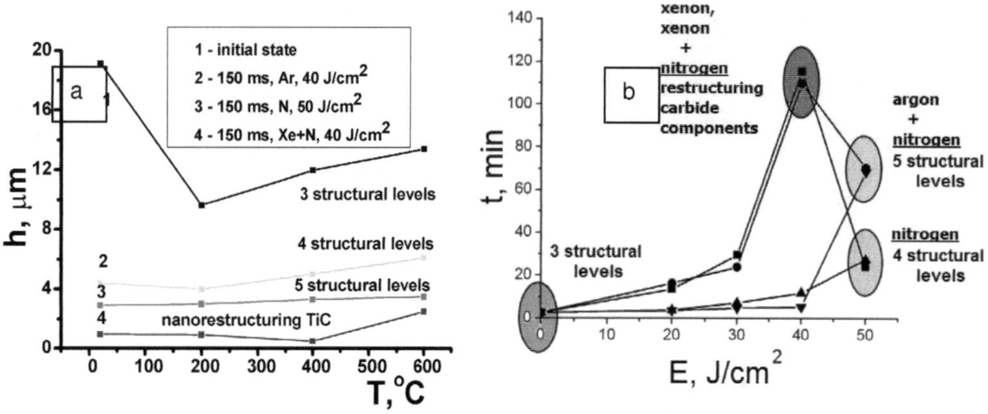

Fig. 7. The temperature dependence of the depth of grooves cut on the surface of samples with the diamond counter-body before and after pulsed electron-beam irradiation of the samples in Ar-, N-, or (Xe+N)-containing gas-discharge plasmas (a) and the durability of hard-alloy plates in steel cutting versus the energy flux density in electron beam during pulsed irradiation of the plates in Ar, N, Xe, and (Xe+N)-based gas-discharge plasmas (b)

The shown temperature dependences of the depth of grooves cut with the diamond counter-body on the sample surface sequentially reflect the influence of additional structural levels formed in the surface layer of hard alloy on the mechanical resistance of the layer during its interaction with the counter-body: with increasing the total number of structural levels the resistance of the hard alloy increases throughout the whole examined range of temperatures, and it reaches a maximal value on restructuring of the hard-alloy carbide component. Here, the mechanical strength of the hard alloy becomes comparable with the mechanical strength of the diamond counter-body. On increasing the total number of structural levels in the surface layer of the material from three levels in the initial state to four levels in the samples irradiated with electrons in Ar-containing gas-discharge plasma, the metal cutting power of the hard alloy increases by a factor of six. An increase of the total number of structural levels in the surface layer of the material from four to five levels in the samples irradiated with electrons in N-containing gas-discharge plasma increases the metal cutting power of the alloy material by a factor of 12. The nanorestructuring of the carbide component of metal-ceramic composition in the surface layer of the metal-ceramic alloy proceeding during the electron-beam irradiation of the alloy in Xe- or (Xe+N)-containing gas-discharge plasma increases the metal cutting power of TiC-(Ni-Cr) hard alloy by more than 20 times.

The data obtained in the present study have a direct bearing on the modification of the phase-structural state of the surface of hard alloys with a high content of ceramic component widely used in mechanical engineering. With the example of T40 hard alloy with composition (75%WC+14%(Ti,Ta,Nb)C)-11%Co it was established that, right on the irradiated surface, their forms a thin (80…140 nm) film of ternary carbide (Ti, W)Cx (Fig. 8, a) that passes into a cellular metal-ceramic structure (Fig. 8, b). The sizes of the cells formed by (Ti, W)Cx carbide particles with cobalt interlayers containing CoxWyC carbide inclusions and/or CoxWy compounds vary in the range 80…200 nm, and those sizes show an increase as we move deeper into the material from the side of its irradiated surface. The cells are separated from each other with cobalt interlayers of thickness 30…50 nm.

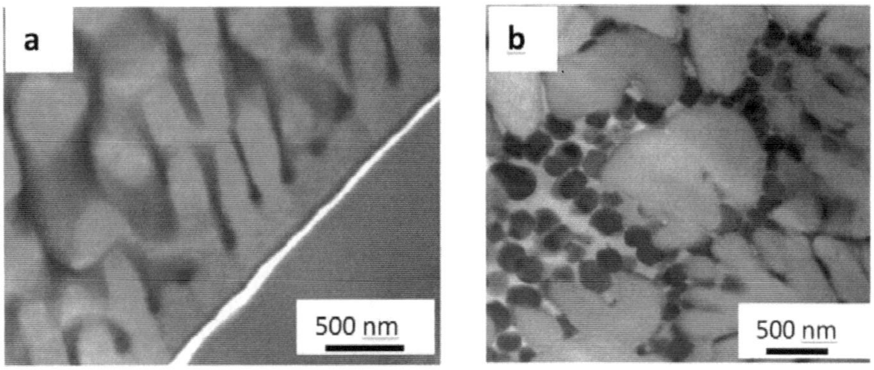

Fig. 8. The microstructure of the surface layer of T40 hard alloy processed with the pulsed electron beam in Ar-containing gas-discharge plasma

The durability of T40 hard alloy in metal cutting depends on the employed regimes of pulsed electron-beam irradiation and on the composition of the gas-discharge plasma (Fig. 9). A maximal improvement of the mechanical strength of T40 metal-cutting plates was achieved in plates irradiated with electrons in Xe-containing gas-discharge plasma.

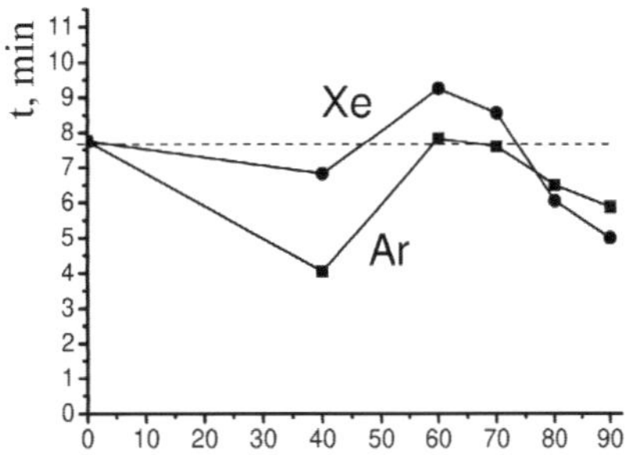

Fig. 9. The durability of T40 hard-alloy plates in steel cutting versus the energy flux density in the electron beam used for irradiation of the samples in Ar- and Xe-containing gas-discharge plasmas.

Conclusions

1. Formation of a multi-level structure in the surface layer of hard alloys offers a well-substantiated concept for increasing the mechanical strength of hard alloys as tool materials.
2. Pulsed (submillisecond time range) electron-beam irradiation offer a promising method for forming a multi-level phase-structural state in the surface layer of hard alloys in nano-size region.
3. The efficiency of pulsed electron-beam irradiation as a method for forming a multi-level phase-structural state in the surface layer of hard alloys depends on the choice of plasma-forming gases. With decreasing the gas ionization energy, the efficiency of the thermal action due to electron beam,

IOP Conf. Series: Materials Science and Engineering **125** (2016) 012008 doi:10.1088/1757-899X/125/1/012008

defining both the rate of phase transformations in the surface layer of hard alloys and the rate of formation of the multi-level phase-structural state in the material, grows in value.

This work performed of the Programme basic scientific research of the state academies of sciences of Russia for 2013-2020.

References

[1] Suryanarayana C, Al-Aqeeli N. Mechanically alloyed nanocomposites // Progress in Materials Science-2013.-V.58-No.4.-P.383-502.
[2] Chawla N, Chawla K K Metal matrix composites. –New York: Springer, 2006. 401p.
[3] Z Zak Fang, Xu Wang, Kyu Sup Hwang, H Y Sohn. Synthesis, sintering and mechanical properties of nanocrystalline cemented tungsten carbide – A review // International Journal of Refractory Metals and Hard Materials.-2009.- V.27, iss. 2.-P. 288-299.
[4] V E Panin (editor), Physical Mesomechanics and Computer-Aided Design of Materials / Nauka, Novosibirsk.- 1995.-Vol.1. – 298 p.
[5] Psakhie S G, Ovcharenko V E, Knyazeva A G, Shil'ko E V Formation of multi-scale structure in surface layers and the strength of metal-ceramic alloys under mechanical loads // Fizicheskaya Mezomekhanika.-2011.-Vol.14.-Iss.6.-P.23-34.

IOP Conf. Series: Materials Science and Engineering **125** (2016) 012009 doi:10.1088/1757-899X/125/1/012009

High-Speed Direct Laser Deposition: Technology, Equipment and Materials

G A Turichin, V V Somonov[1], K D Babkin, E V Zemlyakov, O G Klimova

Peter the Great St.Petersburg Polytechnic University
195251, St. Petersburg, Politechnicheskya st, 29, ph. number (812)5529843

[1]E-mail: vlad@ltc.ru

Abstract. The article presents the principles of direct metal deposition process, types of technological equipment and materials for implementation. The samples of deposited products have been demonstrated.

1. Introduction

Implementation of high production additive technologies, including high-speed laser direct deposition in the production cycle can significantly reduce the cost of manufacturing high-tech products. Reducing the cost by saving material and reducing the time of manufacture is one of its main advantages over machining on CNC machines or technology of casting with further machining. This is due to the fact that the product is created in one step and the original data for it is a computer model of the product. The need to design technological processes, special machining attachments, special tool and the manual labor eliminates and the duration of the time-consuming process of three-dimensional milling of the working profile of die tooling for CNC machines decreases. This method has a high flexibility, allows to consider multivariate technological and design solutions. It allows to combine it with other technologies, including additive, for example, with selective laser sintering (melting). Another important feature of such technologies is the proximity of the producing shape of a product to defined shape. This significantly reduces consumption of materials and production waste. Because of this, the process of high-speed direct laser deposition in comparison with the traditional production technologies has significant potential in terms of reducing costs, saving energy and reducing harmful emissions into the atmosphere. The process is used in various industries [1,2], mainly due to the high performance (0,5-2 kg/h), also research in the field of high-performance laser powder cladding [3-5] allow to forecast the further increase of productivity of the deposition process up to 15 kg/h.

The results of the research high-speed direct laser deposition of metallic products using radiation of a fiber laser demonstrated in this work.

2. Technology of high-speed direct laser deposition

Technology of high-speed direct laser deposition is method of forming metal product due to the partial melting by laser beam of the substrate and the powder transported in a compressed gas-powder jet towards the surface of substrate, coaxially or at an angle to the laser beam [6]. The substrate when melted forms a liquid molten pool of the minimum depth. The metal powder is transported to the substrate and partially melted to form the bead. The solid product is formed by layering on each other. Its geometry is determined by the technological trajectory of the moving head relative to the product.

IOP Conf. Series: Materials Science and Engineering **125** (2016) 012009 doi:10.1088/1757-899X/125/1/012009

There is an opportunity during the process to change the composition of the mixture of feed powders. It provides for the formation of products with graded properties.

During performance technologic experiments in laser deposition varied power of laser radiation, deposition rate of layer, depend up rotation rate of substrate and rate of stepwise rise focusing head for generation next layer, flow rate of powder, spot diameter of laser beam, and angle of incident of powder. As a result of research, the parameters of process have identified. They allow to achieve a maximum material ratio for powder and provide the heterophase mechanism of creation of layer with partial melting of metal particles resulting in an increase in the mechanical properties of the product compared with the casting techniques and selective laser sintering or selective laser melting.

3. Experiment equipment

The experimental studies of the processes high speed direct laser deposition have been carried out at the Institute of Laser and Welding Technologies SPbPU(ILWT) using laboratory bench based on fiber laser LS-5 with power output 5kW (see figure 1). The laboratory bench had five axes to move in the process of deposition of the product. The gas-powder mixture was formed by disk powder feeder Sultzer Metco Twin 10-c. For creation of the samples were applied to the shielding chamber and rotator. The gas protection of the weld metal was carried out by argon, supplied in the shielding chamber.

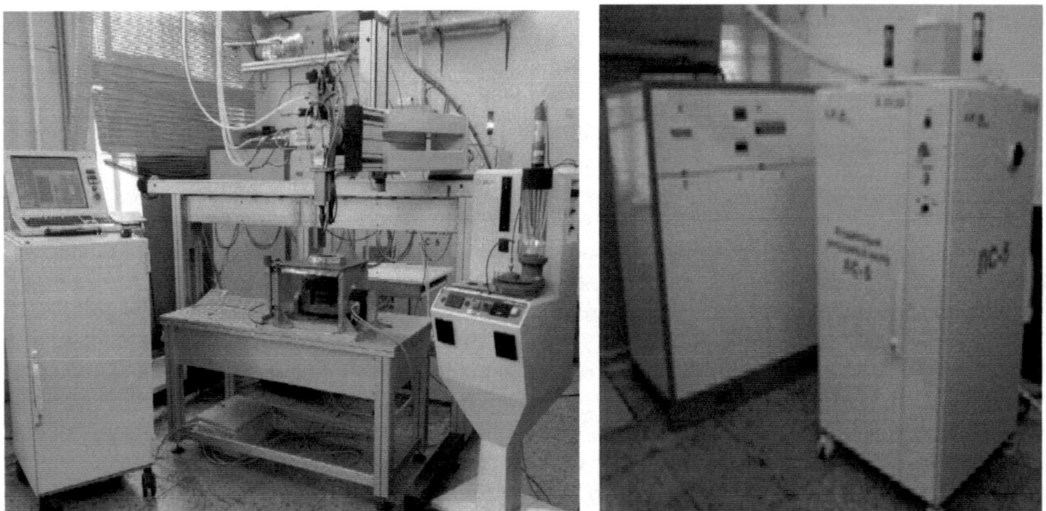

Figure 1. Laboratory bench for studies of the process high-speed direct laser deposition.

The main working part of the equipment for direct laser deposition is a working tool. It is a complex device consisting of a laser head including a focusing system of the laser beam and the cooling system of head, the system of feed of the powder material (nozzle) and the elements of the monitoring and control system of the process of deposition (sensors, video cameras, etc.) [7,8].

The most important part of working tool is the nozzle for feeding of the powder material. It creates the required distribution of powder in the gas powder jet. Research of formation of gas powder jet and gas-dynamic processes of transfer of powder have been carried. Different designs of nozzles were developed and tested (see figure 2).

a) b)

Figure 2. Nozzles for feeding the powder to the zone of laser action: a) Noncoaxial nozzle of circular cross-section with the diameter of outlet port: 1 mm, 1.2 mm, 2 mm; b) Coaxial annular nozzle with the diameter of the ring 14 mm, the width of the slot is 0.4 mm.

The use of noncoaxial nozzles (figure 2a) leads to the dependence of the process of deposition from the direction of motion, because the nozzle has an asymmetry relative to the direction of movement of the working tool. Therefore, the process should be implemented in a single direction relative to the product. The gas-powder jet is simple in structure and symmetric relative to the canal centerline. The using of noncoaxial nozzles efficiently for deposition of products of complex geometry with axial symmetry or bodies of rotation with a minimum diameter of 6mm. The thickness of wall of the deposition product is in the range of from 0.6 to 3 mm with a roughness of surface is not more than 50 μm.

When the flow of powder is transported coaxially to the laser beam, all directions of movement of the substrate are in a plane perpendicular to the direction of powder flow, so a nozzle with a coaxial feed (figure 2b) are independent of the direction of motion. In this case, possible a creation of products with more complex geometry. The geometry of the gas powder jet depends not only on the design of the nozzle, but also on its geometry. The main influence on the performance and stability of the process using a coaxial nozzle are: the angle of convergence, the width of the constriction of jet, the uniformity of distribution of the powder relative to the axis of the laser beam. As a result of the research is the main factor determining the width of the gas powder jet at the outlet of nozzle, is the width of slot of the nozzle. The experimental data show that the jet has a normal distribution in the cross section at a distance of more than double the width of the nozzle (figure 3).

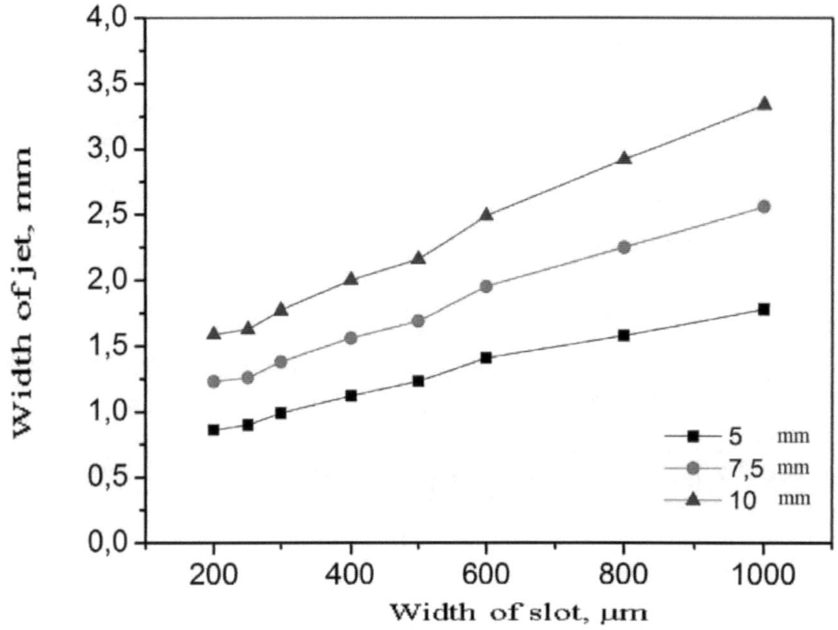

Figure 3. Dependence of the width of jet to the width of slot of nozzle, the size of particles of powder is 50-150 μm.

To implement the process possible to use the following series of technological equipment:

Setups that use linear drive (figure 4a) or robot (figure 4b) to move a work tool and a rotator located in the shielding chamber to control of movement of the substrate;

The equipment is made on the basis of CNC systems for machining, with the replacement of the working tool on a special for direct laser deposition and with the possibility of filling the working chamber with inert gas (figure 4c);

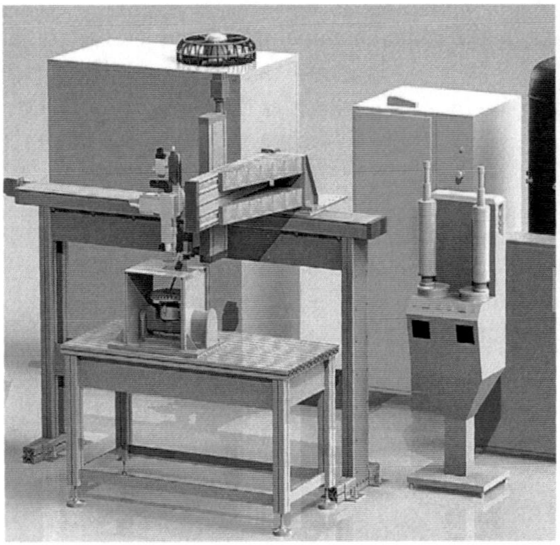

a)

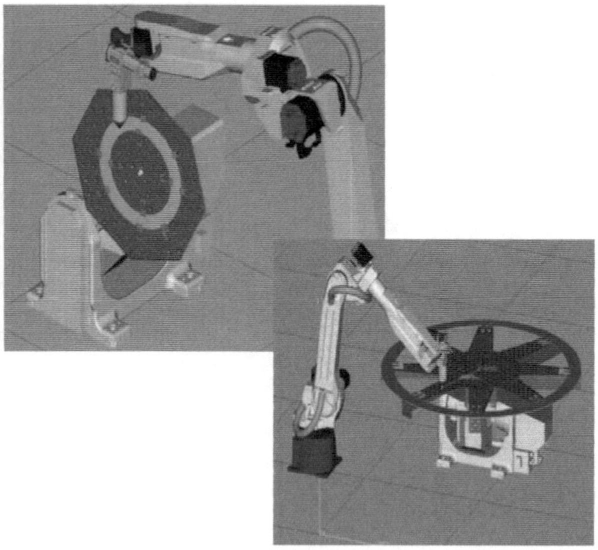

b)

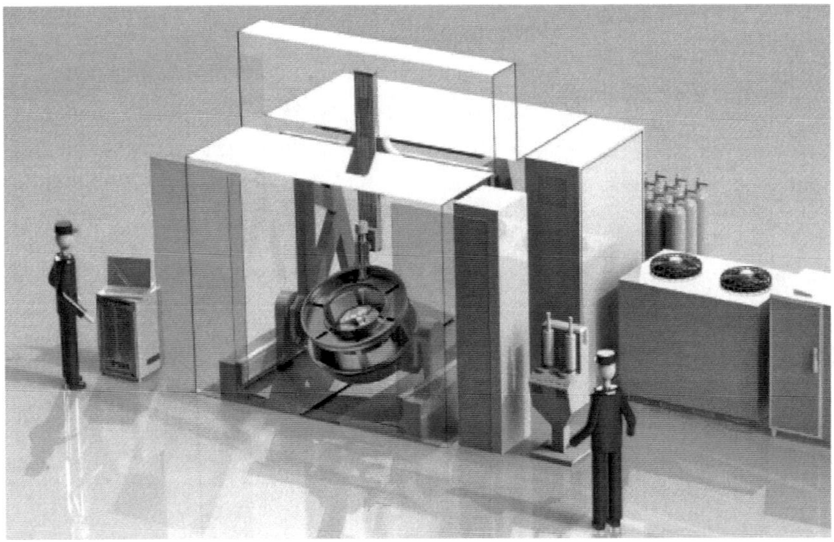

c)

Figure 4. Images of using machines for the process of high speed direct laser deposition of products.

The first two types are created by employees of Institute of laser and welding technologies SPbSPU for the research process high-speed direct laser deposition of products.

4. Materials

Experimental research have been carried out using the metal powder of the base alloys: nickel, including heat resistant (Inconel 625), cobalt, including high strength (Stellite 6), chromium, iron, copper and titanium. The future plans are to include studies of process with powders based on aluminium alloys. The powders had a spherical shape of the particles. This is due, firstly, to the fact that such particles more compactly fit into a certain volume and secondly, to provide a "fluidity" of powder compositions in feed systems with minimal resistance. The original powder have been separated using a vibrating screen 30 GY in divided fractions by sifting of initial powder through a set of sieves with respective cell dimensions. The distribution of powder particles size have been controlled by using a laser analyzer of microparticles LASKA-1K. The choice of this fraction of the powder caused by the economic expediency of its application for the deposition of large products used in industries such as gas turbine engine building and aviation. The using of powder with smaller size of particle allows to obtain products with a smaller roughness of surface, but it increases the cost of the powders and causes difficulty in the formation of stable gas-powder jet of the weight of the flying particle. The material ratio can vary from 20 to 90% depending on the combination of parameters of cultivation. Examples of samples of deposition of products are presented in figure 5.

IOP Conf. Series: Materials Science and Engineering **125** (2016) 012009 doi:10.1088/1757-899X/125/1/012009

a) b)

c) d) e) f)

Figure 5. Samples of the deposition products:
a) Fe alloy: EuTroloy 16316.04; b) Co alloy: EuTroloy 16006.04;
c) Ni alloy : EuTroloy 16496.04; d) Ni alloy (experimental alloy of ODC);
e) Ni alloy: EuTroloy 16625.04; f) Cu alloy: BronzeTec 19868 (80Cu-15Al-5Ni).

Conclusion

The research found out, the technology of high-speed direct laser deposition of products able to replace the current technology of giving multiple increase productivity and save material. The product obtained by this method, do not need next isostatic pressing or heat treatment, in comparison with the technology of selective laser sintering (melting) or casting techniques with next thermal and mechanical treatment. However, this process may be combined with other technologies, further reducing production costs, reducing the time for fabrication, creating products that cannot be obtained by using only this method. During the research the opportunity to create the nozzles of different configurations, to form edges on rotary body and on products with complex profile, to create the objects of variable cross section and with a controlled thickness of wall in the range of from 0.6 to 3 mm and a roughness of surface of not more than 50 μm were confirmed. In the case where this is insufficient to reduce of the roughness to about 10 times the possible use of laser radiation for surface melting of the walls of the deposition products. The dimensions of the resulting product are limited by the size of the shielding chamber.

IOP Conf. Series: Materials Science and Engineering **125** (2016) 012009 doi:10.1088/1757-899X/125/1/012009

5. References

[1] Additive manufacturing: opportunities and constraints. A summary of a roundtable forum held on 23 May 2013 hosted by the Royal Academy of Engineering, Royal Academy of Engineering, November 2013, 21p;

[2] Scott M Thompson, Linkan Bian, Nima Shamsaei, Aref Yadollahi "An overview of Direct Laser Deposition for additive manufacturing; Part I: Transport phenomena, modeling and diagnostics" Additive Manufacturing, Issue 8, 2015, 36-62 pp;

[3] G A Turichin, V.V. Somonov, O.G. Klimova "Investigation and modeling of the process of formation of the pad weld and its microstructure during laser cladding by radiation of high power fiber laser" Applied Mechanics and Materials;

[4] DengZhi Wang, QianWu Hu, YinLan Zheng, Yong Xie, XiaoYan Zeng "Study on deposition rate and laser energy efficiency of Laser-Induction Hybrid Cladding" Optics & Laser Technology, Issue 77, 2015, 16-22 pp;

[5] J C Pereira, J C Zambrano, M J Tobar, A Yañez, V Amigó "High temperature oxidation behavior of laser cladding MCrAlY coatings on austenitic stainless steel" Surface & Coatings Technology 270 (2015) 243–248 pp;

[6] G Turichin, O Klimova, E Zemlyakov, K Babkin, V Somonov, F Shamray "Technological bases of high-speed laser direct growth of products by heterophase powder metallurgy method" Photonika, Issue4, 2015,68-83 pp;

[7] Boisselier, D., Sankaré, S. "Optimization of the laser direct metal deposition process in 5-axis configuration", 32nd International Congress on Applications of Lasers and Electro-Optics, ICALEO 2013, 2013, 326-333 pp;

[8] Didier Boisselier, Simon Sankaré, Thierry Engel "Improvement of the laser direct metal deposition process in 5-axis configuration" Physics Procedia, Volume 56, Issue C, 2014, 239-249 pp.

Physical Nature of the Processes in Structure Forming, Phase and Chemical Composition of pipe Permanent Joints when MMA Welding

D P Il'yaschenko[1,a], D A Chinakhov[1], V I Danilov[1,2], I D Sadykov[1]

[1] Yurga Institute of Technology, TPU affiliate
Leningradskaya 26, Yurga, Russian Federation, 652055
[2] Institute of Strength Physics and Materials Science, Siberian Branch of Russian Academy of Sciences
Akademichesky, 2/4, Tomsk, Russian Federation, 634055

e-mal: [a]mita8@rambler.ru

Abstract. The paper outlines peculiarities of structure formation, phase and chemical composition in regard to heat content in molten electrode metal beads when pipe steel (steel 09G2S) welding using power sources with various energy characteristics. Mathematical calculations indicate an inverter power source provides minor heat content into the bead of electrode metal when welding. Experimental research has pointed at 4 -9 % increase in impact strength of joints produced using an inverter power source in comparison with samples produced applying a diode rectifier. The following factors can possibly give rise to the increasing impact strength: difference in microstructures of weld joints, up to 50% shortening ferritic plates in metal of weld joint, change in dimensions of ferritic grains in the heat-affected zone by as much as 17.5 %, and decrease in the extent of heat-affected zone by 50%.

1. Introduction

The service properties of welded structures, processing strengths of weld metal and heat-affected zone are conditioned by thermal-deformation processes, phase and structural transformations when welding. As the materials used for main pipeline welding are regulated their composition can hardly be changed. However, improvement of service properties is possible provided that various types of allowed welding equipment are used.

In conditions of constantly changing spatial position of a weld pool sufficient dynamic behavior of a current source is required for proper welding of non rotatable joints of pipelines, as well as it necessitates quickly rising voltage from short circuit to arcing, the reasonable rate of current change in order to reduce spattering while beads of melted electrode are transferred into the weld pool; other requirements are to be met too [6].

The authors of this paper deal with studying the influence of heat content into the bead of molten electrode metal on the microstructure and strength properties of a weld joint, which involves application of power sources for arc welding with various energy characteristics.

2. Materials, research procedures and results

For comprehensive analysis we produced pipe welded joint samples Ø159 ×6 of steel 09G2S with electrodes: root - LB-52U (d=2.6 mm), welding current I = 50-60 A, filling - LB-52U (d=3.2 mm), welding current I = 80 – 90 A; power sources are different (diode rectifier VD-306 and inverter Nebula-315).

The set of methods [1, 2] adapted to manual arc welding by means of computer software MatLab was used to calculate the heat content of electrode metal bead according to energy characteristics of welding. Calculation procedure is outlined in Table 1.

Table 1. Procedure to specify the heat content of electrode metal bead when welding with various types of power sources [3]

Procedure to specify the heat content of a bead in software MatLab	Heat content when welding with Nebula 315, Q_{hc}, J,	Heat content when welding with diode rectifier VD 306, Q_{hc}, J,

u1=[u1; u2; u3; u4; u5; u6;.....];
i1=[i1; i2; i3; i4;.....];
[U1,I1]=meshgrid(u1,i1);
t1=t1; t2; t3; t4;.......;
Q_{hc}=sum(trapz(0,2...0,3*U1.*I1.*t1)),
where $u_1, u_2 \ldots u_n$; $i_1, i_2, \ldots i_n$ – instantaneous values of voltage, V; those of current, A per unit of time $t_1, t_2, \ldots t_n$.
	113.25 10^5	125.03 10^5

The difference in heat content is the reason for varying dimensions and mass of transferred beads, which can be estimated by the following formulae:

Mass of transferred beads in conditions of short circuits [4] can be calculated according to equation:

$$m = a \cdot \tau_{k.z.}^{3} \qquad (1)$$

where $\tau_{\text{к.з.}}$ – time a bead is on the electrode tip, s.
a – coefficient $0.3*10^{-4}$ g/s^3.

The radius of transferred electrode metal beads [5]:

$$R = \sqrt[3]{\frac{\frac{-1}{27}\cdot\left(\frac{-\pi\cdot\gamma\cdot\cdot r_0^2}{4a\cdot\tau_0^3\cdot 10^3}\right)^3 + \left(\frac{-3a\cdot\tau_{k.z.}^3\cdot 10^3}{8\cdot\pi\cdot\gamma} + \frac{\pi\cdot\gamma\cdot r_0^2}{24\cdot a\cdot\tau_{k.z.}^3\cdot 10^3}\right) -}{-\sqrt{\frac{2}{27}\left(\frac{-\pi\cdot\gamma\cdot r_0^3}{4a\cdot\tau_{k.z.}^3\cdot 10^3}\right)^3 + \left(\frac{-3a\cdot\tau_{k.z.}^3\cdot 10^3}{8\cdot\pi\cdot\gamma} + \frac{\pi\cdot\gamma\cdot r_0^2}{24a\cdot\tau_{\kappa.z.}^3\cdot 10^3}\right)\cdot\frac{1}{4} - \frac{1}{729}\cdot\left(\frac{-\pi\cdot\gamma\cdot r_0^2}{4a\cdot\tau_{k.z.}^3\cdot 10^3}\right)^6}}} + \frac{\pi\cdot\gamma\cdot r_0^2}{12a\cdot\tau_{k.z.}^3\cdot 10^3}. \qquad (2)$$

where $\tau_{k.z.}$ – time a bead is on the electrode tip, s;
a – coefficient $0.33\cdot 10^{-4}$ g/s^3;
π – mathematic constant, circumference to diameter ratio 3.14;
γ – density of liquid metal, g/mm^3;
R –radius of bead surface curvature, mm;
r_o – core wire radius, mm.

Table 2. Calculated data on transferred bead mass and specific surface area when using power sources with different dynamic characteristics [5]

Power source	Mean values	$\tau_{k.z.}$, 10^{-3}, s	Bead mass m, g	Bead radius R, mm
Inverter	I=100A; U=22V	10…14	$\frac{0.33-0.9}{0.615}$	$\frac{2.16-3.02}{2.59}$
Diode rectifier	I=100A; U=22V	8…12	$\frac{0.168-0.57}{0.369}$	$\frac{1.72-2.89}{2.305}$

As far as it's known [6, 7], downsizing electrode beads cause increase in their total contact surface with the ambient medium. Therefore, liquid metal of beads completely react with this medium (deoxidation, alloying, oxidation, gas dissolution), but the rate of metallurgical reaction is deteriorated provided that the lifetime of beads is shortened. In papers by N.M. Novozhilov [8] the data is presented: specific surface of electrode beads is approximaly 5 – 22 times greater than it of the weld pool, while specific oxidation velocity of electrode bead metal is about 39 times greater than that of weld pool metal. A.A. Erokhin has stated in his paper [9], chemical reactions of welding stop completely at the time of bead formation, beads intensively react with gas and dross. Therefore, chemical composition of weld joint metal can be controlled via affecting chemical composition of electrode metal beads.

Various heat content of molten electrode metal bead, as well as mass and radius of beads transferred from the electrode into the weld pool are the reasons for different heat input into the product, as the consequence, for change in the microstructure, chemical composition and mechanical properties of a weld joint.

Chemical composition and mechanical properties of a product to be welded were experimentally tested in conditions of various energy impact of power sources when manual arc welding. The analysis of the data presented in Table 3 and 4 reveals the increase in mass concentration of alloying elements: (Mn – 0.02 to 0.28%, Si – 0.24 to 0.48%, Cr – 0.04 to 0.5%) in deposited metal and reduction of oxide concentration (SiO_2, MnO) in a dross phase if inverter is used instead of diode rectifier. Various heat content in electrode metal beads can be a probable reason of it [6] provided that different power sources are used, as one can see in Table 4. This fact is confirmed by the data in paper [10].

Table 3. Chemical composition of weld joint metal, produced from pipe Ø159×6 (steel 09G2S) with electrodes LB 52U [11]

Power source	Element concentration, mass %							
	C	Si	Mn	S	P	Cr	Ni	Cu
Diode rectifier	0.10± 0.012	0.52± 0.03	1.03± 0.05	0.010	0.014	0.03±0.01	0.05 ±0.01	0.03
Inverter	).09±0.00 5	0.60±0.03	1.23±0.05	0.010	0.014	0.03±0.01	0.06 ±0.01	0.03

Table 4. Typical chemical composition of dross [11]

Electrodes	Power source	CaO, %	SiO_2, %	TiO_2, %	NbO, %	MnO, %	Fe_2O_3, %	Cr_2O_3, %	Al_2O_3, %
LB-52U	Diode rectifier	38.66	25.37	9.57	0.10	7.21	18.31	0.17	3.61
	Inverter	36.27	24.187	8.74	0.05	7.48	13.89	0.15	3.66

The results presented above are the consequences of heterogeneous reactions of equilibrium, which dross MnO, SiO_2 and base metal enter into, they depend on heat content of beads and their temperature (they are different for various sources of power Table 3 and 4):

Interacting dross and metal components in the course of oxidation-reduction reaction on the interface can be both viewed as a heterogeneous reaction or electrochemical process [12]:

Dross:	*(MnO)*	*(FeO)*	
	$\updownarrow$	$\updownarrow$	(3)
Metal:	$[MnO] + Fe \leftrightarrow [Mn] + [FeO]$		

Dross:	*(SiO₂)*	*2(FeO)*	
	$\updownarrow$	$\updownarrow$	(4)
Metal:	$[SiO_2] + Fe \leftrightarrow [Si] + [FeO]$		

$$\Delta G^0 = \Delta H - T\Delta S \quad (5)$$

where ΔG^0- Gibbs energy, kJ/ mole;

ΔH – enthalpy change, kJ/ mole;

ΔS – entropy change, kJ/ mole;

T – temperature of electrode metal bead heating, ^{0}K. As it is specified in paper [5] the data are as follows: for inverter T= 2213^0 K, for diode rectifier T=2283^0 K.

$$\Delta H = \Delta H_{FeO} - \Delta H_{MnO} \quad (6)$$

$$\Delta S = \Delta S_{FeO} - \Delta S_{MnO} \quad (7)$$

We specify Gibbs energy (5) given in Table 5 making use of reference data [13, 14] ΔH_{FeO}= -265 kJ/ mole; ΔH_{MnO}= -385.1 kJ/ mole; ΔS_{FeO}= 60.8 kJ/ mole K^{-1}; ΔS_{MnO}= 61.5 kJ/ mole; ΔH_{SiO2}= -908 kJ/ mole; ΔS_{SiO2}= 42.7 kJ/ mole K^{-1} in formulae 6, 7.

Table 5. Calculated Gibbs energy of chemical reactions in metallurgical processes when welding

	ΔG^0 for reaction 6, kJ/ mole	ΔG^0 for reaction 7, kJ/ mole
Inverter	121.649	180.129
Diode rectifier	121.708	219.415

In the area affected by high temperatures there is equilibrium direction shift of processes (3,4) towards oxidation of Si, Mn, moreover, the process velocity increases as the temperature goes up, the data given in Tables 3 – 4 confirm this fact.

The change in chemical composition of deposited metal is to cause changing structure and phase composition, as the consequence, operating properties of deposited metal.

Table 6. Mechanical properties of weld joints produced of pipe Ø159×6 (steel 09G2S) with electrodes LB 52U [11]

Power source	Limit of temporary tensile strength σ_B, MPa	Angle of root, inward, and edgewise bend, grad.	Impact strength KCU, J/cm^2 (cut along the center of a weld)			
			+20°C	0°C	-20°C	-40°C
Diode rectifier	$\frac{541-543}{542}$	$\frac{120-120}{120}$	$\frac{201-220}{210}$	$\frac{212-223}{216}$	$\frac{200-233}{219}$	$\frac{143-230}{182}$
Inverter	$\frac{550-560}{555}$	$\frac{120-120}{120}$	$\frac{208-226}{217}$	$\frac{215-254}{235}$	$\frac{224-250}{237}$	$\frac{150-258}{193}$

The analysis of data given in Table 6 reveals 4 -9 % growth of impact strength of joints produced using an inverter power source in comparison with the samples produced by diode rectifier at various temperatures. The following factors can possibly give rise to the increasing impact strength: difference in microstructures of weld joints (Figure 1), up to 50% shortening ferritic plates in metal of weld joint, change in the dimensions of ferritic grains in the heat-affected zone by as much as 17.5 %, and decrease in the extent of heat-affected zone by 50%.

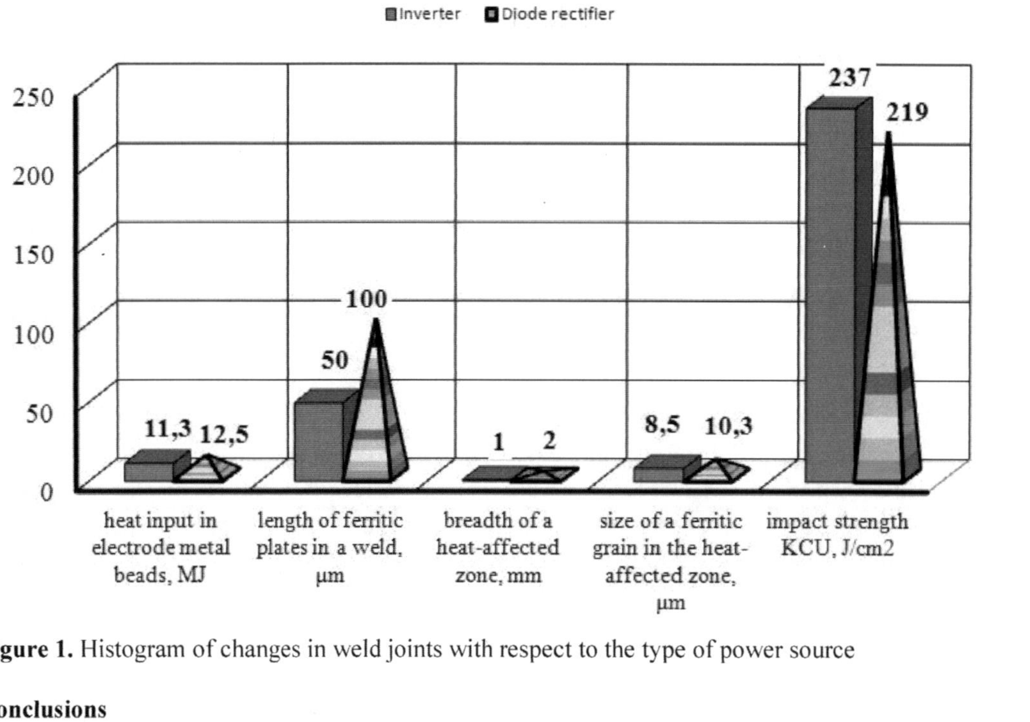

Figure 1. Histogram of changes in weld joints with respect to the type of power source

Conclusions

The experiments have revealed an inverter power source provides minor heat content into the bead of electrode metal when MMA welding as compared with the diode rectifier. The mass concentration of alloying elements goes up: Mn 0.02 to 0.28%, Si 0.24 to 0.48%, Cr 0.04 to 0.5%) in deposited metal, whereas the share of oxides (SiO_2, MnO) is reduced in the doss phase when inverter using instead of a diode rectifier. Impact strength of joints produced using an inverter power source increases by 4 -9 %

in comparison with the samples produced by diode rectifier due to difference in microstructures of weld joints, up to 50% shortening ferritic plates in metal of weld joint, change in the dimensions of ferritic grains in the heat-affected zone by as much as 17.5 %, and decrease in the extent of heat-affected zone by 50%.

References

[1] Solodsky S A, Brunov O G, Il'yaschenko D P; 2012 Power sources for arc welding (Yurga Institute of Technology – Tomsk: Publishing of Tomsk Polytechnic University) p 165

[2] Brunov O G 2007 Mechanized pulsed wire feed welding in active shielding gases (Tomsk: Publishing of Tomsk Polytechnic University) p 137.

[3] Ilyashchenko D P and Chinakhov D A 2011 Materials Science Forum Intestigating the Influence of the Power Supply the Weld Joints Properties and Health Characteristics of the Manual Arc Welding 12 pp 704-705.

[4] Makarenko V D, Shatilo S P; 199 Welding production Calculation of electrode bead kinetic characteristics transferred through the arc space when coated electrode welding 12 pp.6 -10.

[5] Il'yashchenko D P, Chinakhov D A and Gotovshchik Yu M Applied Mechanics and Materials Calculation of the Heat Content of the Electrode Metal Droplet When Applying Power Supplies for Manual Arc Welding With Different Volt-Ampere Characteristic. 756 pp 101-104.

[6] Fedko V T, Chipalyuk A S; 2003 Welding production Melting and transfer of electrode metal when coated electrode welding 2 pp.3-10.

[7] Bagryansky K V,. Dobrotina Z A, Khrenov K K; 1976 Theory of welding processes. (Publishing: «Higher school»). 424 p.

[8] Novozhilov N M; (1979) Fundamentals of arc welding metallurgy in gases(M. Machinebuilding) 231 p.

[9] Erokhin A A;1973 Fundamentals of fusion welding. Physical and chemical regularities. (M. Machinebuilding) 448p.

[10] Ignatova A M 2013 Nauchno Tekhnicheskij Vestnik Povolzhja Mineral formation in particles of solid component in welding fumes at high temperatures and short pyrogenic welding processes 5 pp 166-173. (in Russian).

[11] Il'yaschenko D P, Chinakhov D A, Danilov V I, Schlyakhova G V, Gotovshchik Yu M; 2015 IOP Conf. Series: Materials Science and Engineering Physical Nature of the Processes in Forming Structures, Phase and Chemical Compositions of Medium-Carbon Steel Welds 91 012006

[12] Frolov V V 1988 Theory of welding processes (Moscow: Higher school) 559 p.

[13] Garkushin I K, Istomova M A; 2012 Reference book on chemical analysis (Samara: Samara State Technical University) 237 p.

[14] Kryukov R E, Kozyrev N A, Galevsky G V, Bendre Y V, Goryushkin V F, Valuev D V; 2015 Materials Science and Engineering. J. IOP Conference Series Some aspects of oxidation-reduction reactions under carbon-bearing flux welding,: 91 p. 1-6.

IOP Conf. Series: Materials Science and Engineering **125** (2016) 012011 doi:10.1088/1757-899X/125/1/012011

On Organizing Quick Change-Over Mass Production

S I Petrushin[a], R H Gubaidulina[a], S V Gruby[b], Sh C Nosirsoda[a]

[a]Yurga Institute of Technology of National Research Tomsk Polytechnic University Affiliate, Tomsk, Russia
[b]Moscow State Technical University by N.E. Bauman, Moscow, Russia

E-mail: [a]victory_28@mail.ru, [b] gru@bmstu.ru.

Abstract. The terms "type of production" and "coefficient of assigning operations" are analyzed. A new method of calculating the optimum production plan based on profit projections is suggested. We recommend using the cycle time values as initial data for designing and developing technology. On the basis of existing techniques used to convert productions we suggest a new approach to production change-over with the service life of manufacturing facilities equal to the time to product's obsolescence. The factors to maximize profits using this change-over method are indicated, with maximum profits being a condition for the organization of quick change-change mass production.

Introduction

In accordance with the existing practices as to process designing, the concept of production planning in terms of volume requirements (single-unit, medium series and mass production) is served as a specific criterion for choosing a particular type of technology and equipment configuration to be used. According to the Russian Standards, a type of production is chosen on the basis of the so-called coefficient of assignment of operations representing a ratio of the number of different operations to be made per a month to the number of workplaces involved into performing those operations. This approach, in our opinion, has a significant week point, namely, at the start of engineering processes an engineer does not know a number of operations in the process to be designed or the number of workplaces. These data should be specified in the course of the subsequent engineering process, so there is considerable uncertainty as to defining a type of production that is essential for making future critical decisions.

Results and Discussion

Determining an optimum solution to annual production planning

When developing a technology intended for producing machinery, a product design and the so-called annual output program in units appear to be basic initial data. If the design documentation for machinery should be developed in the previous stage of product life cycle (PLC), the issue of determining the optimal size of a series of machinery or run of production items remains open. In the case of planned economy existed in Russia before, production plans were administered with respect to government-controlled needs and interests of the entire population. With the transition to the market – based economy in a competitive environment, this task becomes particularly imperative, since, on one hand, the number of made and sold products must refund the costs of production and, on the other hand, generate some profits to the owner of an enterprise without overproduction of these products. Therefore, errors in specifying numbers in production plans causing underproduction or overproduction can lead to negative consequences to the extent of bankruptcy of a company.

Analyze a production plan of a machine-engineering company on the basis of the so-called break-even charts [1,4-7], which are specified with respect to the coordinate system "production plan - value" (Figure 1). The total cost to produce a certain type of products S_i are made up of the initial capital investment K_i required to organize a new machine production and operating expenses (cost) associated with manufacturing products in accordance with the formula:

$$S_i = K_i + C_i N, c.c.u., \tag{1}$$

where C_i is the cost of a product in conditional currency units (c.c.u.); N is the current production plan in units.

Dependency (1) corresponds to the line of total costs (see Figure 1), which is a straight line when the value is constant.

On the other hand, the total revenue of the enterprise D received from the product sales is directly proportional to their numbers according to the production plan:

$$D = Z_i \cdot N \tag{2}$$

where Z_i is the sale price of a product in c.c.u.

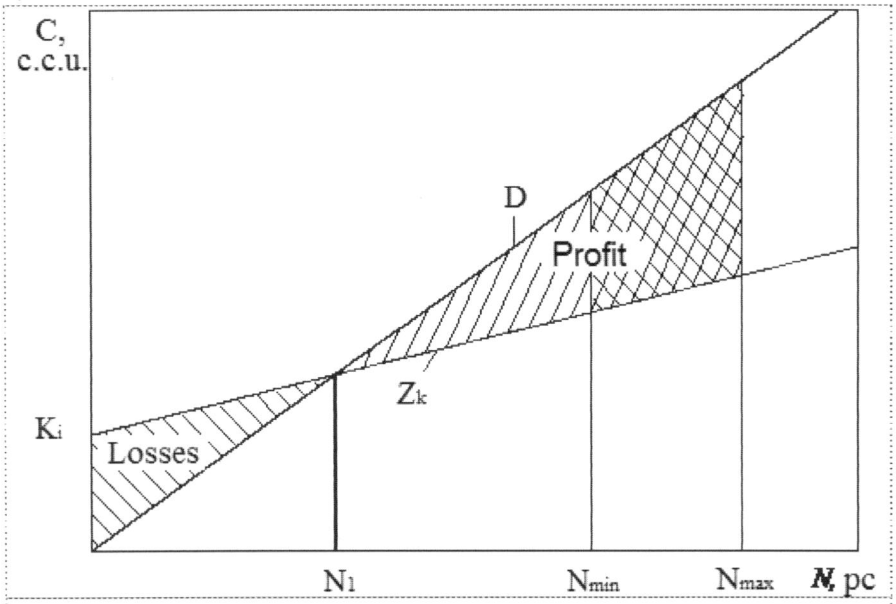

Figure 1. Break-even chart relative to the production plan.

With a certain volume of products made and sold, the line of total costs intersects the line of total revenue (see Figure 1). The point of intersection on the x-axis has the following coordinates:

$$N_1 = \frac{K_i}{Z_i - C_i}. \tag{3}$$

Below the value N_1 the company incurs losses due to the necessity of covering the total expenditure made by that time, and when $N \geq N_1$, it begins receiving the current profit. It should be noted that

IOP Conf. Series: Materials Science and Engineering **125** (2016) 012011 doi:10.1088/1757-899X/125/1/012011

the company can balance absolute total costs and absolute total revenue only when it will manufacture and sell more than the number of products N_1, which we designate as $N_{\min}$ (profitable production plan). In the case of linear dependence S_i and D upon the production plan, we have the following formula $N_{\min} = 2N_1$ or taking into account expression (3):

$$N_{\min} = \frac{2K_i}{Z_i - C_i}. \quad (4)$$

If the company under examination is able to make a net income (absolute profit) starting from $N_{\min}$, then this value is considered as a lower limit in the context of the determination of its production plan.

The analysis of expression (4) shows that the plan profitability is determined by the difference between the price of the product and its cost or what is the same, the income received from the sale of its product. The larger the difference, the less $N_{\min}$ is, and conversely, when the product's price is close to the cost of its production, the cost-effectiveness of production sharply increases.

The value $N_{\min}$ defines the quantity of products that makes possible to cover the total costs of their production after selling them on the market. The company starts getting the absolute profit when $N \geq N_{\min}$; and generating the maximum profit is a global goal in the conditions of the capitalist mode of production. On this account, there is a crucial task of determining the maximum output value $N_{\max}$ within the production plan; after reaching this value by the company the product is discontinued. The so-called "moral depreciation" of a machine should be accepted as a key to this task. Karl Marx was apparently the first to introduce this term. In his work [2], he wrote: "But in addition to the material wear and tear, a machine also undergoes, what we may call a moral depreciation. It loses exchange-value, either by machines of the same sort being produced cheaper than it, or by better machines entering into competition with it.64 In both cases, be the machine ever so young and full of life, its value is no longer determined by the labour actually materialised in it, but by the labour-time requisite to reproduce either it or the better machine. It has, therefore, lost value more or less. The shorter the period taken to reproduce its total value, the less is the danger of moral depreciation...".

It should be noted that the terms "obsolescence" or "moral depreciation" as well as all the basic concepts in political economy are not clearly defined. Therefore, in future we assume that obsolescence is a calendar period from the moment when the first machine of a new model is manufactured to the moment this model is discontinued.

If we use T_m to designate this period of product's obsolescence in conditional time units (c.t.u.), then the maximum production plan can be expressed by the following ratio:

$$N_{\max} = \frac{T_m}{t_d}, \quad (5)$$

where t_d is the interval products are made (cycle time), c.t.u.

Figure 1 shows the top line $N_{\max}$ to limit the area of absolute profit and the bottom line has a zero limit along the line $N_{\min}$ as it follows from the above. If these values are known, then in the context of linear dependences the size of raised profits P corresponds to the cross hatched area (see Figure 1). After integration of this area in view of expression (4) we obtain the following formula to calculate the absolute value of profit:

$$P = N_{\max}\left[0.5N_{\max}(Z_i\text{-}C_i)\text{-}K_i\right]. \quad (6)$$

The analysis of expression (6) proves that the company's profit will grow by increasing $N_{\max}$ and the difference between the product price and the product cost, as well as by the decreasing the initial capital invested in organizing the process to make these products. With this, the value $N_{\max}$ is the most effective means for increasing the absolute profit as it is raised to the second power in formula (6).

At the same time, as it follows from expression (5), the value $N_{\max}$ can be increased by either extending the period of product obsolescence or reducing the cycle time for the products. Currently, the market is largely saturated with products having different levels of quality, and increasing the value T_m is difficult under these conditions. On the contrary, competition leads to a reduction of "moral" service life of a particular product design. This trend has already been noted by professor G.A.Shaumyan [3].

We substitute (5) into formula (6):

$$P = \frac{T_M}{t_d}\left[0.5\frac{T_M}{t_d}(Z_i - C_i) - K_i\right]. \qquad (7)$$

From expression (7) it follows that for a specified time of product obsolescence the company's profit will be inversely related to the product time. Thus, the main focus should be laid on the full reduction of cycle time required for manufacturing a product in order to achieve the maximum possible profit within the specified limits. With the focus on the goal of gaining the maximum profit, values of $N_{\max}$ corresponding to the cycle time is accepted the estimated production plan for product output.

On the other hand, unlimited decreasing a cycle time can lead to the overproduction of a certain model of products causing severe consequences for the manufacturer. From our point of view, the lower limit of cycle time must be associated with a profit target, and relative to its amount the manufacturer maintains an advantage not only over its competitors in the same sector of market but also other manufacturers in various industries with the aim to prevent the drain of capitals there.

Suppose the company's management set a profit target Δ as ratio of profit value to sales revenue [6] that is $\Delta = P / D$. Then in view of (6) and (2) the maximum acceptable production plan will be equal to

$$N_{\max} = \frac{2(\Delta C_i + K_i)}{Z_i - C_i} .pc.$$

The optimal (also as the minimum) value of cycle time associated with this plan will be determined by the following expression using formula (5):

$$t_{\partial.opt.} = \frac{T_M(Z_i - C_i)}{2(K_i + Z_i\Delta)}, c.t.u. \qquad (8)$$

This value should be the ground for organizing production of a specified machinery product. For the conditions of running a particular machine-engineering enterprise a value of cycle time according to formula (8) should be corrected taking into account the time required for the organizational process, which are not included in the fixed labour costs.

Rationale for the organization of quick change – over mass production

The problem of "painless" changeover to start a new engineering product is a major problem for manufacturers, because costs for production facilities that are idle and a company's associated financial losses are fully dependent on how efficient this conversion process is.

Currently, there are several ways to change productions to run new machine models [4]: a full stop of production for a period of reconstruction, a parallel changeover approach, a "non-stop" approach, reengineering in conditions of a flexible production. These processes used to convert main production facilities to run a new product have one common disadvantage: the intention to organize a new production where the previous industrial base is located. Karl Marx wrote about changing the means of production [2]: " The instruments of labour are largely modified all the time by the progress of industry. Hence they are not replaced in their original, but in their modified form. On the one hand the mass of the fixed capital invested in a certain bodily form and endowed in that form with a certain average life constitutes one reason for the only gradual pace of the introduction of new machinery, etc., and therefore an obstacle to the rapid general introduction of improved instruments of labour. On the other hand competition compels the replacement of the old instruments of labour by new ones before the expiration of their natural life, especially when decisive changes occur. Such premature renewals of factory equipment on a rather large social scale are mainly enforced by catastrophes or crises. "

This statement has already been discussed in detail by Professor G.A. Shaumyan [3]. "The service life of machinery is not always determined by its physical durability. They are most often defined by the quality of products, the requirements for which are constantly growing, forcing to upgrade or replace all manufactured products ... We can say that the quality of manufacturing equipment is determined by how to use it to produce more, better and cheaper. As we can see, the quality and mass nature are in conflict, since the first requires continuous improvement of products (quick-change production), and the second the stability of product quality... Consequently, the quick changeover of production is a feature inherent to not only small-scale production, as it is commonly supposed, but the particular feature of modern technology development... it is necessary to observe the birth of a new kind of production with the following characteristics: quick change-over mass production ... (emphasis added). This task of huge difficulty must be solved to prevent that the automation from being the instrument of technological progress turns into its brake. "

In our view, it is this situation relating to the world's mechanical engineering industry we have observed in recent years. Refusal to introduce automation of core operations as well as transition to the principles of flexible technology led to the reduction of profit rates in the engineering industry in comparison with other industries and, consequently, to capital outflow to other industries and even to withdrawal of funds from manufacturing. This situation can be improved in terms of attracting investments to mechanical technology only by dramatically increasing profitability of the machine-building industry.

From p.1 it follows that the changeover time required to run a new product must be the same as the time to obsolescence T_m of the product. Determining a T_m value is an important task in terms of market research, and this value should be known before the start of production of any given goods, since it is taken as the base for estimating an optimal production plan and expected profit [1,4,7].

What prevents production facilities of the company from upgrading to the full extent and the simultaneous discontinuation of the product? The answer is obvious meaning different economically justified service lives of equipment that usually exceed the time to obsolescence of a product. This implies a very important conclusion for the stage of a product manufacture: that optimum service life of production facilities, including equipment, tooling and other means of technological equipment should be equal to the time to obsolescence of products. If we accept this statement as a basis, then the suggested in [1] approach relating to the usage of the machine and the optimization of its production process is fully applicable. There is only one difference indicating that in this work the machine is understood as a complete range of casting, forging, pressing, metal-cutting, assembly, material handling and other auxiliary facilities and tooling used for making products of a certain design.

Based on the above, the specified costs at the production stage must be calculated according to formula [1]:

$$S_p = \mathrm{C}_p \tau + \frac{\mathrm{K}_i}{\tau + 1}; \quad \frac{\text{c.c.u.}}{\text{c.t.u.}}, \tag{9}$$

where K_i is the capital investments in the organization and development of production of a new machine (c.c.u.); C_p is the rate of operating expenses associated with the support of running the production. $\text{c.c.u.}/(\text{c.t.u.})^2$

The minimum expenses according to expression (9) must correspond to the time to obsolescence relating to a product design. Then using formula (9) we obtain the following expressions for:

- Value to estimate obsolescence

$$T_м = \sqrt{\frac{K_i}{C_p}}; \quad (10)$$

- Optimal amount of capital investments in production

$$K_{i.opt} = C_p T_m^2. \quad (11)$$

If we put (11) into formula (7), we obtain an expression for calculating profit of a machine engineering company that operates in accordance with the stated above principle of optimal changeover to run a new machine (complete replacement of manufacturing facilities at the moment of product shift)

$$P = \frac{T_м}{t_d}\left[0.5\frac{T_м}{t_d}(z_i - C_i) - C_p \cdot T_м^2\right]. \quad (12)$$

Figure 2 shows the results of calculating profit (12) depending on the value of a product obsolescence and its cycle time based on the following conditions: the product price is 1000 c.c.u.; the product cost is 800 c.c.u.; the ratio of operating expenses C_p is 10 c.c.u./(c.t.u.)^2. On Figure 2 it is seen that there is a maximum profit and an optimal time relating to product's obsolescence for each value of cycle time. If we differentiate T_m from expression (12) and equate it to zero, we get

$$T_{m.opt} = \frac{z_i - C_i}{S \cdot C_p \cdot t_d}, c.t.u. \quad (13)$$

In other words, the optimal value of obsolescence with respect to a particular design of products is inversely proportional to their cycle time, with other conditions being equal. When the value of T_m is assumed as an initial data, it is possible to calculate the optimal value of cycle time that ensures maximum profit:

$$t_{d.opt} = \frac{Z_i - C_i}{3 \cdot C_p \cdot T_{м.opt}}, c.t.u. \quad (14)$$

Note, however, that the smaller $T_{m.opt}$ is, the less maximum profit a given company can raise (see the dashed line on Figure 2).

When substituting (13) into (12), we have a formula for calculating the maximum profit:

$$P_{max} = \frac{(Z_i - C_i)^3}{54 \cdot C_p^2 \cdot t_d^4}, c.c.u. \quad (15)$$

In expression (15) when we compare with (12), the maximum profit P_{max} is inversely proportional to the cycle time as well, and not to the second power but the fourth power. From (15) it follows that to maximize profits it is necessary to increase the difference between the product price and its cost, to cut operating expenses of production, to reduce the cycle time for production. If the first two approaches are traditional and also practically assured among businessmen, the third approach in the context of increasing profit requires careful analysis due to a number of reasons.

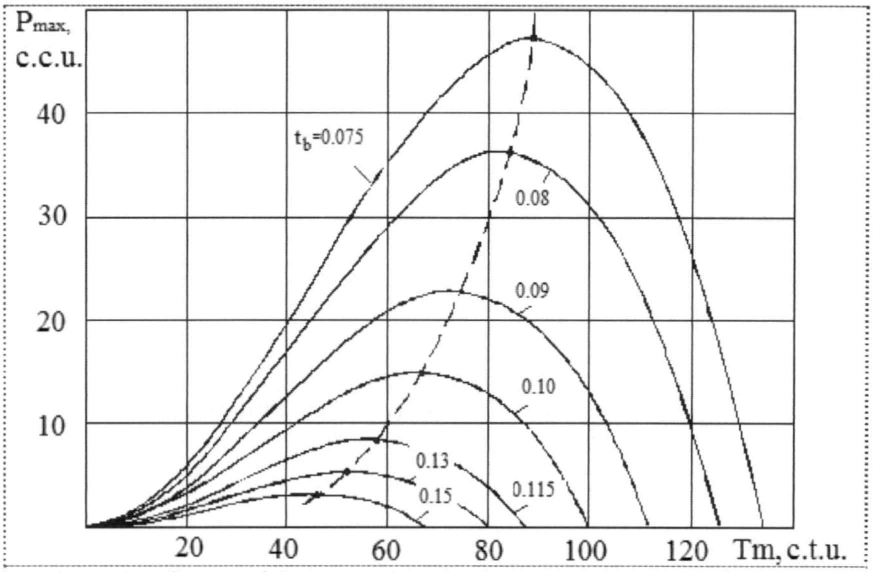

Figure 2. Influence of time to product obsolescence and the cycle time on the company's maximum profit:
$Z_i = 1000 c.c.u.; C_i = 800 c.c.u.; C_p = 10 c.c.u./(c.t.u.)^2$

First, the relationship between the cycle time and the maximum profit is the most strong (to the fourth power), and this proves that cutting cycle time to the possible extent is the most effective way to improving company's profitability. According to Figure 2, when t_d decreases two times (from 0.15 c.t.u to 0.075. c.t.u.), the maximum profit increases almost 16 times (from 2.9 mln c.c.u. to 46.8 mn. c.c.u.).

Secondly, according to expression (13), decreasing t_d automatically leads to increasing the optimal value of time as to machine's obsolescence, which is primarily determined by performance and competitiveness of the product design.

Thus, to maximize profits (see Figure 2), both conditions (13) and (14) must be observed at the same time, i.e. it is required to optimize the design engineering phase using $T_{m.opt}$ and the production phase using t_d. The conclusion is: The design of a machine and technology for its manufacture are closely connected with each other by means of the maximum possible profit.

It should be noted that the above stated is true only for a production created on the principle of optimal product shifting. How can this changeover be implemented in practice to enable a new machine production running without significant loss? In our view, production capacities of a machine engineering company should be divided into two equal parts, with one part being in operation at a particular moment and producing conventional products. At the same time the second part is used for setting up and gearing the production for making a new product with a predetermined manufacturing cycle time. At the moment when the product made using the first part capacities becomes obsolete, the second production facilities are quickly changed over to start run a new product, and the first production facilities are subject to a full dismantling for further production organization of more promising goods. To put in other words, an enterprise is supposed to have two parallel production lines at the same time. This approach allows solving successfully a problem relative to quick-change mass production detected by Professor A.G. Shaumyan.

For sustainable operation of main production facilities it is required to set up a sufficient material base for auxiliary production. The auxiliary production is mainly focused on designing and manufacturing equipment and tooling to support main processes engaged in making a new machine. And this work must be carried out systematically on an ongoing basis to ensure optimal changeover to production of more sophisticated goods. Since this type of production is characterized as a single-made and small-scale industry, the extensive application of principles aiming at using flexible technology is appropriate. Strong tool and prototype productions should serve as the third cornerstone of a modern machine engineering company, thereby ensuring sustainability and high profitability of the mechanical engineering industry.

Conclusions

The quantity of products to be manufactured is calculated on the basis of profit projections and competitive level of profitability of a manufacturer.

The optimum amount of cycle time instead of the production type approach is regarded as basis for designing and developing technological processes.

For a quick and loss-free changeover to production of a new machine, the service life of manufacturing facilities must be equal to the time to product obsolescence.

A product design and its technology are closely connected. To have the maximum possible profit for a company it is required to ensure the cycle time needed for producing a machine corresponds to the time to its obsolescence.

References

[1] Petrushin S, Gubaidulina R New prinziples of mechanical engineering organization

[2] The7thinternational Forum on Strategic Technology IFOST 2012 September 17-21, 2012 Tomsk polytechnic University.VOLUME II s/129-133.:http://www.tpu.ru.

[3] Karl Marx. Capital. A Critique of Political Economy. Volume 1. The Process of Production of Capital. - L. State. Publ.,Polit. Literature, 1949. - 794 p.

[4] Shaumyan, G A Complete Integrated Automation of Production Processes. - M.: Mashinostroeniye (Eng: Engineering), 1973. - 640 p.

[5] Gubaidulina R H, Petrushin S I, Galeeva A A, Selecting an Economical Variant of the Manufacturing Method of Engineering Product Fabrication under Current Conditions Applied Mechanics and Materials 682 2014 p. 613-616.

[6] N A Saprykina, A A Saprykin, Influence of Layer-by-layer Laser Sintering Modes on the Thickness of Sintered Layer of Cobalt-chromium-molybdenum Powder, Applied Mechanics and Materials. 1040. (2014). p.808-811.

[7] Belomestnykh V N, Soboleva E G Behavior of Poisson's ratio in the crystal Cu2O // Applied Mechanics and Materials. - 2014 - Vol. 682. - p. 170-173

[8] Petrushin, S I, Gubaidulina R H Optimizing the change over to the production of new engineering products. // Vestnic Mashinostrieniya (Eng; Journal of Mechanical Engineering). - 2011. – No. 12. – P.1.

IOP Conf. Series: Materials Science and Engineering **125** (2016) 012012 doi:10.1088/1757-899X/125/1/012012

Cutting Edge Geometry Effect on Plastic Deformation of Titanium Alloy

G I Korovin[1], A V Filippov [2,3,*], A V Proskokov[3] and V V Gorbatenko[2]

[1] Prod. Company MION, Tomsk, Verchinina str. 46/5, Russia
[2] Institute of Strenght Physics and Materials Sciencs SB RAS, Tomsk, per. Academicheskii 2/4, Russia
[3] Tomsk Polytechnic University, Tomsk, Lenina av. 30, Russia

e-mail: andrey.v.filippov@yandex.ru

Abstract. The paper presents experimental studies of OT4 titanium alloy machining with cutting edges of various geometry parameters. Experiments were performed at a low speed by the scheme of free cutting. Intensity of plastic shear strain was set for defining of cutting edge geometry effect on machining. Images of chip formed are shown. Estimation of strain magnitude was accomplished with digital image correlation method. Effect of rake angle and cutting edge angle has been studied. Depth of deformed layer and the area of the plastic strain is determine. Results showed that increasing the angle of the cutting edge inclination results in a change the mechanism of chip formation.

Introduction

Cutting_operation_for many years is one of the most effective ways of metal shaping. Complexity of the processes, occurring in the cutting zone, makes it difficult to study inter-coupling of various physical characteristics of cutting operations. Workpiece material strain is one of key physical components of cutting operations. The most common ways to examine cutting strains are micro-sections and use of a grade grid. These methods require fragments of the cutting zone with intact chips, which can be achieved by use special tools for momentary pause of the operation. The tool wear resistance, the performance and stability of the cutting process, the quality of the machined surface depend on the cutting edge geometry, the cutting conditions, the rigidity of the technological system and the materials of tool-workpiece pairs [1-8].

It is known that different materials and cutting conditions are characterized by different chip formation mechanism The copper chip forming is realized by continuous mechanism in a wide range of cutting conditions and the cutting edge geometry. The copper chip forming is realized by continuous mechanism in a wide range of cutting conditions and the cutting edge geometry.The element discontinuous and continuous chips can be made under machining titanium alloys.

* Corresponding author: A V Filippov - andrey.v.filippov@yandex.ru.

Application of digital image correlation (DIC) for cutting operation research is described in [9, 10]. DIC method summary consists in registering of cutting operation with a digital camera with simultaneous cutting zone highlighting with monochromatic laser. The laser light creates a contrasting image on the sample surface. Block-by-block search using correlation algorithms allows computing a vectorial field of displacements [11, 12].The main advantage of DIC is the ability to study operation recording, using different computation algorithms. In addition, DIC method does not require micro-sections and metallographic studies. Thus the complexity of experimental studies is significantly reduced.

We investigated the peculiarities of plastic deformation in polycrystalline copper cutting using tools with different rake angles and cutting edge inclination angles. The result revealed that the increase of the rake angle and the cutting edge inclination angle reduces the intensity of the plastic shear strain and sizes of plastic deformation area in chip formation zone [12].

The work's aim is the experimental research of the cutting edge geometry effects on the plastic deformation region formation in the OT4 titanium alloy samples.

Research method

This work is a continuation of a series of experimental studies of in-cut strain processes. In this paper we consider the effect of cutting edge geometry on plastic strain in the sample made of OT4 titanium alloy. Corrosion-resistant OT4 titanium alloy is used for manufacture of aircraft parts, operating at temperatures up to 350°C. OT4 relates to a Ti-Al-Mn system (Ti-4.2Al-1.6Mn) with pseudo α-structure and a small amount of β-phase stabilized with manganese.

Tool material is R6M5K5 (SKH 55). Cutting speed is limited to 13 mm/min, due to characteristics of recording equipment (video recording speed is 27 fps). Experiments were performed by the scheme of free cutting (see fig.1). While oblique cutting, incline plane of the cutting edge was out of sight of the camera.

In the study two sets of cutters were used. The first set, with the rake angles (γ) +4°, 0° and -4°, rear angle α=7.5°, cutting edge angle λ=0°. The second, with cutting edge angles 7.5°, 15°, 30° and 35° if α=7.5° and γ=15°.

Video recording was processed using special software developed in ISPMS SB RAS. Correlation analysis revealed a magnitude and direction of displacement of strained material. A schematic image of vectorial field of displacements is shown in Figure 1.

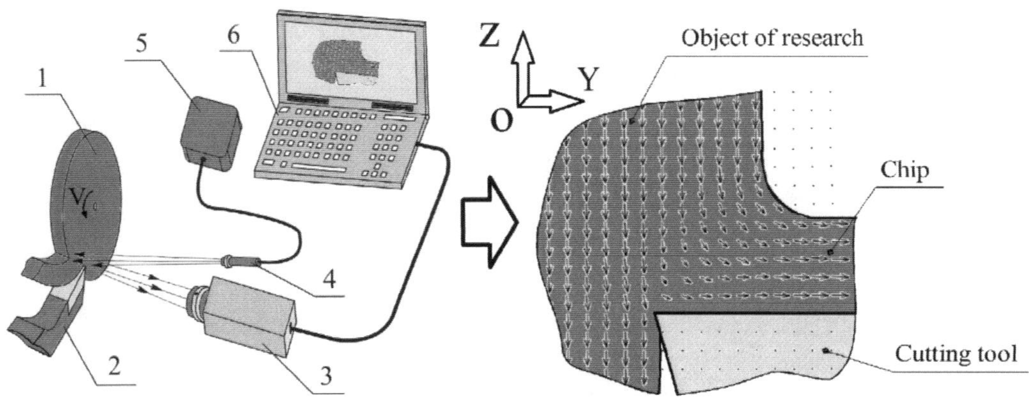

Figure 1. The experimental scheme and vectorial strain field in cutting zone

During experiments a side face of the sample was observed, therefore lateral strain was not determined. Further calculations were performed for plane-strain state. Further on computations were performed for plane-strain state. According to general concepts of the plastic strain theory [13] strain components can be determined by known displacements:

$$\varepsilon_z = \frac{\partial V_z}{\partial z};\ \varepsilon_y = \frac{\partial U_y}{\partial y};\ \gamma_{zy} = \frac{1}{2}\left(\frac{\partial V_z}{\partial y} + \frac{\partial U_y}{\partial z}\right). \qquad (1)$$

where V_x is a displacement vector projection on vertical axis OZ,

V_y is a displacement vector projection on vertical axis OY.

Intensity of shear strain for plane-strain state [13]:

$$e_i = \sqrt{2}/3\sqrt{\left(\varepsilon_z - \varepsilon_y\right)^2 + \varepsilon_z^2 + \varepsilon_y^2 + \frac{3}{2}\cdot\gamma_{zy}^2}. \qquad (2)$$

Computation by the formulas (1) - (2) gives value e_i matrix within a cutting zone with 0.12 mm pitch along axes OY and OZ.

Based on the data obtained, topograms with e_i-distribution were constructed in Mathcad. For illustrative purposes contours of a tool and a test sample are located on the topograms. Plastic strain area (s) and deformed layer depth (Δ) are defined as quantitative assessments of tool geometry effect on formation of plastic strain. Relative strain magnitude upwards of 0.01 is taken as a criterion of plastic strain. The values s and Δ are found by a graphic method considering a scale factor.

Results and discussions

Topograms of intensive plastic strain when cutting titanium alloy OT4 with a tool with rake angles +4°, 0° and -4°, if α=7.5°, λ=0° are shown in Figure 2. Schematic images of formed and detached chip elements are shown too. The greatest magnitude of e_i is observed at the tool top, in contact area of the tool clearance face with the sample and on the sample surface near the shearing range. In our researches formation of a narrow zone with great strain intensity is observed. This zone has form and arrangement similar to a conventional plane of share [6].

In this case, chips are formed elementally: by accumulation of strain in the material, with its localization in the narrow zone. When strain reaches a certain critical value, chip elements displace along the shearing range (see Fig. 1). Shearing begins from the tool top and progresses along the chain of zones with high values of e_i.

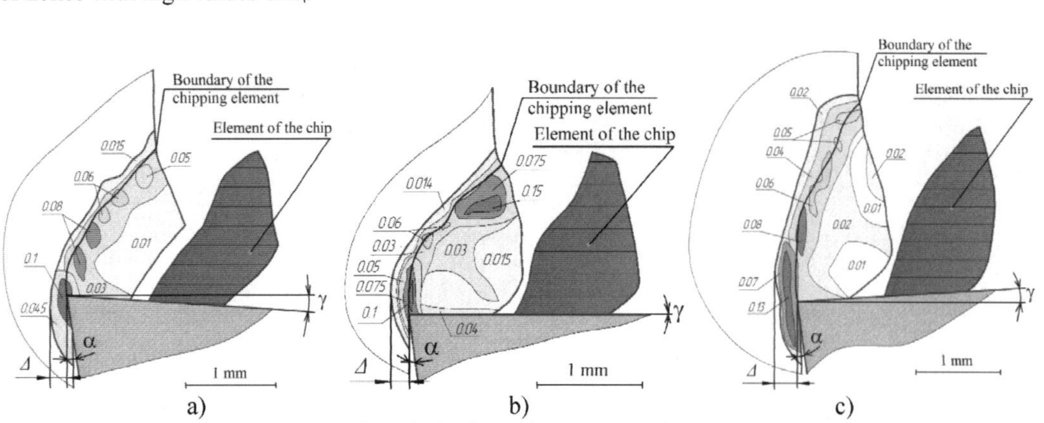

Figure 2. Intensity of plastic strain a) γ= +4°, b) γ=0°, c) γ= -4°

Figure 3 shows images of chip elements produced with cutting tool with angle γ = 0 °, obtained with a laser scanning microscope Olympus LEXT OLS 4100. The element has a shape similar to a triangle. The base is a flat, formed as a result of pressing of the front tool face on the metal. Compression of the material results in broadening of elements near the contact area of chips with the tool (see. Figure 3b), consistent with a high strain intensity e_i=0.04 near the front tool face (see Figure 2b).

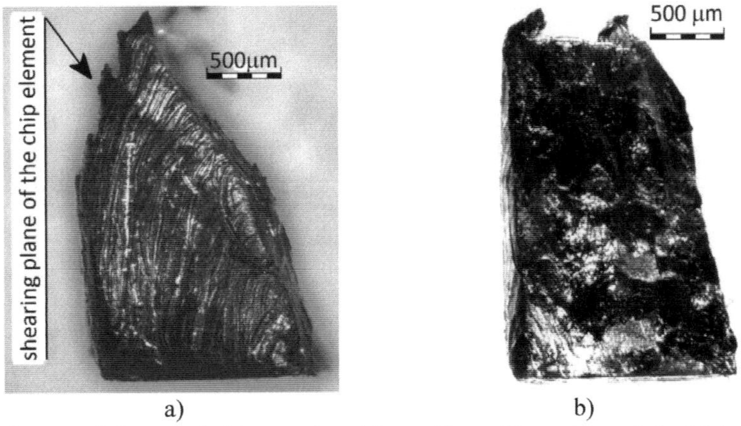

Figure 3. Image of elemental chips produced by cutting with angle γ = 0 °: a) lateral surface, b) shearing plane

Figure 4 shows the topograms of intensive plastic strain when cutting of OT4 titanium alloy with a tool with blade angles 7.5°; 15°; 30° and 35° if α=7.5° and γ=15°.

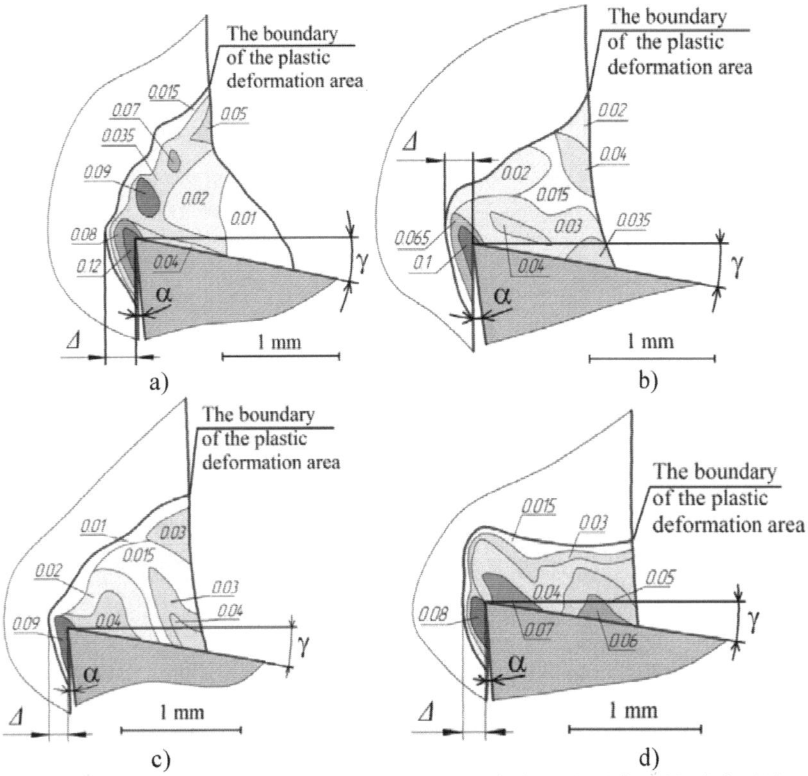

Figure 4. Intensity of plastic strain a) λ=7.5°, b) λ=15°, c) λ=30°, d) λ=35°

IOP Conf. Series: Materials Science and Engineering **125** (2016) 012012 doi:10.1088/1757-899X/125/1/012012

Chips flowing from angled front face are formed as a continuous helical spiral. When cutting with a tool with angles λ (15°, 30° and 35°) chip formation represents a continuous flow, that is, a continuous plastic deformation of the sample material occurs without chipping and separation into elemental chips. When cutting with the angle λ=7.5° chips are formed as connected segments, without chipping.

The highest strain intensity when cutting with a tool with the angle λ=7.5° is localized within a narrow zone, the same is when cutting with tools with λ=0° if γ +4°, 0° and -4°. With increase λ angle strain is less localized. e_i value increases from the strain zone boundary to the front tool face. In all cases, larger e_i values are observed near the tool blade tip.

Figure 5 shows an image of a fragment of the helical chip produced by cutting with the angle λ=35°. Chips are undivided (without segments) in the form of helical spiral with a pitch of 3mm and a diameter of 4mm. Increasing of the blade angle changes shear direction due to additional shear in transverse direction. This is the main cause of chip twisting.

Figure 5. View of twisted chips resulting in cutting with angle λ=35°.

Figure 6 shows a curve of the area of plastic strain (s, mm2) and depth of the deformed layer (Δ, mm) for various values of the cutting edge angle (λ) and the rake angle (γ).

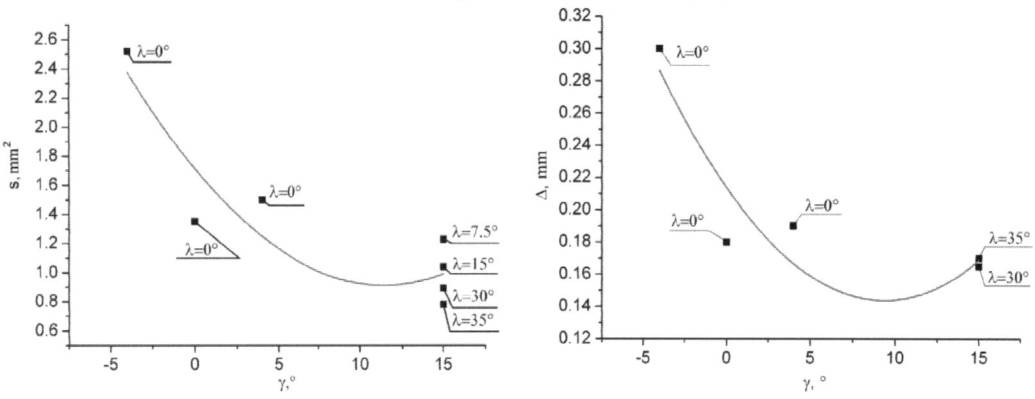

a) b)

Figure 6. Effect of cutting edge geometry on plastic strain area (a) and depth of deformed layer (b)

According to the curve (see Figure 6a), plastic strain area decreases with increasing of the cutting edge angle and the rake angle of the tool. The depth of the deformed layer also decreases with increasing of the rake angle of the tool (see Figure 6b). Accordingly, a severity of strain effect exposure on the workpiece face layer decreases. In turn, this reduces strain hardening and residual stresses in the face layer.

Conclusion

The findings on the effects of tool geometry on plastic strain of OT4 titanium alloy make it possible to draw the following conclusions:

1) Cutting with the angle $\lambda = 0$ ° causes compression and shear of the material sample which occur in the plane ZOY (see Figure 2) Cutting edge angle increase changes strain state from flat to volumetric and chip formation from elemental to continuous.

2) Rake angle increase and cutting edge angle increase diminish the area of plastic strain.

3) Intensity of plastic shear strain e_i subsides with increase of the cutting edge angle and the rake angle of the tool.

4) With increase of angle λ, chip formation changes from elemental to continuous making cutting operation more stable.

5) With increase of angles γ and λ the depth of deformed layer diminishes which as well is a positive aspect of tool geometry effect on cutting operations.

References

[1] Komanduri R and Von Turkovich B F 1981 Wear 69 179-188
[2] Komanduri R 1982 Wear 76 15-35
[3] Komanduri R and Brown R H J. 1981 Eng. Ind. 103 33-51
[4] Merchant M E 1945 J. Appl. Phys. 16 267
[5] Merchant M E 1945 J. Appl. Phys.. 16, 318
[6] Zorev N N 1966 Metal cutting mechanics (Oxford: Pergamon Press)
[7] Oxley P L B Int. 1961 J. Mach. T. Des. Res. 1 89-97.
[8] Bobrov V F 1962 Influence of the inclination angle of the main cutting edge upon the process of metal cutting (Moscow: Mashgiz)
[9] Filippov A V and Proskokov A V 2014 Bull. of MSTU N.E. Bauman. Ser. "Mashinostroyeniye" 2 100-113
[10] Filippov A V and Gorbatenko V V 2014 Appl. Mech. and Mat. 682 525-529
[11] Zuev L B, Gorbatenko V V and Pavlichev K V 2010 J. Measur. Sci. Technol. 21 90-93.
[12] Polykov S N, Gorbatenko V V, Lopaev E L and Zuev L B 2003 J. Avtometriya. 39 102-111.
[13] Thomsen E G, Yang C T and Kobayashi S 1965 Mechanics of plastic deformation in metal processing (New York: MacMillan Pub. Com.)

Influence of Welding With Two-Jet Gas Shielding On the Shaping of a Welding Joint

D A Chinakhov[1], E D Chinakhova[2], S V Grichin[1], Y M Gotovschik[1]

[1]Yurga Institute of Technology of National Research Tomsk Polytechnic University, 652055, Leningradskaya st., 26, Yurga, Kemerovskaya region, Russia
[2]Siberian State Medical University, 634055, Moskovsky trakt 2, Tomsk, Russia

e-mail: chinakhov@tpu.ru

Abstract. The author considers gas-dynamic influence upon microhardness and weld configuration of single-pass welds from steel 30HGSA when welding with consumable electrode under double-jet shielding. The relations to the chosen controlled welding parameters (Q, L, I) are developed and the controlling influence of the gas-dynamic affect of dynamic shield gas jet over formation of welds from alloy-treated steel 30HGSA is determined.

1. Introduction

Alloyed steels possessing a wide range of operational properties are used in manufacturing essential welded constructions [1, 2]. They provide a high strength of a construction and simultaneously reduce its specific quantity of metal. However, under the thermal cycle of welding in the heat-affected area of joint welds quenched structures are formed. The latter show high values of hardness at low values of toughness. As the consequence, the maximum quantity of low-temperature cracks when welding is formed in the heat-affected area [3].

There are various methods for improving operational properties of alloyed steel joint welds and their resistance to forming low-temperature cracks when welding. For instance, the level of high-temperature chemical micro-inhomogeneity, structural and mechanical inhomogeneity in the fusion area and in the weld metal can be decreased by means of intense mixing of electrode and base metals [4]. Electrode and base metals are mixed better as the time of weld pool metal being liquid increases, but at the same time, it causes the saturation of the weld metal with hydrogen and results in its embrittlement [1]. We can reduce the time of the weld pool metal being liquid and simultaneously increase the rate of its mixing by the impulse-dynamic impact, for example, by controlling the electrode metal transfer into the weld pool [5–7] or by controlling the impact of gas shielding medium [8–10].

The paper is aimed to determine the influence of the parameters of consumable electrode welding under double-jet gas shielding upon microhardness and weld configuration in single-pass joint welds from steel 30HGSA.

2. Methodology

The mechanized single-pass welding of plates from steel 30HGSA 150x300 mm in size and 8 mm thick was completed in CO_2 with Sv-08G2S welding wire 1.2mm in diameter and the stationary arc under double-jet gas shielding without preheating and following-up heat treatment. The controlled parameters were varied according to the matrix of planning (Tabl. 1), arc voltage U = 26…27 V, welding rate V = 25…26 cm/min.

Table 1

The matrix for planning of a full factorial experiment

Controlled welding parameter	№ experiment							
	1	2	3	4	5	6	7	8
1. Shielding gas consumption Q, l/min	15	20	15	20	15	20	15	20
2. Electrode wire extension L, mm	8	8	14	14	8	8	14	14
3. Welding current I, A	170	170	170	170	200	200	200	200

3. Results and Discussion

The dispersion of microhardness in the cross-section of the welded samples was measured by microhardness tester Duramin-5 (Fig. 1, 2) in accordance to lines located 2, 4, 6 mm from the upper surface of welded samples.

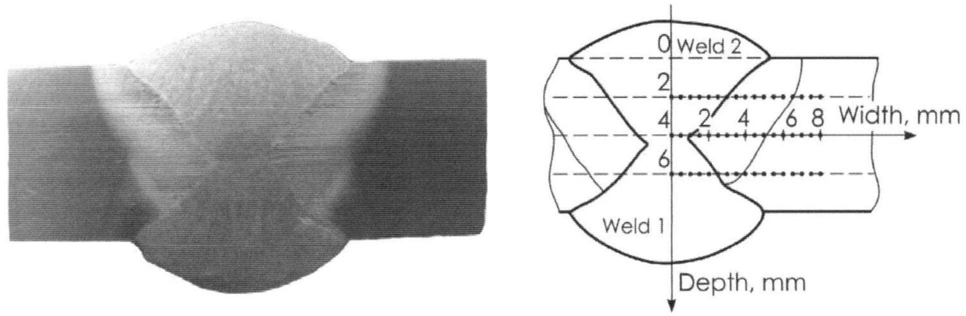

Figure 1. Macrosection and diagram for measuring microhardness of welded samples

The analysis of the dispersion of microhardness in the cross-sections of joint welds showed the increase in microhardness of the heat-affected area, particularly beyond Y=2 mm from the upper surface of the welded samples. The value of microhardness in the centre of the weld metal of all welded samples is nearly equal and amounts to 300-330 N, which is approximately congruent to the microhardness of the base metal 270-300 N.

To determine the relation of microhardness in the heat-affected area of single-pass joint welds from steel 30HGSA to the controlled parameters (Q, L, I) maximum values of microhardness on each side of the weld axis in line Y=2 mm were implemented (the percentage error doesn't exceed 8%):

$$H\mathrm{max\_2} = 1428 + 25.94\cdot Q - 78.28\cdot L - 6.49\cdot I - 2.358\cdot Q\cdot L + 0.59\cdot L\cdot I \quad (1)$$

The results of research of microhardness dispersion in the cross-sections of joint welds from steel 30HGSA demonstrated the dependence of the maximum value of microhardness upon the electrode extension. When decreasing the electrode extension there is an increase in microhardness peak, which is possible due to the increase in the density and outflow rate of the shielding gas (when its consumption held equal). It causes the growth of the rate of metal cooling-down under the welding nozzle and, consequently, leads to the increase in microhardness in the heat-affected area [11]. The maximum value of microhardness also rises as the consumption of the shielding gas is increased and falls down as the welding current rises.

The microstructure of joint weld areas formed according to welding modes 1 and 8 of experiments from the matrix for panning a full factorial experiment (Tabl. 1) is shown in Figure 3. The analysis of the microstructure combined with the analysis of microhardness dispersion revealed the significant influence of controlled welding parameters (Q, L, I) upon the microstructure of a weld and both of a heat-affected area and its length.

IOP Conf. Series: Materials Science and Engineering **125** (2016) 012013 doi:10.1088/1757-899X/125/1/012013

No. EXPERIMENT	MICROHARDNESS, N
1	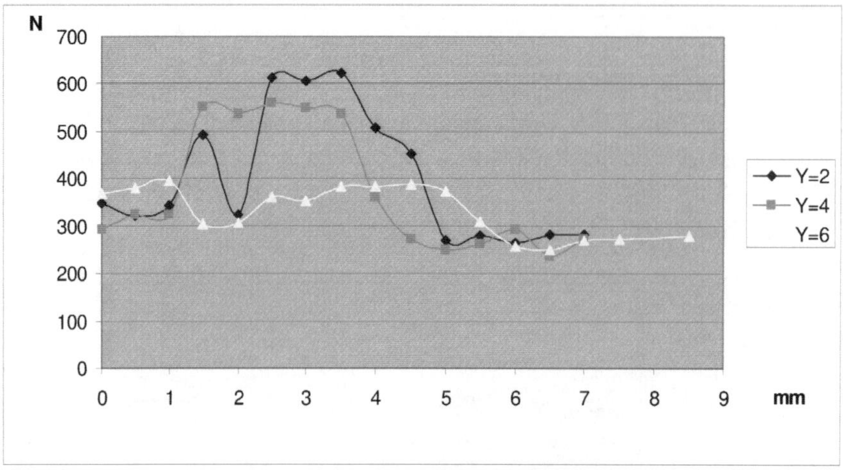
8	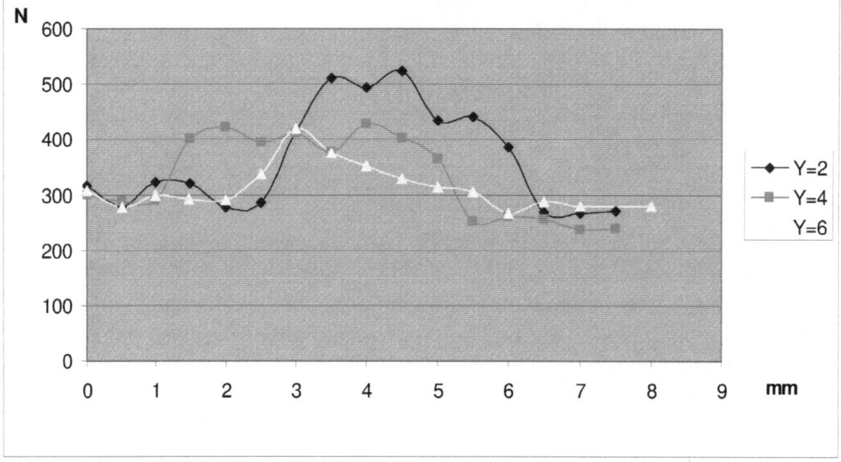

Figure 2. The dispersion of microhardness in the cross-sections of joint welds formed according to 1 and 8 welding modes from the matrix of a full factorial experiment

The formed welded samples were used for carrying out measurements of weld dimensions (E – weld width, g – weld reinforcement, h – weld penetration). According to the results of research the multi-factorial relations of single-pass weld dimensions (E, g, h) to the controlled parameters of welding under double-jet gas shielding were developed, the percentage error doesn't exceed 10%:

1. The relation of weld width (mm) to the controlled parameters.

$$E = 64.4 - 3.8\cdot Q - 4.18\cdot L - 0.317\cdot I + 0.3\cdot Q\cdot L + 0.02\cdot Q\cdot I + \\ + 0.02\cdot L\cdot I - 0.00156 Q\cdot L\cdot I \qquad (2)$$

2. The relation of weld reinforcement (mm) to the controlled parameters.

$$g = 0.925\cdot Q + 0.099\cdot I - 0.005\cdot Q\cdot I - 15.8 \qquad (3)$$

3. The relation of weld penetration (mm) to the controlled parameters.

IOP Conf. Series: Materials Science and Engineering **125** (2016) 012013 doi:10.1088/1757-899X/125/1/012013

$$h = 54.3 - 3.01 \cdot Q - 4.605 \cdot L - 0.245 \cdot I + 0.266 \cdot Q \cdot L + 0.01463 Q \cdot I + \\ + 0.023 \cdot L \cdot I - 0.00133 Q \cdot L \cdot I \quad (4)$$

Area	1	8
Weld	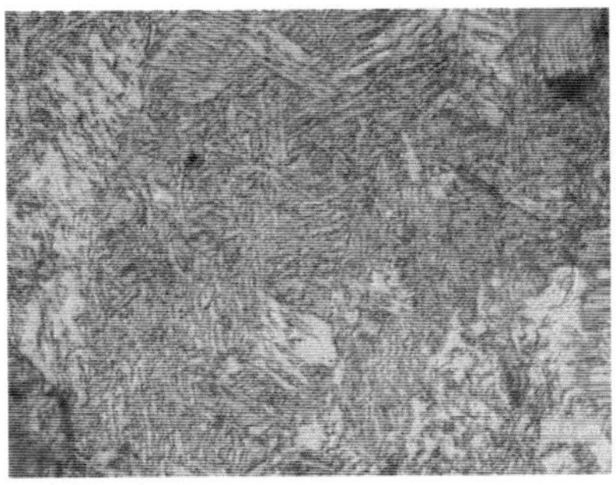	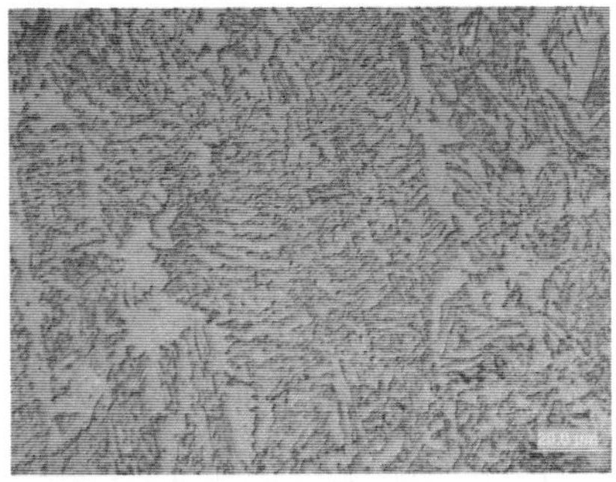
Heat-affected area, 3 mm from weld axis	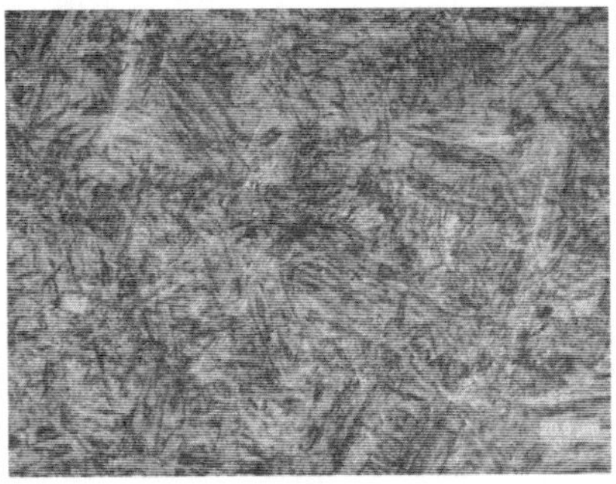	
Heat-affected area, 4 mm from weld axis	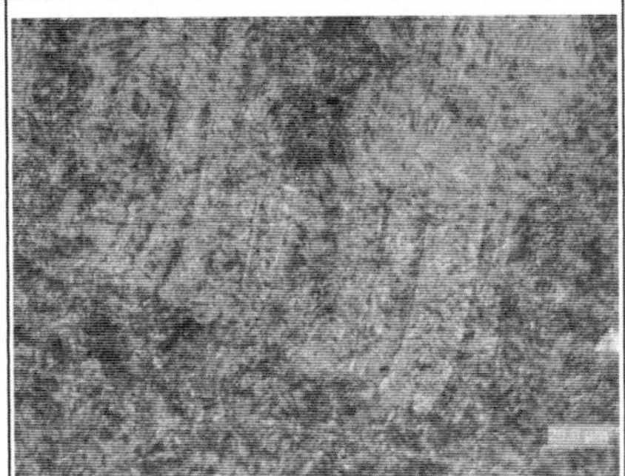	
Heat-affected area, 6 mm from weld axis	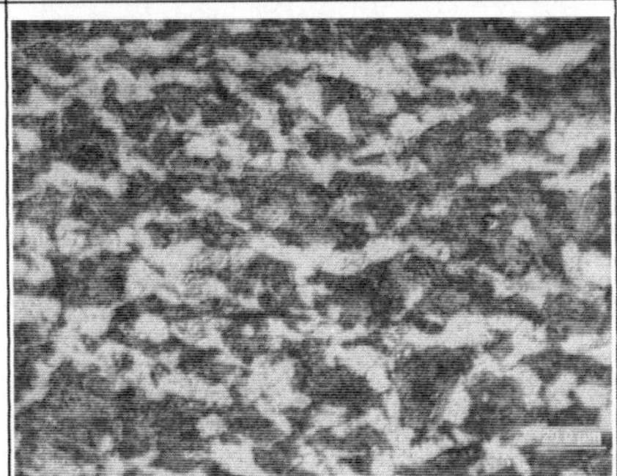	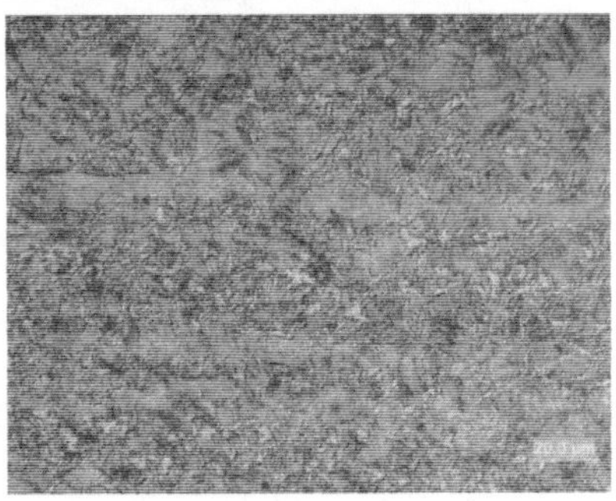

Figure 3. Microstructure in the joint weld areas, formed according to welding modes 1 and 8 of experiments from the matrix for planning a full factorial experiment, in line Y=2mm

IOP Conf. Series: Materials Science and Engineering **125** (2016) 012013 doi:10.1088/1757-899X/125/1/012013

All the dimensions of single-pass welds (E, g, h) of joint welds from steel 30HGSA (in conditions of the experiment) are affected both by the welding current, the shielding gas consumption, separately and combined, and by the three factors (Q, L, I), acting simultaneously. There is also a considerable influence of the electrode extension combined with the shielding gas consumption and welding current upon the weld width and penetration.

Conclusion

It's determined that when double-jet gas shielded welding the gas-dynamic impact influences the structural and phase condition, the dispersion of microhardness in the cross-section of joint welds and weld configuration, in other words, it allows controlling operational properties of joint welds from alloyed hardenable steels.

References

[1] Lakhtin Yu.M., Leontyeva V.P. Material science: a study guide for higher technical schools. The 3d edition revised and completed. Maschinostroyenie. 1990. 528 p.
[2] Welding and welded materials: In 3 v. V.1. Weldability of materials: reference book / edited by E.L. Makarov. Metallurgy, 1991. 528 p.
[3] Chinakhov, D.A., Saraev, Yu.N. Production technology of the multi-layered welded joints from 30KhGSA steel without heating. Svarochnoe Proizvodstvo. 2002. № 7. P. 18-20.
[4] Elagin V.P., Snisar V.V., Savitsky M.M., Gordan G.N., Vasilyev V.G., Doroshenko L.K. Chemical and structural inhomogeneity in fusion area of low-carbon steels by austenitic welds when gas shielded welding / Automated welding. 2001. № 4. P. 8-13.
[5] Lenivkin V.A., Dyurgerov N.G., Sagirov H.N. Technological properties of a welding arc in shielding gases. Machinebuilding. 1989. 264 p.
[6] Paton B.E., Voropay N.M., Buchinkky V.N. and others. Controlling the process of arc welding through programming the rate of electrode wire feed. Automated welding. 1977. № 1. P. 1–5.
[7] Brunov O.G., Fed'ko V.T., Solodskii S.A. Transfer of electrode metal in welding with the pulsed feed of welding wire. Welding International. 2007. T. 21. № 1. C. 50–54.
[8] Fedko V.T., Kiyanov S.S., Shmatchenko V.S., Sapozhkov S.B. Applying double-jet nozzles for gas shielded welding / Automation and contemporary technologies. 2003. № 3. P. 12-18.
[9] Shtrickman M.M., Pavlov A.S., Sabantzev A.N., Egorov V.N. Arc welding with impact of directed gas ejection flows on a weld pool / Welding production. 1999. № 12. P. 3-6.
[10] Chinakhov D.A. Gas Dynamic Control of Properties of Welded Joints from High Strength Alloyed Steels. China Welding. Vol. 23 No.3 (2014) pp. 27-31.
[11] Chinakhov D.A. A gas-dynamic impact on properties and configuration of single-pass joint welds from steel 30HGSA when welding under double-jet gas shielding. Heavy engineering. 2012. No 3. P. 29-34.

IOP Conf. Series: Materials Science and Engineering **125** (2016) 012014 doi:10.1088/1757-899X/125/1/012014

Mathematical Model of Motion of Modifying Powder Particles in the Shielding (Carrier) Gas

M A Kuznetsov [a], A V Kryukov [b], A G Krampit [c], E A Zernin [d], I D Sadykov

Yurga Institute of Technology (Affiliate) of Federal State Autonomous Educational Institution of Higher Professional Education "National Research Tomsk Polytechnic University", Yurga, Russia (652050, Yurga, 26, Leningradskaya str.)

e-mail: [a]kyznechik_85@mail.ru, [b]earendill@mail.ru, [c]akrampit@mail.ru, [d]yuti_sp@bk.ru

Abstract: The paper gives some consideration to the model of including ultradisperse modifiers into the surface layer of material through the shielding (carrier) gas. It has been revealed this technique of inclusion necessitates ultradisperse particles to be 100 – 150 nm for their efficient transfer by the shielding (carrier) gas. Particles of smaller dimensions can't get in a jet of the shielding (carrier) gas. Probability of such defect as "impurity" occurs provided that particles are over 150 nm.

1. Introduction

Low stability of equipment against intercrystalline corrosion, especially in zones affected by high temperature in the process of manufacturing is a vital problem of pharmaceutical, chemical, power, nuclear and food industries. Therefore, in recent years researchers have been focusing their attention on the issues of increase in corrosion resistance by means of modifying layers. Obtaining special properties of the surface via modifying is inseparably linked with solution of the problems how to produce, test modifying powders and choose the method of their inclusion into the layer to be processed.

Authors of the paper deal with the technique of ultradisperse powder inclusion through the shielding (carrier) gas; this technique makes it possible to control concentration of ultradisperse particles in the shielding (carrier) gas [1]. The authors decided to specify the dimensions of modifier particles by means of mathematical simulation.

Mathematical simulation has been developed successfully and now it is considered a powerful tool to study non-linear problems in various fields of science. If natural experiment is impossible, complicated or cost intensive mathematical simulation is used [2,3,4,5,6,7,8].

Development of computer engineering provides excellent perspectives for application of complex models displaying multiple-factor character and interconnection of different physical processes and phenomena [9].

Software package CONSOL "Multiphysics" intended for simulation of various physical processes, systems and their interconnection is a powerful simulation tool. "Consol" is a flexible platform allowing adapting model to real conditions, supplementing or modifying some physical processes described both by standard and set up by the user interfaces based on the system of differential equations in partial derivatives. As the consequence, a model maximally resembling the reality can be obtained.

2. Mathematical statement of the problem

Software package “CONSOL Multiphysics” was used to develop a computer model in the course of work. We calculated particle distribution in the volume and their mechanical trajectory on the base of gas jet rate distribution and data on particles inside the mixer (Fig. 1).

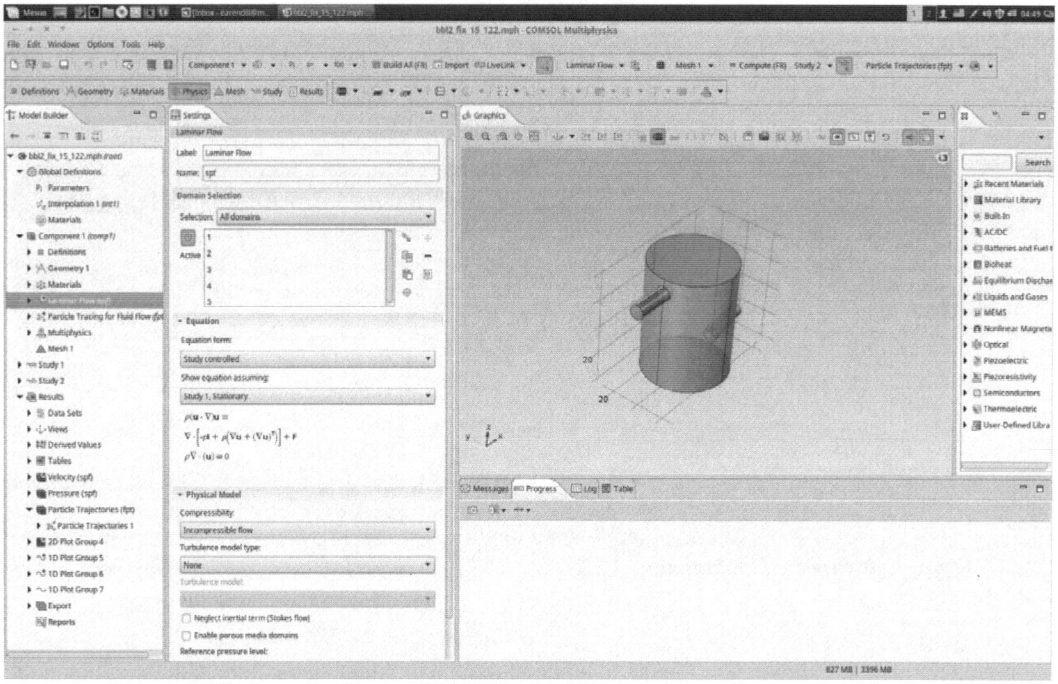

Figure 1 Window of software “CONSOL Multiphysics” to calculate the trajectory

The aim defined in the paper was attained in two steps:

1. Simulating argon supply into the mixers;
2. Simulating the trajectory of particles in the gas jet.

1. Argon supply into the mixers was simulated on the base of differential equation of laminar flow (1).

$$\rho(u \cdot \nabla)u = \nabla \cdot [-\rho i + \mu(\nabla u + (\nabla u)^T] + F \tag{1}$$

where $\rho\nabla \cdot (u) = 0$, ρ – gas density; u – gas rate component; ∇ – inverted delta, μ – dynamic viscosity; T – ambient temperature.

The following boundary conditions were determined for modelling: inlet conditions include $u = -U_0 n$, where U_0 specifying according to consumed shielding gas. Outlet conditions correspond to zero pressure. The rate distribution of shielding (carrier) gas jet is depicted in Figure 2.

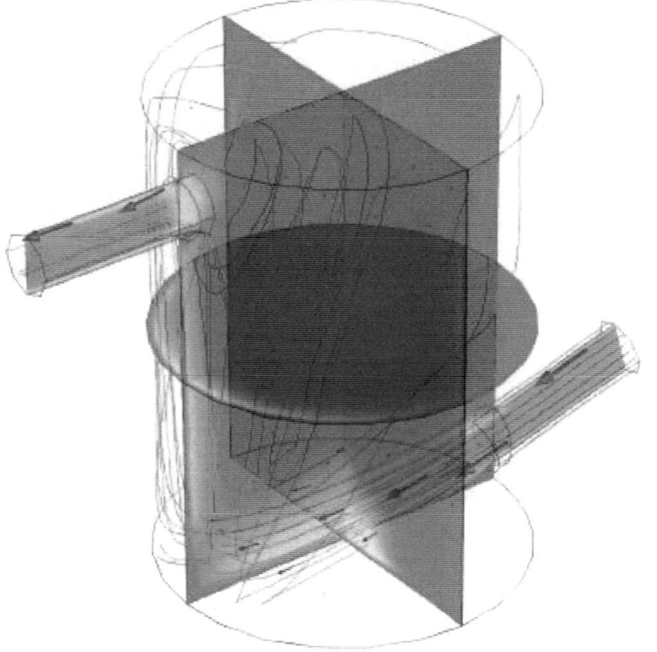

Figure 2 Distribution of gas jet rate

2. Simulating the trajectory of particles in the gas jet.

The motion of particles in the jet of shielding (carrier) gas can be described by equation (2)

$$\frac{d(m_p v)}{dt} = F_t \quad (2)$$

The following boundary conditions were specified for modelling: initialization of particles at the zero second in terms of finite element mesh; driving force $F = \frac{1}{\tau_p} m_p (u - v)$; $\tau_p = \frac{\rho_p d_p^2}{18\mu}$

where mp – particle mass; u – particle rate component; v – jet rate component, ρp – density of a particle; dp – diameter of a particle.

The motion of particles in shielding (carrier) gas is presented in Figure 3.

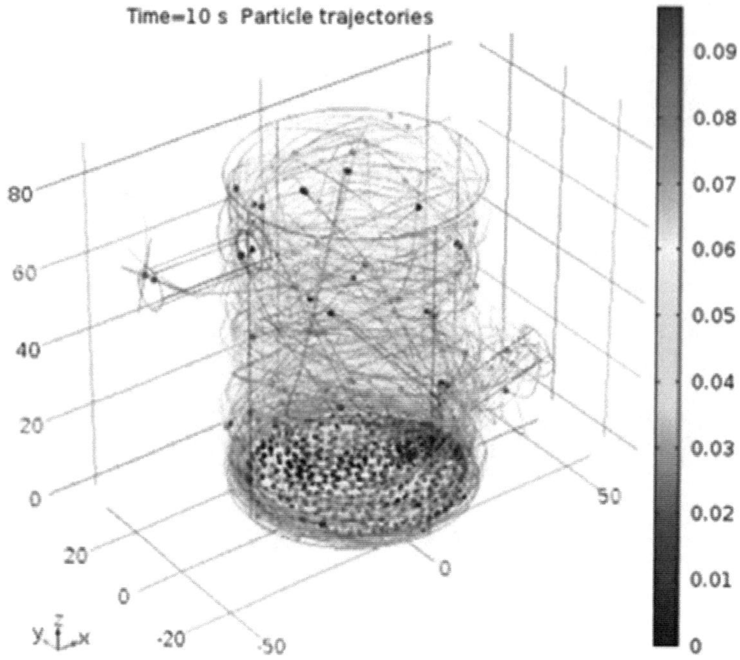

Figure 3 Motion of particles in the jet of shielding (carrier) gas

3. Results and Discussion

The obtained model was applied to study 4 sizes of moving particles (below 100 nm, 100-150 nm, 200-500 nm, and over 500) of modifying powders (tungsten, molybdenum, aluminium oxy-hydroxide).

In view of calculations the total amount of inlet initialized particles was estimated, as well as the percentage of particles transferred by jet to the outlet. Figure 4 presents the dependence of particles transferred by jet on their size. As the result of obtained data analysis, 100 to 150 nm particles are required for their efficient transfer.

The outcomes of mathematical simulation point to the following fact: maximum 5 of 1000 initialized in the mesh particles get into the jet of shielding (carrier) gas for 10 seconds provided that their size does not exceed 100 nm. If the particles are 100 to 150 nm, 100 of them get into the jet of shielding gas, if the sizes are 200 to 500 nm, their number in the shielding gas goes up to 50, and such defect as "impurity" is probable, particles 500 nm and above are not transferred into the jet of shielding (carrier) gas because of their mass.

Then, we studied the range of particle dimensions 100 to 150 nm to determine the efficient size of particles of the following modifying powders: tungsten, molybdenum, aluminium oxy-hydroxide). Table 1 provides main results of based on the above ideas calculations of particle motion of modifying powders in the jet of shielding (carrier) gas in the mixer according to the dimensions of particles.

Table 1 – Efficient technological parameters of modifying powder particles

Modifying powder	Size, nm	Mean density, g/cm^3	Consumption, mg/m
Tungsten	122	15	4
Molybdenum	127	4	4
Aluminium oxy-hydroxide	150	0.4	4

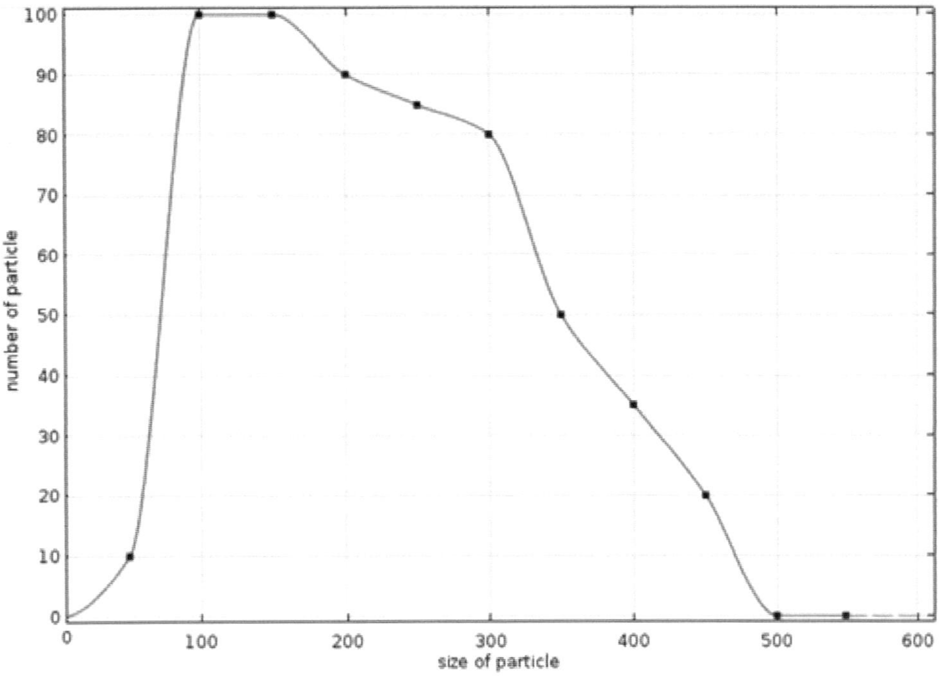

Figure 4 The number of particles in shielding (carrier) gas determined by the size of a particle

Production of powders by electric explosion of conductors relied on the carried out simulation. This method allows varying dispersibility of the powder by changing both physical (diameter, length of the wire to be exploded, gas pressure in the facility) and electro-physical parameters (operating voltage, capacity, circuit inductance), as the consequence, energy applied to the conductor under electric explosion [10].

For the purpose of work we used tungsten (W) and molybdenum (Mo) powders produced by electric explosion of conductor and aluminium oxy-hydroxide AlO(OH) produced by electric explosion of conductor and subjected to thermal hydrolysis.

Experiments pointed at the possibility of significant influence on grain dimensions and corrosion resistance of the arc-processed surface via changing powder concentration in the carrier gas [11,12,13].

The surface layer, as we have identified, has a laminated structure: a sub-layer of fine polyhedral austenite grains with the spots of non-oriented dendrites adjacent to the free surface, a sub-layer of short and thick non-oriented dendrites, and a sub-layer of extended oriented dendrites. The thickness of sub-layers depends on powder modifier (Figure 5).

We have found out modifying the surface layer by ultra-disperse powders AlO(OH), W and Mo at the rate of 4 mg per meter of the layer has the most pronounced effect of modifying the surface layer in the system Fe-C-Cr-Ni-Ti due to arising in the melt additional centres of crystallization – inoculators in crystallizing metal. Mean thickness of dendrite gets reduced by the factor of 1.5 - 2 (30-40%), its width is decreased by the factor of 2 - 3 (45-55%) (Figure 6).

IOP Conf. Series: Materials Science and Engineering **125** (2016) 012014 doi:10.1088/1757-899X/125/1/012014

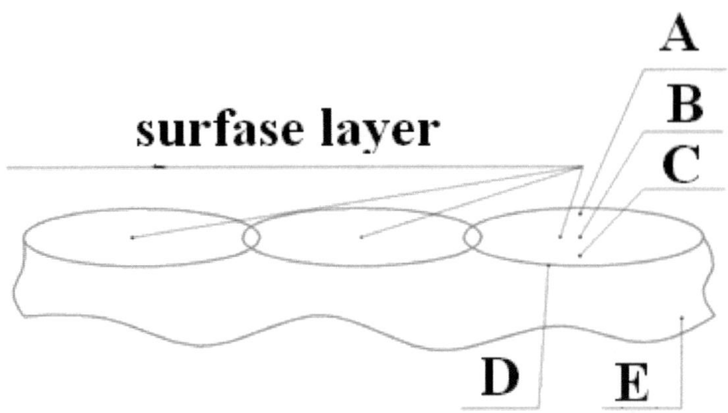

Figure 5 – A layout chart of zones in the microstructure of the surface layer to be examined: A – a sub-layer of fine polyhedral austenite grains with the spots of non-oriented dendrites, B – a sub-layer of short and thick non-oriented dendrites, C – a sub-layer of extended dendrites, D – a surface – base layer transition zone, E – base metal

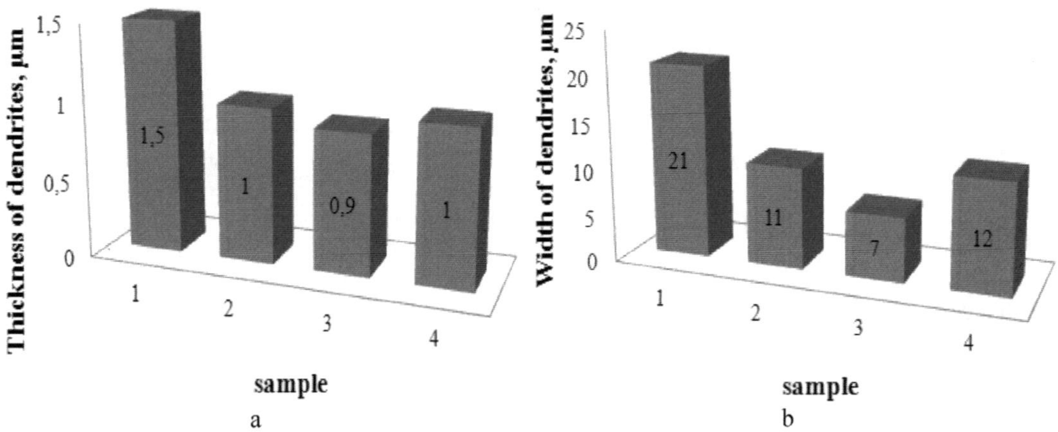

Figure 6 Dendrite dimensions: a – thickness of dendrites, b – width of dendrites
1 – without powder; 2 – with powder W; 3 – with powder AlO(OH);
4 – with powder Mo

Polyhedral grain boundaries are clear and their width did not increase after corrosion resistance tests. The structure of the surface layer modified by ultradisperse powders did not change somehow both near the melting border and in the bulk.

Therefore, high resistance to intercrystalline corrosion of the surface layer modified by ultradisperse powders was determined when applying standard metallographic methods. Atomic force microscopy tests of the surface layer structure confirm the results of metallography: the surface layer does not tend towards intercrystalline corrosion.

Grain boundaries do not really change. Etching was carried out not on grain boundaries but on their whole surface. The surface of grains was etched less or more according to their orientation. As the consequence, grooves formed between the grains (Figure 7).

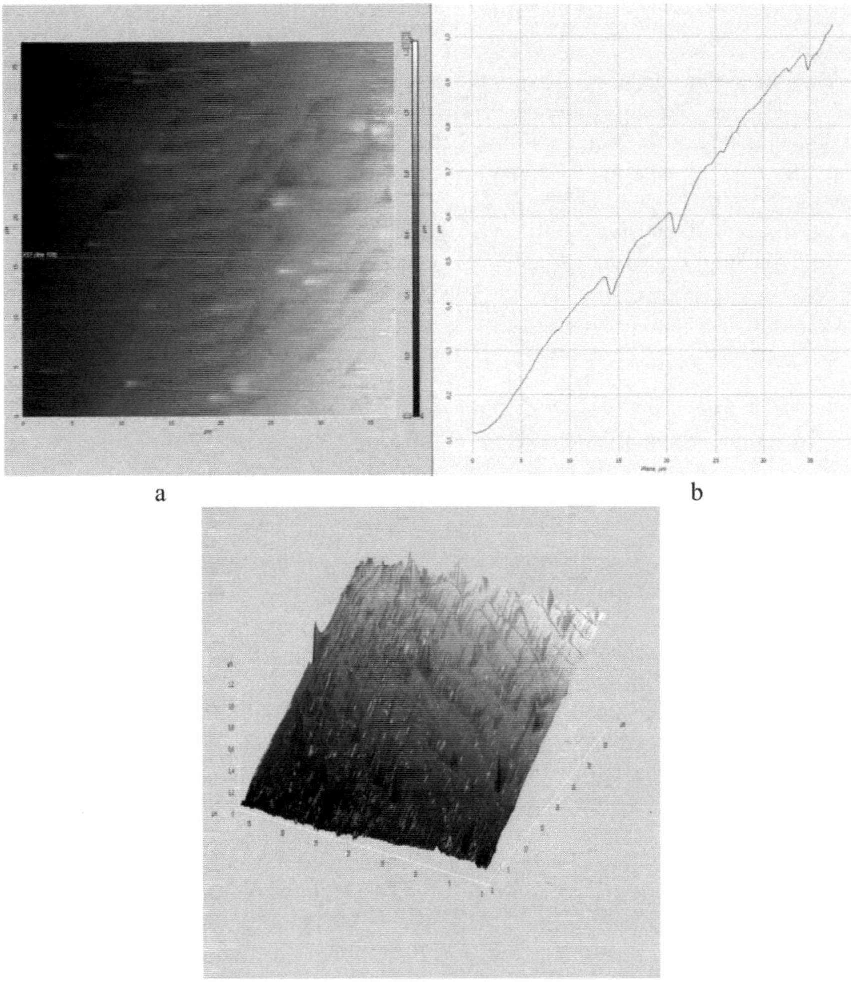

a b

c

Figure 7: a – AFM 2D image of welded metal,
b – profilogram on the line marked in the image:
c – AFM 3D image of the same zone

The sample modified by ultradisperse tungsten powder had the lowest jog; that is why; it was less etched under corrosion tests.

Conclusions

1. The mathematical model of motion of modifying powder particles in a streamline flow of a carrier gas has been proposed.

2. Efficient dimensions of particles for transferring a modifier by gas jet were specified on the base of proposed model. The minimum size of particles is 120 nm as smaller particles are not transferred into the jet of a carrier gas. If the size of particles exceeds 150 nm such defect as "impurity" is probable. Therefore, particles are to be 120 to 150 nm.

3. Rational technological parameters (dimensions of particles, their mean bulk density) of three type modifiers for austenite steels were determined according to the number of particles transferred by the gas jet: tungsten (W) – 122 nm, 15 g/cm3; molybdenum (Mo) – 127 nm, 4 g/cm3; aluminium oxy-

hydroxide (AlO(OH)) – 150 nm, 0.4 g/cm3. That is, the number of particles transferred by the gas jet is to tend towards a maximum.

4. Modifying the surface layer by ultra-disperse powders AlO(OH), W and Mo at the rate of 4 mg per meter of the layer has the most pronounced effect of modifying the surface layer in the system Fe-C-Cr-Ni-Ti due to arising in the melt additional centres of crystallization – inoculators in crystallizing metal. Mean thickness of dendrite gets reduced by the factor of 1.5 - 2 (30-40%), its width is decreased by the factor of 2 - 3 (45-55%).

5. It has been revealed, key factors for determining the resistance to intercrystalline corrosion of permanent joins are composition and structure of the heat-affected zone, its stability under operational influence as well. Samples produced by inclusion of nano-structured powders into the welded metal are the most corrosion resistant in view of structure and metal stability in the heat-affected zone.

References

[1] Kuznetsov M.A., Barannikova S.A., Zernin E.A., Filonov A.V., Kartsev D.S. 2015 Applied Mechanics and Materials Methods for defining the concentration of nanostructured powders in protective gas and its effect on the microstructure of deposit metal 770 pp. 28-33.

[2] Mazhukin V.I., Shapranov A.V., Perezhigin V.E. 2012 Mathematical simulation of thermophysical properties, heating and melting processes of metals by the method of molecular dynamics // X International Seminar "Mathematical Models & Modelling in Laser-Plasma Processes & Advanced Science Technologies, Vol XXIV pp. 47-65.

[3] Mazhukin V.I., Shapranov A.V. 2012 Mathematical simulation of metal heating and melting. Part I. Model and Computational algorithm // Preprints of Keldysh Institute of Applied Mathematics 31 – 27 p.

[4] Chinakhov D.A., Vorobjev A.V., Tomchik A.A. 2013 Materials Science Forum Simulation of Active Shielding Gas Impact on Heat Distribution in the Weld Zone 762 pp 717-721.

[5] Chinakhov D.A., Vorobjev A.V., Gotovshchik Y.M. 2014 Applied Mechanics and Materials Simulation of wind Influence on the thermal processes in gas-shielded welding 682 pp. 91-95.

[6] Krampit M.A., Krampit N.Y. 2015 Applied Mechanics and Materials Electronic circuit design of power supplies for welding in ltspice lv program // 770 pp. 323-327.

[7] Temlyantsev M.V., Starikov V.S., Kondratiev V.G. 2000 Proceedings of the higher educational institutions. Iron and steel modeling of temperature fields and resistance to deformation in cylindrical workpieces by heating with a hot tenements for rolling 6. pp. 51-54.

[8] Temlyantsev M.V., Starikov V.S. 2000 Proceedings of the higher educational institutions. Iron and steel calculation of temperature fields in prismatic workpieces during thermal cycling with 2. pp. 42-45.

[9] Makhnenko V.I. 2002 Perspectives of mathematical simulation and information technologies in welding and related processes // Mathematical simulation and information technologies in welding and related processes: Proceedings of International Conference; Paton Welding Institute Kiev pp. 3-11.

[10] Yavorovsky N.A. 1996 Proceedings of Higher Educational Institutions. Physics. Production of ultradisperse powders by electric explosion 4 pp. 114 – 136.

[11] Kuznetsov M.A., Zhuravkov S.P, Zernin E.A., Kolmogorov D.E., Yavorovsky N.A. 2014 Advanced Materials Research Influence of Nanostructured Powder Modifiers on the Structure of a Welding Bead 872 pp. 118-122.

[12] Kuznetsov M.A., Zernin E.A., Zhuravkov S.P., Yavorovsky N.A. 2014 Advanced Materials Research Influence of nanopowders on corrosion resistance of welted joints 1040 pp. 837-844.

[13] Kuznetsov M.A., Zernin E.A., Danilov V.I., Kolmogorov D.E., Zoubenko L.N. 2015 IOP Conference Series: Materials Science and Engineering Corrosion and mechanical properties of austenic steel weld joints 91 pp. 012010 doi:10.1088/1757-899X/91/1/012010.

IOP Conf. Series: Materials Science and Engineering **125** (2016) 012015 doi:10.1088/1757-899X/125/1/012015

The Structure and Principles of Work Intellectual Processing Control System on CNC

R Y Nekrasov, U S Putilova, A I Starikov, I V Soloviev and O V Bekareva

Tyumen State Oil and Gas University, 38, Volodarskogo street, Tyumen, 625000, Russia

e-mail: aragorn89_89@mail.ru

Abstract. The high rate of technological progress imposes stringent quality demands of engineering production. The main requirement is dimensional accuracy of metal-cutting equipment with a minimal needed operating costs during the period of operation. The article considers the problem of increasing the accuracy of processing engineering products on CNC. The authors have proposed a solution of the problem by providing entering operational adjustment to the control of the trajectory of the executive bodies of the machine. The necessity of creation of mathematical models of processes in an automated process system operations (APSO). Based on the research, authors had proposed that a generalized scheme of diagnosis and operation input of corrections and approximate mathematical models of the individual processes of diagnosis.

Introduction

Currently, the processing engineering products implemented on equipment that uses computer numerical control (CNC), has been some marginal level of precision machining, so the minimum discrete movement of the executive bodies of most modern machines, is 1 micron, and some of them are able to carry out moving in quantities less than one micron. Further increases the accuracy of processing implemented on CNC equipment, is due to a decrease minimum possible discrete movement of executive bodies is not possible to restrictions imposed by both the physics of metal cutting on the micron and submicron level and the actual condition of the equipment. For example, in the micron and submicron level becomes critical influence of errors caused by the influence of the equipment: power loads, the heat flux and wear of the cutting tools that change the shape and spatial position as the executive bodies of the machine as workpiece.

In almost all cases the development of control programs for CNC equipment is carried out either with the use of computer-aided design (CAD), or with built-in simulation directly on the CNC control panel. Should be noted that in this case, and CAD and CNC panel working with a mathematical model of the workpiece, cutting tools and equipment, and any mathematical model is a kind of idealized object affected by a greater or lesser degree of simplification. In this case, simplified physical and mechanical properties of objects which appear to CAD as absolutely rigid, not exposed to heat fluxes not undergo wear. Therefore, errors occur in the technological system in the implementation of control programs (CP), created by means of CAD or CNC control panel, will be inherited by the workpiece. Prevent the occurrence of errors in the processing system, at realization of CP is impossible due to the physical nature of the objects. In this case it is necessary to implement conversion of the control trajectory (CT), in a way that errors of processing system will be compensated.

2. Methods of Research

Widespread CNC lathes in modern machining production leads to appearance another problem - reducing the quality of products manufactured by the technological system during exploitation, that requires a decision. The point is that any material object is exposed to the phenomenon of ageing, i.e., gradual deterioration of its original characteristics. Technological system is no exception to this rule, it is also susceptible to deterioration in their operational characteristics of over time, and this process as faster, as more intense use of technological possibilities. One of the major operational characteristics of technological system is its ability to ensure the desired dimensional accuracy and surface quality at minimum production costs. The deterioration of the ability of technological to ensure the required dimensional accuracy, is connected with many physical phenomena occurring during exploitation.

The main factors causing the aging of technological system the following:

– wear in use friction surfaces of the elements technological equipment
– error of the spatial position of the individual elements of technological equipment;
– thermal deformation ;
– changes in the geometric parameters of the cutting tool;
– various kinds of breakdowns.

These factors do not influence individually and independently of each other, but operate simultaneously, thereby significantly degrading the performance of technological system.

With an ever increasing requirements, placed on scientific and technical progress to the quality of product engineering production, it is necessary to ensure the minimization of the impact, exerted by the unfavorable factors in the technological system. In spite of the revealed problem of reducing the dimensional accuracy during the operation of equipment, CNC lathes are an essential part of a technological system of modern engineering production. Therefore it is necessary look for ways to solve this problem. In order to improve the dimensional accuracy of processing on CNC lathes, it is proposed to provide operational input corrections in CT movement of executive bodies of the machine, which allows to compensate the system error appearing in work. From the definition of the technological system implies that a complex of functionally related technological equipment, the production of objects and performers to execute regulated conditions of production. This definition gives us a generalized representation of what is a technological system, to the same from the definition is not entirely clear degree of automation, performed by the system processes or operations. So it is important to differentiate definitions technological systems, where used universal equipment and where used CNC lathes.

We introduce the concept of an automated system of technological operations (ASTO). ASTO – it is a complex of functionally related technological equipment, the production of objects, executors and intellectual control system processing on CNC lathes to execute in regulated production conditions specified of technological processes of or operations in the shortest possible machining time and at the least cost of material and human resources.

The general form of structure automated ASTO can be represented as a neural graphic pattern illustrated in figure 1.

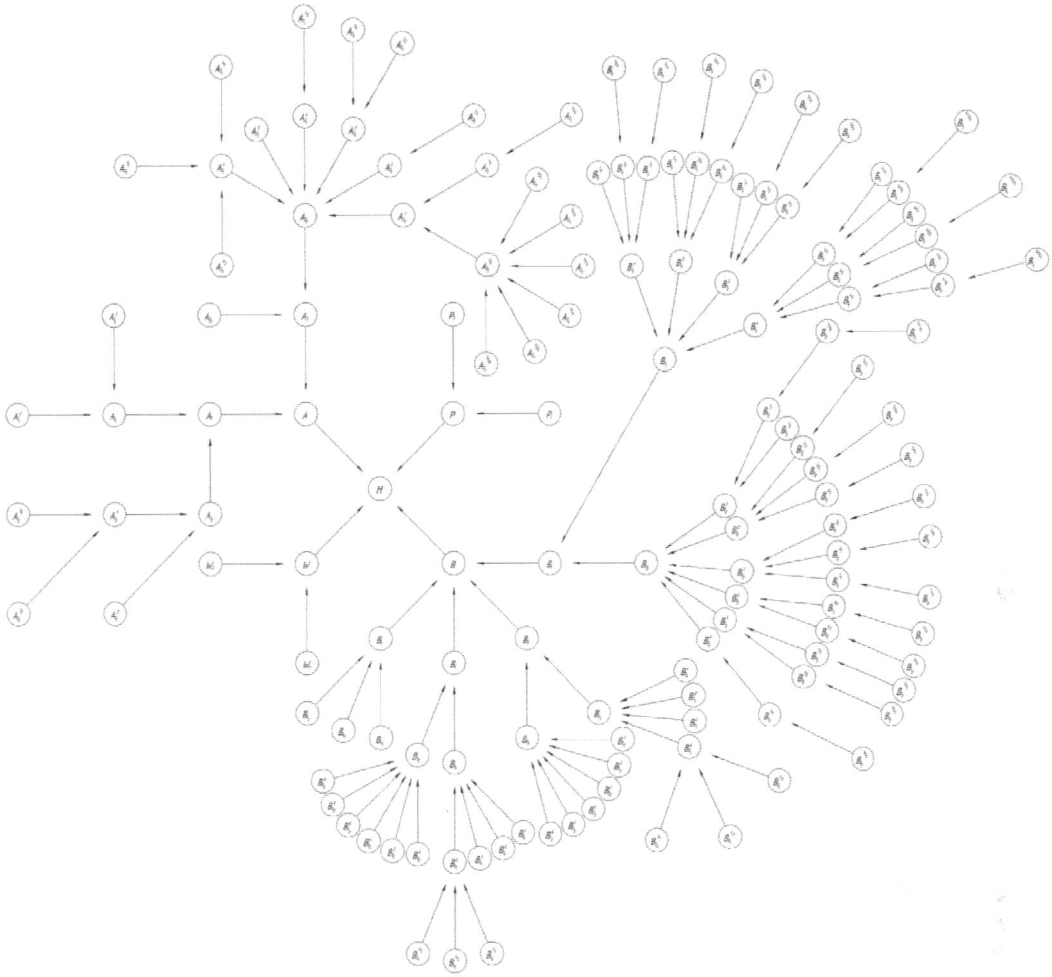

Figure 1. The neural representation graphic pattern intellectual control system processing on CNC lathes.

The proposed structure of ASTO is not a full and complete (elements that make up the ASTO shown in Figure 1), but enough to approach the question of mathematical modeling and the development of mathematical models of various processes occurring in it and necessary for the development of intellectual control system processing on CNC lathes.

Table 1. The elements constituting ASTO.

Symbol	Name of element ASTO
M	Automated system of technological operations (ASTO – ISDU)
P	Subject of production (labor)
P_1	Variety of physical and mechanical properties of the material object of production (labor)
P_2	Variety geometrical parameters of the object of production (labor)
W	Workers on engineering process
W_1	Mechanical technician of engineering process

Symbol	Name of element ASTO
W_2	Technological equipment operator
A	Variety of technological equipment
A_1	Technological equipment
A_{1_1}	Variety of technological equipment
A_{1_2}	Variety of tool support
A_2	Technological equipment
A_{2_1}	Cutting machine
A_{2_2}	Variety means of maintenance of working capacity tool
B	Technical equipment an automated system control of engineering process (TEASC EP)
B_1	System diagnostic aids
B_{1_1}	System of aids preliminary diagnosis of subsystems and elements of technological system
B_{1_2}	System of aids operative diagnosis of subsystems and elements of technological system
B_2	System processing diagnostics data
B_{2_1}	System processing data of the provisional diagnosis of the technological system
B_{2_2}	System processing of operative diagnostic technology system
B_3	System determining the correction parameters
B_{3_1}	System determining the correction parameters according to the preliminary diagnosis
B_4	System of numerical control machine (CNC)
B_{4_1}	System of formation and operational input of corrections
B_{4_2}	System operational data during processing
B_{4_3}	System of numerical control of lathe (PCNC)

As part of the study on an experimental setup shown in Figure 2, the possibility of creating intellectual control system processing on CNC machines, exercise power loading of the executive bodies to simulate the errors occurring in the technological system from the impact of the cutting forces, leading to the rejection of the surfaces to be treated details reversal of the turret and a change in spatial position the caliper.

It was found, that the results obtained during the research of deviation of the elements of technological system leading to the formation of error processing, are correlated with applicable cutting forces.

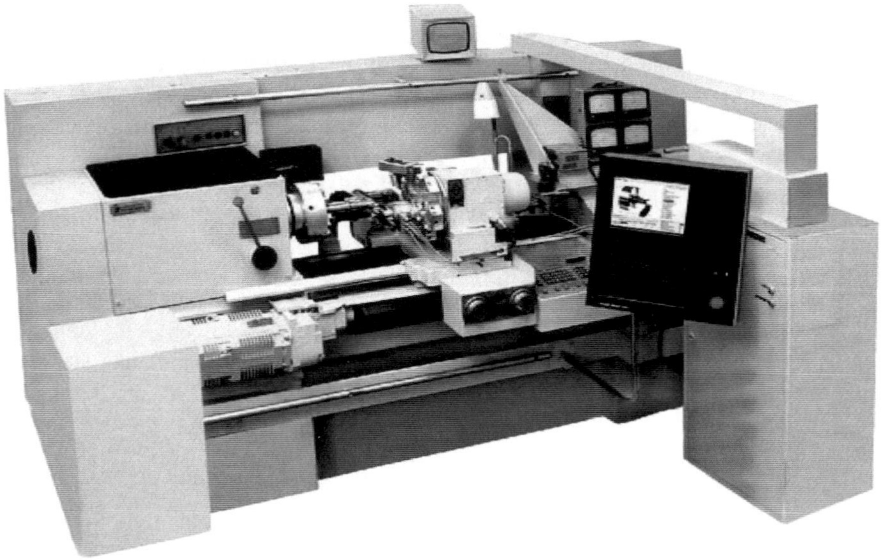

Figure 2. Experimental CNC lathe – based unit.

The study also noted that the processes of diagnosing the state of technological systems and drives, it is advisable to divide the preliminary and operational, that reduces the complexity of setting up specific technology systems and meets the requirements for the implementation of flexible processing technologies.

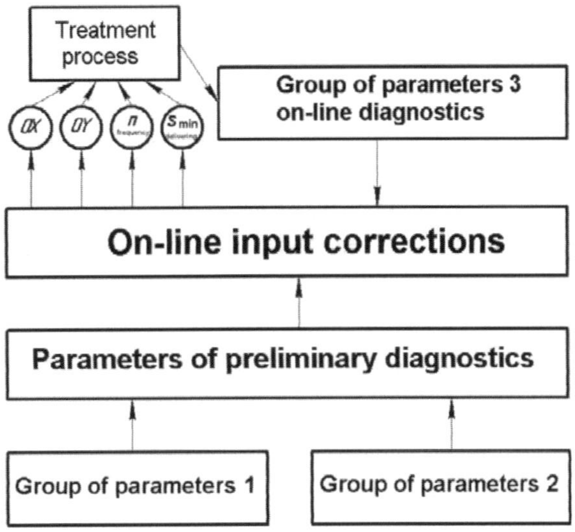

Figure 3. The relationship of process parameters of diagnosis and management.

From this graphic can be seen that the three basic groups of parameters stand apart:

- parameter group 1, a preliminary diagnosis of the parameters used to define values of operative correction in the processing for all the various parts and tools.

- parameter group 2, a preliminary diagnosis of the parameters used to define values of operative correction in the processing of specific parts and tools.
- parameter group 3, a current operational parameters of the diagnosis used to define the values of the input sum and surgical correction in the processing of the path of the tool and cutting conditions.

This generalized scheme of diagnosing the state of the technological system allows to come close to the issue of mathematical modeling of diagnostics.

Mathematical model of automating the process of diagnosis is not adequate diagnostic process performed by a man. The mathematical model contains features inherent in the machine process, and saves the basic features of the process performed by a man. This article presents mathematical models and their verbal description, diagnosing the state of the technological system, written using the language of mathematical logic and predicate logic. In this article, as an example, the mathematical model (1) diagnosis of deviations of surfaces of structural elements of a part under the influence of the cutting forces arising in ASTO during the processing of written with the help of mathematical logic and its section of predicate logic.

$$\begin{gathered} \forall M \exists B_{1_2}{}^{2} \in \left(\left(\left(P \subset A_{1_1}{}^{1} \right) \in \left(B_{1_2}{}^{2_2} \supset B_{1_2}{}^{2_{2_1}} \right) \right) \in \left(B_{1_2}{}^{2_1} \supset B_{1_2}{}^{2_{1_1}} \right) \right) \to \left(B_{1_2}{}^{2_3} \in B_{1_2}{}^{2_{3_1}} \right) \\ M = A \supset P \supset W \supset B \\ B_{1_2}{}^{3_{2_1}} = \left\{ x \in N, F\left(x\right) \right\}; \\ F\left(x\right) = x_1 \cdot q^{n-1}; \end{gathered} \tag{1}$$

Read: for a variety of ASTO (M), including multiple of technological equipment (A), a multiple of items of production (labor) (P), a lot of workers (W) and a multiple of means of automated process control systems (B) - there are many systems diagnostics changes in the spatial position.

($B_{1_2}{}^{2}$)the subject of work (P),established in technological equipment ($A_{1_2}{}^{1}$),loading using a device ($B_{1_2}{}^{2_2}$) a variety of weights ($B_{1_2}{}^{2_{2_1}}$), measured by device ($B_{1_2}{}^{2_1}$),including feeler ($B_{1_2}{}^{2_{1_1}}$), registered device ($B_{1_2}{}^{2_3}$) a lot of deviations ($B_{1_2}{}^{2_{3_1}}$) spatial position.

Or, from the perspective of an engineer: diagnostics automated technological system operations consisting of technological equipment, the subject of production (labor), contractors and equipment of automated process control system, by definition, changes in the spatial position of the workpiece relative to the OZ axis when it is loading component of cutting force PZ is as follows: on the blank mounted in the fixture affected by a device that simulates the action of cutting force component PZ, the measuring device disposed at the opposite side, comprising a probe, determines the amount of deflection recorded recording unit.

The proposed approach to the mathematical modeling of processes in ATCO allow to solve the problem of intellectual control system processing CNC and, consequently, reduce the dimensional accuracy machining with CNC machines as a result of the phenomenon of aging. The solution to this problem provides a reduction in production costs associated with the manufacture of technological engineering products.

References

[1] G S Andreev, The stress state of the cutting part of the cutter with periodic cutting, Durability of the cutting tool, Moscow, 1969, pp. 28-37.
[2] A I Betaneli, Communication method of calculating the strength of the cutting part of the tool, Bulletin of engineering. 2 (1965) 16-18.
[3] V F Bobrov, Determination of stresses in the cutting part of the cutting tools, High-performance cutting in mechanical engineering, Moscow, 1966, pp. 228-233.

[4] A M Geniatulin, Holographic method research strained elements of assorted cutters, Optical-geometrical methods in the study of deformations and stresses and their standardization, Gorky, 1982, pp. 83-84.

[5] M B Gordon, Methodology and some results of a study of patterns of distribution of the friction forces and tangential stresses on the contact length of the chip from the front surface of the cutter, Issues in the application of cutting fluids in machining metals, Ivanovo, 1965, pp. 41 - 44.

[6] V A Kuznezhov, Technological processes in mechanical engineering textbook for use in the learning process of educational institutions implementing programs of vocational education - Moscow, 2009, Ser. Secondary vocational education. Welding production.

[7] O S Lomova, Mathematical modeling of structural changes in the surfaces of the workpieces at thermal perturbations in the process of sanding, Omsk scientific bulletin. 2(120) (2013), pp. 95-98.

[8] S I Petrushin, I M Bobrovich, M A Korchuganova, Optimal design of the shape of the cutting blade tools, Tomsk, 1999.

[9] M F Poletika, M H Uteshev, Study of the processes of cutting the polarization-optical method, Bulletin of the Tomsk Polytechnic Institute, 1964, pp. 114-115.

[10] S I Takhman, Patterns of wear and forecast basis the wear resistance of tools from the standard hard alloys. Mechanics and physics of processes on the surface and in the contact of solid bodies, parts of technological equipment, S. I. Tahman. 3(2010), pp. 64-72.

[11] V G Shalamov, Mathematical modeling in machining of metals, Chelyabinsk, 2007.

[12] A S Yanushkin, S O Safonov, D V Lobanov, Improvement of technological processes of machine-building productions, Bratsk, 2006.

[13] A I Afonasov, A A Lasukov, Elementary Chip Formation in Metal Cutting, Russian Engineering Research. 3 (2014), pp. 152–155.

IOP Conf. Series: Materials Science and Engineering **125** (2016) 012016 doi:10.1088/1757-899X/125/1/012016

Influence of Shielding Gas and Mechanical Activation of Metal Powders on the Quality of Surface Sintered Layers

N A Saprykina, A A Saprykin, D A Arkhipova
Yurga Institute of Technology of National Research Tomsk Polytechnic University Affiliate, Tomsk, Russia
The study was performed by a grant from the Russian Science Foundation (project №15-19-00191).

e-mail: saprikina@tpu.ru

Abstract. The thesis analyses the influence of argon shielding gas and mechanical activation of PMS-1 copper powder and DSK-F75 cobalt chrome molybdenum powder on the surface sintered layer quality under various sintering conditions. Factors affecting the quality of the sintered surface and internal structure are studied. The obtained results prove positive impact of the shielding gas and mechanical activation. Sintering PMS-1 copper powder in argon shielding gas after mechanical activation leads to reduced internal stresses and roughness, as well as improved strength characteristics of the sintered surface. Analysis of sintered samples of mechanically activated DSK-F75 cobalt chrome molybdenum powder shows that the strength of the sintered surface grows porosity and coagulation changes.

Introduction

Layered synthesis technologies are a rapidly progressing trend in the development of modern industry. A separate line of this development is layered laser sintering of metal powder compositions based on 3D CAD-model allowing applicative products. As far as the layered synthesis of an applicative product basically consists in shaping a single layer, a problem resides in strains which arise in a single sintered layer and thus hamper a uniform application of the next powder layer and deform the product. For a product obtained by laser layered sintering to perform its function, the required quality characteristics must be provided. The main quality parameters of a sintered product are precision, surface layer condition, physical and mathematical properties, durability. A considerable problem in providing quality of the surface layer of a sintered product consists in strains which prevent a uniform application of the following powder layer and distort a product shape. To solve this problem it is necessary to research the influence of physical and mechanical properties of powder materials, laser sintering modes and process conditions of powder layer application [1]. The purpose of this research is to study factors effecting quality and inner structure of a sintered surface as well as to analyze the influence of argon shielding gas, mechanical activation of copper powder PMS-1 and cobalt chrome molybdenum powder DSK-F75 on the quality of a sintered surface layer under different process conditions. Thickness and roughness of a sintered surface and strength of sintered samples have been examined.

Result and Discussion

The sintered process starts with assigning a powder material which directly determines sintering conditions. Important properties of metal powders include their shape, structure, character of interparticle interaction. Particles can be of the following shapes: fibers, flakes, plates, cubes, spheres etc. The size of porous intervals depends on the particle shape and their just proportions. To minimize pores spherical particle are combined with particles of a similar shape and dendritic particles fuse with flaky ones [2]. It also should be kept in mind that powder materials melting temperature decreases as degree of dispersion grows [3]. Packaging density is influenced by particle surface roughness. Mechanical interlocking is one of the types of particle connection [4]. Friction between the particles results from adhesive interaction, that's why great attention is paid to activating powder compositions [5]. When one component powder material is sintered with materials with similar melting temperature under continuous laser impact drops are formed [6]. It should be taken into consideration that thermal physical and chemical properties of powder materials differ from reference ones of solid materials.

Sintering process conditions greatly influence the surface layer quality. Intensity of laser radiation depends on powder material melting temperature, heat conductivity index as well as particle shape and size. For heat-resistant materials laser radiation power must be increased and the speed of laser ray movement must be reduced. When the intensity is improper a powder material is lighted [7].

Increase in powder density leads to decrease in depth of sintering. It can be increased by raising laser radiation power but it causes energy loss.

The following factors considerably influence a prototype quality: coagulation (fusion of molten metal into separate drops), pore formation, internal strains and deformations.

Study of the influence of argon shielding gas on the sintered surface layer quality was carried out with copper powder PMS-1 and cobalt chrome molybdenum powder DSK-F75 under different process conditions. Argon provides a shielding environment which eliminates interaction of powder products with oxygen and nitrogen and also strengthen the product surface. Copper powder PMS-1 being sintered in argon, the sintered surface changed its color, it became goldish, the samples had a harder surface without any cracks [8, 9]. Figures 1-2 shows samples sintered in the air and in argon shielding gas compared. Sintered layer thickness increases from 115mkm up to 200mkm, the sample strength grows.

Figure 1. External appearance of PMS-1 surface (x2). Process conditions: P=30 W, V=3000 mm/min, S=0.3 mm, t=26 ^{0}C, in the air.

Figure 2. External appearance of PMS-1 surface (x2). Process conditions: P=30 W, V=3000 mm/min, S=0.3 mm, t=26 ^{0}C, in argon

Figures 3-4 shows samples sintered in argon under the following process conditions: P=15 W, V=200 mm/min, S=0.3 mm, t=200 ^{0}C. Cross cracks and longitudinal cracks are not found, sintered layer thickness changes insignificantly from 1675 when sintered in the air to 1735 when sintered in argon.

Figure 3. External appearance of PMS-1 surface (x2). Process conditions: P=15 W, V=200 mm/min, S=0.3 mm, t=200 ^{0}C, in the air.

Figure 4. External appearance of PMS-1 surface (x2). Process conditions: P=15 W, V=200 mm/min, S=0.3 mm, t=200 ^{0}C, in argon

IOP Conf. Series: Materials Science and Engineering **125** (2016) 012016 doi:10.1088/1757-899X/125/1/012016

Figures 5-6 shows samples sintered under the following conditions P=30 W, V=200 mm/min, S=0.1 mm, t=200 ^{0}C, in the air and in argon. Surface layer quality is dramatically changed. Roughness changes from 525 to 115 mkm, thickness of the sintered layer changes insignificantly from 850 to 915 mkm. The sample sintered in argon lacks cross and longitudinal cracks.

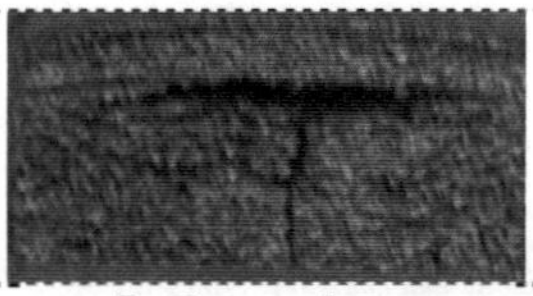

Figure 5. External appearance of PMS-1 surface (x2). Process conditions: P=30 W, V=200 mm/min, S=0.1 mm, t=200 ^{0}C, in the air.

Figure 6. External appearance of PMS-1 surface (x2). Process conditions: P=30 W, V=200 mm/min, S=0.1 mm, t=200 ^{0}C, in argon

Sintering in argon shielding gas improves the quality of the surface layer produced of copper powder PMS-1. Roughness is reduced, defects are eliminated.

The influence of mechanical activation on the sintered surface layer quality was studied through sintering copper powder PMS-1 and cobalt chrome molybdenum powder DSK-F75 under different process conditions. Mechanical processing of the powder was carried out in a centrifugal mill AGO-2, in steel mill shells filled with steel balls 6mm in diameter with total mass 600gr and loaded with 30gr of the powder. Mechanical activation increases degree of powder dispersion, aggravates particles crystal lattice imperfection [10] which leads to quick oxidation and allows sintering at unusually low temperatures [11]. Intensive milling of particles increases their sum surface, rises surplus energy of the powder and enlarges surface layer thickness. The mechanical processing was carried out during one, three, five and seven minutes. Figures 4-8 show pictures of sintered cobalt chrome molybdenum surfaces obtained of non-activated and variously activated powders under different processing conditions. In figures 7-11 P=10 W V=100 mm/min, S=0.1 mm t=200 ^{0}C for non-activated sintered surface R_Z=525 mkm, after 1 minute activation Rz=545 mkm after 3 minutes activation Rz=540 mkm, after 5 minutes activation Rz= 615 mkm, coagulation increases, after 7 minutes activation Rz=545 mkm [12].

Figure 7. External appearance of sintered surface DSK-F75 (x2). Processing modes: P=10 W, V=100 mm/min, S=0.1 mm, t=200 ^{0}C, non-activated.

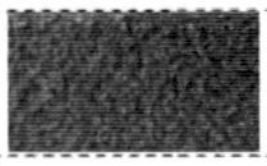

Figure 8. External appearance of sintered surface DSK-F75 (x2). Processing modes: P=10 W, V=100 mm/min, S=0.1 mm, t=200 ^{0}C, after 1 minute activation.

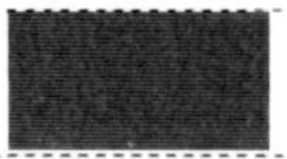

Figure 9. External appearance of sintered surface DSK-F75 (x2). Processing modes: P=10 W, V=100 mm/min, S=0.1 mm, t=200 ^{0}C, 3 minutes activation.

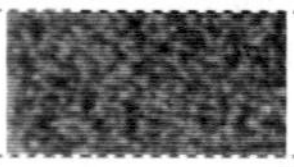

Figure 10. External appearance of sintered surface DSK-F75 (x2). Processing modes: P=10 W, V=100 mm/min, S=0.1 mm, t=200 ^{0}C, after 5 minutes activation.

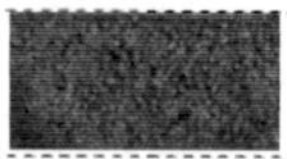

Figure 11. External appearance of sintered surface DSK-F75 (x2). Processing modes: P=10 W, V=100 mm/min, S=0.1 mm, t=200 ^{0}C, after 7 minutes activation.

Figures 12-16 shows that roughness of the sintered surface changes insignificantly, strength increases. The sintered sample after seven minutes of activation has reduced porosity and solid spots on the surface.

IOP Conf. Series: Materials Science and Engineering **125** (2016) 012016 doi:10.1088/1757-899X/125/1/012016

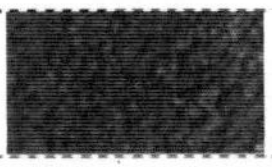

Figure 12. External appearance of sintered surface DSK-F75 (x2). Processing modes: P=20 W, V=100 mm/min, S=0.1 mm t=200 ^{0}C, non-activated.

Figure 13. External appearance of sintered surface DSK-F75 (x2). Processing modes: P=20 W, V=100 mm/min, S=0.1 mm t=200 ^{0}C, after 1 minute activation.

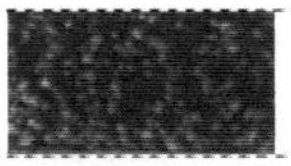

Figure 14. External appearance of sintered surface DSK-F75(x2). Processing modes: P=20 W, V=100 mm/min, S=0.1 mm t=200 ^{0}C, 3 minutes activation.

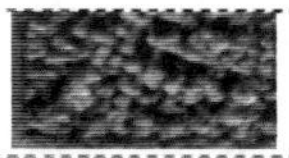

Figure 15. External appearance of sintered surface DSK-F75(x2). Processing modes: P=20 W, V=100 mm/min, S=0.1 mm t=200 ^{0}C, after 5 minutes activation.

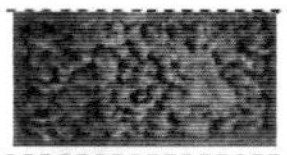

Figure 16. External appearance of sintered surface DSK-F75(x2). Processing modes: P=20 W, V=100 mm/min, S=0.1 mm t=200 ^{0}C, after 7 minutes activation.

Increase in activation time leads to reduced diameters of coagulated particles and their uniform distribution on the sintered surface. The strength of the sintered surface increases.

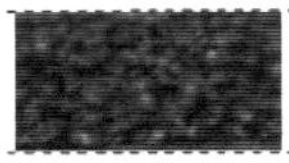

Figure 17. External appearance of sintered surface DSK-F75 (x2). Processing modes: P=20 W, V=300 mm/min, S=0.1 mm t=26 ^{0}C, non-activated.

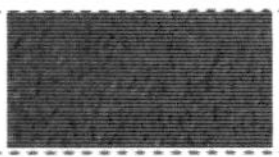

Figure 18. External appearance of sintered surface DSK-F75 (x2). Processing modes: P=20 W, V=300 mm/min, S=0.1 mm t=26 ^{0}C, after 1 minute activation.

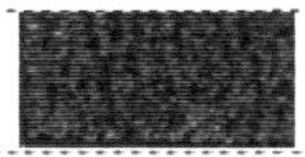

Figure 19. External appearance of sintered surface DSK-F75 (x2). Processing modes: P=20 W, V=300 mm/min, S=0.1 mm t=26 ^{0}C, 3 minutes activation.

Figure 20. External appearance of sintered surface DSK-F75 (x2). Processing modes: P=20 W, V=300 mm/min, S=0.1 mm t=26 ^{0}C, after 5 minutes activation.

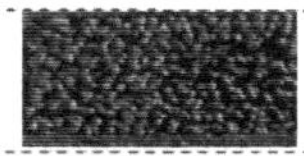

Figure 21. External appearance of sintered surface DSK-F75 (x2). Processing modes: P=20 W, V=300 mm/min, S=0.1 mm t=26 ^{0}C, after 7 minutes activation.

Figure 22. External appearance of sintered surface DSK-F75 (x2). Processing modes: P=10 W, V=300 mm/min, S=0.15 mm, t=26 ^{0}C, non-activated.

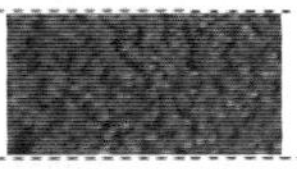

Figure 23. External appearance of sintered surface DSK-F75 (x2). Processing modes: P=10 W, V=300 mm/min, S=0.15 mm t=26 ^{0}C, after 1 minute activation.

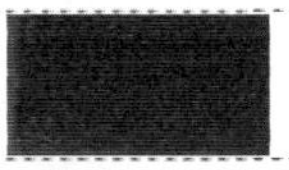

Figure 24. External appearance of sintered surface DSK-F75 (x2). Processing modes: P=10 W, V=300 mm/min, S=0.15 mm, t=26^{0}C, 3 minutes activation.

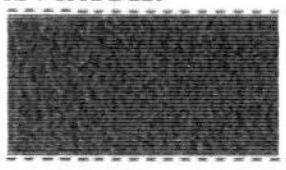

Figure 25. External appearance of sintered surface DSK-F75 (x2). Processing modes: P=10 W V=300 mm/min, S=0.15 mm t=26 ^{0}C, after 5 minutes activation.

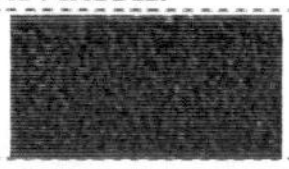

Figure 26. External appearance of sintered surface DSK-F75 (x2). Processing modes: P=10 W, V=300 mm/min, S=0.15mm t=26 ^{0}C, after 7 minutes activation.

Figure 27. External appearance of sintered surface DSK-F75 (x2). Processing modes: P=20 W, V=300 mm/min, S=0.15 mm, t=26 ^{0}C, non-activated.

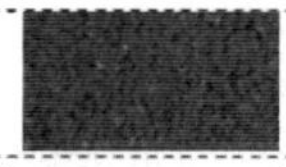

Figure 28. External appearance of sintered surface DSK-F75 (x2). Processing modes: P=20 W, V=300 mm/min, S=0.15 mm, t=26 ^{0}C, after 1 minute activation.

Figure 29. External appearance of sintered surface DSK-F75 (x2). Processing modes: P=20 W, V=300 mm/min, S=0.15 mm t=26 ^{0}C, 3 minutes activation.

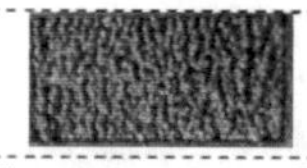

Figure 30. External appearance of sintered surface DSK-F75 (x2). Processing modes: P=20 W, V=300 mm/min, S=0.15 mm, t=26^{0}C, after 5 minutes activation.

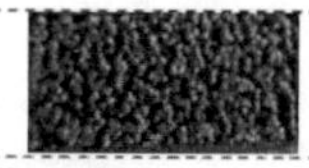

Figure 31. External appearance of sintered surface DSK-F75 (x2). Processing modes: P=20 W, V=300 mm/min, S=0.15 mm, t=26 ^{0}C, after 7 minutes activation.

Figures 32-34 shows diffraction patterns of cobalt chrome molybdenum powder. Y-axis shows diffraction peaks intensity, X-axis shows angular characteristic with maximum shooting angle 2 theta.

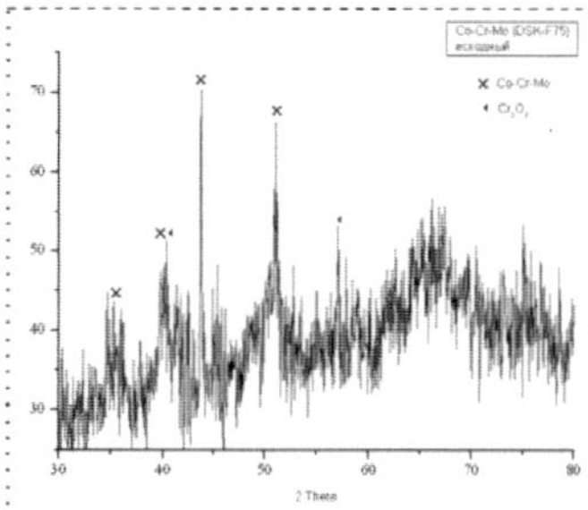

Figure 32. Diffraction pattern of cobalt chrome molybdenum powder, non-activated.

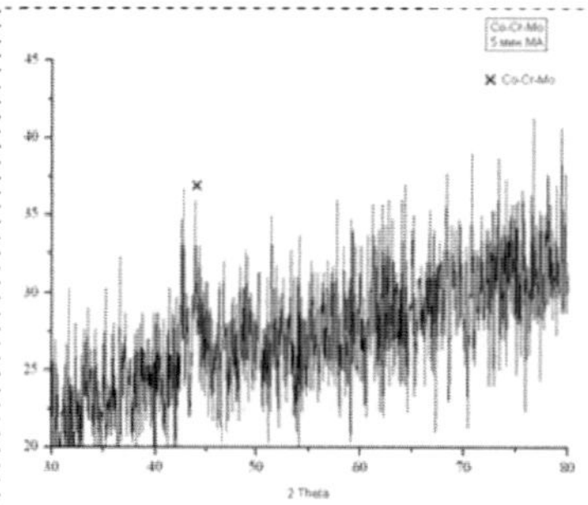

Figure 33. Diffraction pattern of cobalt chrome molybdenum powder, after 5 minutes activation.

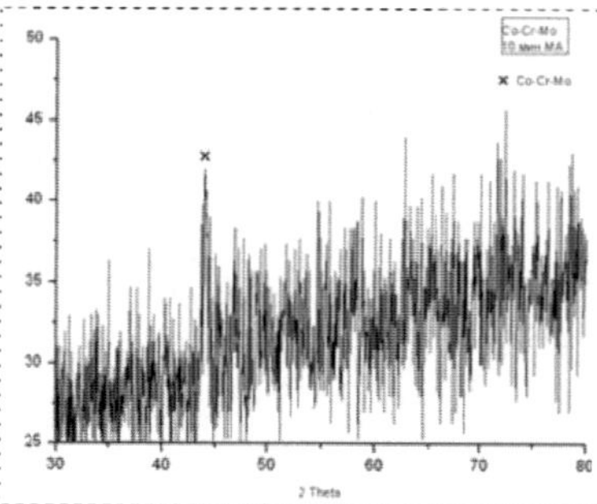

Figure 34. Diffraction pattern of cobalt chrome molybdenum powder, after 10 minutes activation.

Comparison of diffraction patterns of non-activated (Fig.9a) and mechanically activated (Fig.9b) cobalt chrome molybdenum powder shows that after mechanical activation chromic oxide vanishes. Activation time does not considerably influence the diffraction pattern. After activation peaks arise only in one position.

Figures 35-54 show comparative pictures of sintered layers of copper powder composition PMS-1 obtained under different process conditions as well as of non-activated and variously activated powders. Three minutes activation results in increasing strength of a sintered sample, figures 35-37, Rz changes slightly from 625mkm to 630mkm, sintered layer thickness grows from 820mkm up to 950mkm.

Figure 35. External appearance of PMS-1 surface (x2). Process conditions: P=30 W, V=3000 mm/min, S=0.3 mm, t=26 ^{0}C, non-activated.

Figure 36. External appearance of PMS-1 surface (x2). Process conditions: P=30 W, V=3000 mm/min, S=0.3 mm, t=26 ^{0}C, after 1 minute activation

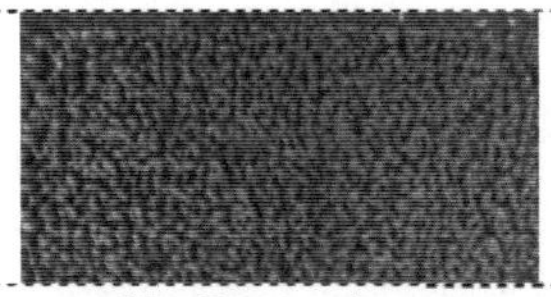

Figure 37. External appearance of PMS-1 surface (x2). Process conditions: P=30 Wt, V=3000 mm/min, S=0.3 mm, t=26 ^{0}C, after 3 minutes activation.

Comparison of two samples figures 38-39 shows improved quality of the surface layer, cracks vanish, roughness is reduced from 975 mkm to 715 mkm.

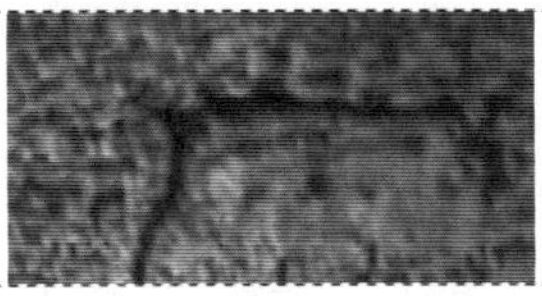

Figure 38. External appearance of sintered surface PMS-1 (x2). Process conditions: P=30 W, V=200 mm/min, S=0.3 mm, t=26 ^{0}C, non-activated.

Figure 39. External appearance of sintered surface PMS-1 (x2). Process conditions: P=30 W, V=200 mm/min, S=0.3 mm, t=26 ^{0}C, after 3 minutes activation.

In figures 40-42 surface roughness of the sintered samples of mechanically activated powder materials decreases from 700mkm to 426mkm.

Figure 40. External appearance of PMS-1 surface (x2). Process conditions: P=15 W, V=200 mm/min, S=0.3 mm, t=26 ^{0}C, non-activated.

Figure 41. External appearance of PMS-1 surface (x2). Process conditions: P=15 W, V=200 mm/min, S=0.3 mm, t=26 ^{0}C, after 1 minute activation

Figure 42. External appearance of PMS-1 surface (x2). Process conditions: P=15 W, V=200 mm/min, S=0.3 mm, t=26 ^{0}C, after 3 minutes activation.

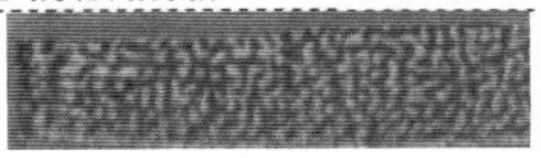

Figure 43. External appearance of PMS-1 surface (x2). Process conditions: P=30 W, V=3000 mm/min, S=0.1 mm, t=200 ^{0}C, non-activated.

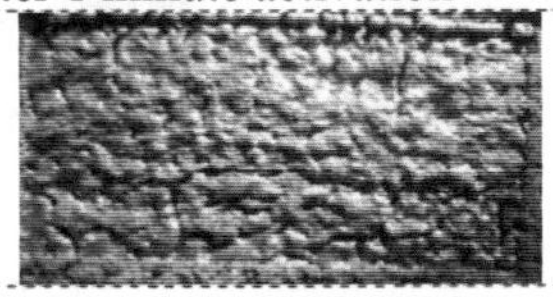

Figure 44. External appearance of PMS-1 surface (x2). Process conditions: P=30 W, V=3000 mm/min, S=0.1 mm, t=200 ^{0}C, after 1 minute activation

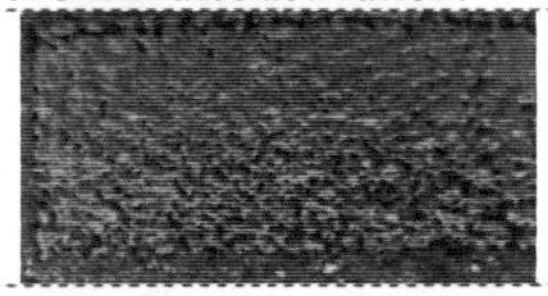

Figure 45. External appearance of PMS-1 surface (x2). Process conditions: P=30 W, V=3000 mm/min, S=0.1 mm, t=200 ^{0}C, after 3 minutes activation.

Figure 46. External appearance of PMS-1 surface (x2). Process conditions: P=15 W, V=3000 mm/min, S=0.1 mm, t=200 ^{0}C, non-activated.

Figure 47. External appearance of PMS-1 surface (x2). Process conditions: P=15 W, V=3000 mm/min, S=0.1 mm, t=200 ^{0}C, after 1 minute activation

Figure 48. External appearance of PMS-1 surface (x2). Process conditions: P=15 W, V=3000 mm/min, S=0.1 mm, t=200 ^{0}C, after 3 minutes activation.

Figure 49. External appearance of sintered surface PMS-1 (x2). Process conditions: P=30W, V=200 mm/min, S=0.1 mm, t=26 ^{0}C, non-activated.

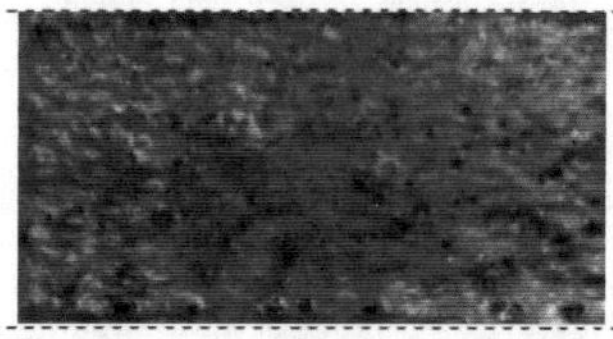

Figure 50. External appearance of sintered surface PMS-1 (x2). Process conditions: P=30 W, V=200 mm/min, S=0.1 mm, t=26 ^{0}C, after 3 minutes activation.

Figure 51. External appearance of sintered surface PMS-1 (x2). Process conditions: P=15 W, V=200 mm/min, S=0.1 mm, t=26 ^{0}C, non-activated.

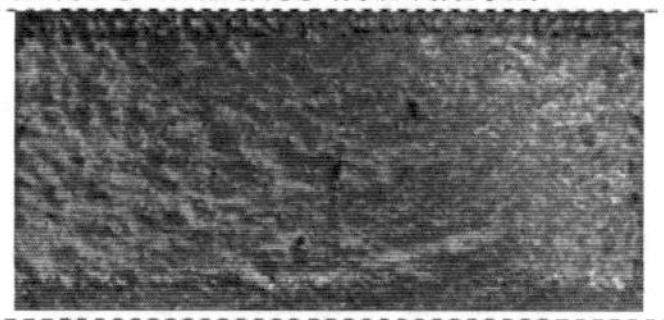

Figure 52. External appearance of sintered surface PMS-1 (x2). Process conditions: P=15 W, V=200 mm/min, S=0.1 mm, t=26 ^{0}C, after 3 minutes activation.

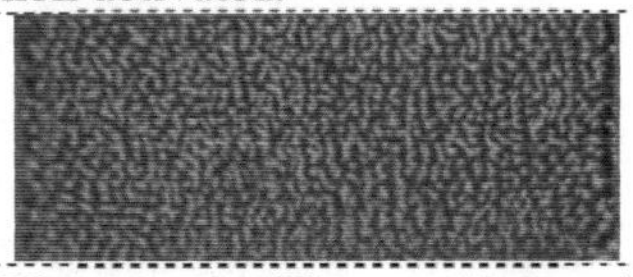

Figure 53. External appearance of sintered surface PMS-1 (x2). Process conditions: P=22 W, V=160 mm/min, S=0.2 mm, t=26 ^{0}C, non-activated.

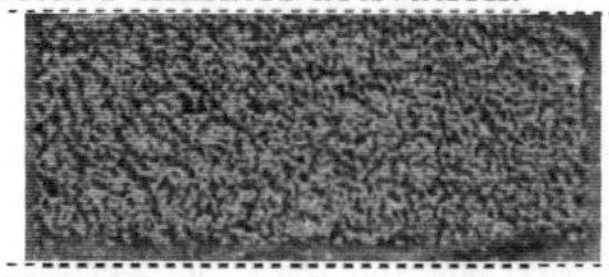

Figure 54. External appearance of sintered surface PMS-1 (x2). Process conditions: P=22 W, V=160 mm/min, S=0.2 mm, t=26 ^{0}C, after 3 minutes activation.

Thus the picture of samples compared clearly show the influence of mechanical activation on the surface quality: it leads to reduced coagulation and roughness.

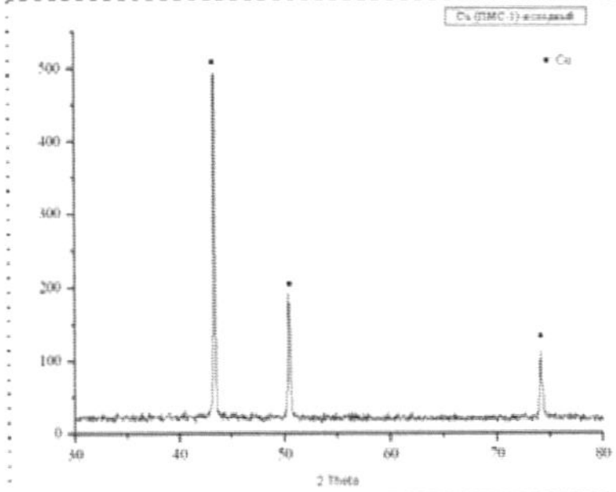

Figure 55. Diffraction pattern of copper powder PMS-1, non-activated.

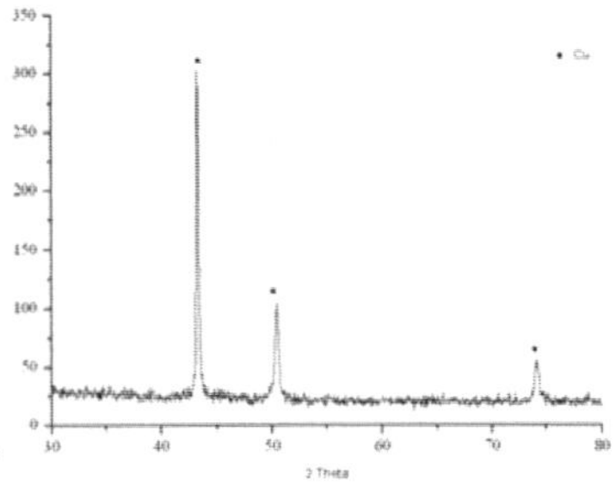

Figure 56. Diffraction pattern of copper powder PMS-1, after 3 minutes activation.

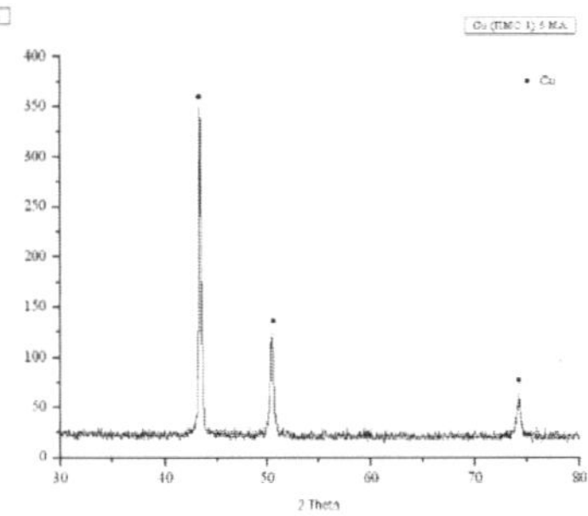

Figure 57. Diffraction pattern of copper powder PMS-1, after 5 minutes activation.

Conclusion

The research in properties of a sintered layer of activated and non-activated powder materials showed that a preliminary mechanical treatment effects the sintering process and provides an improved surface quality: the diameter of coagulated particles decreases, roughness is reduced. Internal structure and strength characteristics are improved.

Positive influence of a shielding gas and mechanical activation of metal powders on the quality of surface sintered layers is proved. In order to reduce roughness and improve the sintered layer quality it is recommended to sinter metal powders subjected to one and three minutes of activation in argon shielding environment which reduces roughness, improves internal structure and strength characteristics.

References

[1] Schmid M, Amado F, Levy G, Wegener K, Flowability of powders for selective laser sintering (SLS) investigated by round robin test, High Value Manufacturing: Advanced Research in Virtual and Rapid Prototyping: Proceedings of the 6th International Conference on Advanced Research in Virtual and Rapid Prototyping Leiria 2013.

[2] Panchenko V Ya, Laser technologies of machining: contemporary problems of fundamental research and applied designs, monograph M Phismatlit 2009

[3] Belomestnykh V N, Soboleva E G, Acoustic analogs of elasticity theory ratios for determining anisotropic Poisson ratios of cubic monocrystals 7th International Forum on Strategic Technology (IFOST - 2012) Tomsk: TPU Press 1 2012 p 499-502

[4] Gubaidulina R H, Petrushin S I, Galeeva A A, Selecting an Economical Variant of the Manufacturing Method of Engineering Product Fabrication under Current Conditions Applied Mechanics and Materials 682 2014 p 613-616.

[5] Arkhipov P V, Yanyushkin A S, Lobanov D V, The effect of diamond tool performance capability on the quality of processed surface Applied Mechanics and Materials 379 2013 p 124-130

[6] Chinakhov D A, Dependence of Silicon and Manganese Content in the Weld Metal on the Welding Current and Method of Gas Shielding Applied Mechanics and Materials 756 2015 p 92-96

[7] Saprykin A A, Saprykina N A, Dudikhin D V, Emelyanenko S M, Influence of layer-by-layer laser sintering modes on the thickness of sintered layer of cobalt-chromium-molybdenum powder, Advanced materials research 1040 2014 p 805-808

[8] Lychagin D V, Tarasov S Yu, Chumaevskii A V, Alfyorova E A, Macrosegmentation and strain hardening stages in copper single crystals under compression International Journal of Plasticity. 69 2015 p 36-53

[9] Babakova E V, Gradoboev A V, Saprykin A A, Ibragimov E A, Yakovlev V I, Sobachkin A V,

Comparison of Activation Technologies Powder ECP-1 for the Synthesis of Products Using SLS Applied Mechanics and Materials 756 2015 p 220-224
[10] Saprykina N A, Saprykin A A, Shigaev D A, Investigation of the factors affecting the quality of the surface obtained by laser sintering Metal 4 2011 p 78-82
[11] Saprykina N A, Saprykin A A, Matrunchik M S, Formation of Surface Layer of Cobalt Chrome Molybdenum Powder Products with Differentiation of Laser Sintering Modes Applied Mechanics and Materials 682 2014 p 294-298
[12] Saprykin A A, Ibragimov E A, Yakovlev V I, Influence of mechanical activation of powder on SLS process Applied Mechanics and Materials 682 2014 p 143-147

IOP Conf. Series: Materials Science and Engineering **125** (2016) 012017 doi:10.1088/1757-899X/125/1/012017

Surface Quality Multicriteria Optimization of Flat Parts from 06Cr14Ni6Cu2MoVaTi-SH Steel While Grinding by Varigrain High Porosity CBN Wheels

Y I Soler[1] and M T Nguyen[2]

[1]Soler Yakov Iosifovich, candidate of Technical Sciences, associate professor, Irkutsk National Research Technical University, Russian Federation
[2]Nguyen Manh Tiem, postgraduate, Irkutsk National Research Technical University, Russian Federation

e-mail: solera@istu.irk.ru;

Abstract. The grinding of flat parts from 06Cr14Ni6Cu2MoVaTi-SH high-strength corrosion-resistant steel was made by the highly porous wheel (HPW) of CBN30 (B76, B107, B126, V151) 100 OVK27-KF40 (GOST R 53922-2010, GOST R 53923-2010). As input variables for fuzzy logic modeling in the *Matlab* the following description of the surface quality is chosen: the microrelief parameters (GOST 25142-82) – R_{a1}, R_{max1} in the cross-feed direction, S_{m2} in the length feed direction, flatness deviation (GOST 24642-81) are introduced with the EFE_{max}, EFE_a and EFE_q; microhardness *HV*. Every parameter at the model input is presented with position measures (medians) and scattering measures (quartile latitude). The *Matlab* modeling has shown that the best quality of the part surfaces is provided with HPW CBN30 B151 during the multicriteria optimization.

1. Introduction

The grinding is used at the final stage of the manufacturing of high load critical parts. Its effectiveness is estimated by a great number of factors, among which there is roughness, form error, microhardness and etc. relating to surface engineering [1]. Most of them are introduced with some correlated specified attributes. These attributes are difficult to be estimated qualitatively and with high accuracy. Thus, it seems to be impossible to describe physical analogs and their experimental ones are of particular character having a limited area of application. That is why it is settled to use fuzzy logic making it possible to analyze a large number of grinding variables and cover a wide range of process specification.

The fuzzy logic is a perspective trend of the cybernetics development suggested and elaborated by *L. A. Zadeh* [2]. He suggested using words and phrases for linguistic variables and inaccurate reasoning. As a result, the content, meaning of the information and processing logic is passed in the form of probability for problem solving that cannot be described exactly, specifically in engineering, like automatic control, diagnostics, artificial intelligence and etc. In this regard the fuzzy logic is equivalent to the fuzzy-set theory A_l, i.e. the classes with fuzzy boundaries, presented with the ordered couple package made of the elements y_l of the universal sets $\{y_{lv}\}$ and corresponding grade of membership $\mu_A(y_l)$:

$$A_l = \{(y_l, \mu_A(y_l)) | y_l \in \{y_{lv}\}\}, \ v = \overline{1,n}.$$

The purpose of the present article is to study the interrelation between roughness, microhardness and flatness deviation of the surface of corrosion-resistant steel parts 06Cr14Ni6Cu2MoVaTi-SH (EP817-SH) with the help of the theory of fuzzy reasoning and the search for optimization parameters improving the quality of the part surfaces with pendulum grinding of the highly porous wheel (HPW) from the grains of cubic boron nitride (CBN).

2. Experimental technique

To implement the concept of fuzzy logic some grinding experimental data are used. The verification nature tests are carried out under the following constant conditions: flat surface grinding machine 3E711V; 1A1 wheels dimensioned of 200×20×76×5 mm (GOST R 53923-2010). The process-dependent parameters are cutting speed $v_w = 28$ m/s, longitudinal feed $s_l = 6$ m/min, cross-feed $s_c = 4$ mm/double stroke, cutting depth t=0.01 mm, operational allowance z = 0.1 mm. The lubricoolant 5% Akvol-6 emulsion (TU 0258-024-0014842-98), is supplied flowing on the workpiece in the amount of 7-10 l/min; the number of duplicating tests is in (1) – n = 30. The test subject is forms of steel 06Cr14Ni6Cu2MoVaTi-SH with mechanical properties: σ_{UST}=1310-1400 MPa, $\sigma_{0.2}$=1210-1240 MPa, δ=12-14%, ψ=57-60% - and the dimension of B×L×H =60×60×60 mm, ground butt end B×L. The index $l = \overline{1;4}$ reflects the characteristics of HPW: 1-CBN30 B76 100 OVK27-KF40; 2-CBN30 B107 100 OVK27-KF40; 3-CBN30 B126 100 OVK27-KF40; 4-CBN30 B151 100 OVK27-KF40 (GOST R 53922-2010, GOST R 53923-2010), in which only the size of CBN grains (grain) from B76 (200/230) up to B151 (100/120), where in the brackets the grain is indicated in mesh. The parameters of surface roughness (GOST 25142-82): R_a, R_z, R_{max}, S и S_m are measured with the profilograph-profilometer of 252 type produced by «Caliber» in two mutually orthogonal directions $i = \overline{1;2}$: correspondingly according to the vectors s_l (R_{a1}, R_{z1}, R_{max1} and etc.) and s_c (R_{a2}, R_{z2}, R_{max2} and etc.). The measuring of microhardness was made with the device PMT-3 under the condition of loading 1,96 H, the indenter lowering rate of 0,15 mm/s, the persistence time under load of 10-15s. The methods of searching for the attributes of flatness deviation (GOST 24642-81) such as the maximum EFE_{max}, arithmetic mean EFE_a, quadratic mean EFE_q are discussed in the article [4].

The specificity of the grinding is that the cutting capacity of the HPW cannot be possibly imagined as a determined value during the edge cutting machining. The above mentioned is conditioned by the fact that the abrasive grain in the tools have an arbitrary shape, chaotic arrangement in the bonded, the difference in height in radial axis and a different number of operating grain and cutting edges per unit area of its contact with the entry into the workpiece. This makes it possible to examine the observation of random variables (RV) and estimate their behaviour on the base of probability-theoretic approaches. In this case the experimental data presentation is supposed to be given in the form of independent sets $l = \overline{1;k}$:

$$\{y_{lv}\}, \ v = \overline{1;30}, \tag{1}$$

where v is a number of replicate tests.

The statistical methods are divided into two groups: parametric and nonparametric, specifically rank ones. Each of them has «its own field» [3] for the effective use. In the first case it is necessary to ensure the fulfillment of the two constraints imposed on RV: homogeneity of variance of deviation and normalcy of distribution. The discussed grinding requirements are often violated to any extent that can be accompanied by a considerable bias of estimator, confidence bounds and factors [3]. In such a situation it is reasonable to use a nonparametric method that is not connected with a certain family of distributions and does not use its properties. RV are estimated by the following univariate frequency allocation [5-6], GOST R 5725-2-2012:

- position measures (reference values)

averages $\bar{y_l} = y_{l\bullet}$, (2)

medians $\tilde{y_l}$; (3)

- scattering measures (precision)

deviation standards SD_l, (4)

the swing $R_l = (y_{max} - y_{min})_l$, (5)

quartile latitudes $QL_l = (y_{0,75} - y_{0,25})_l$. (6)

From the theoretical statistics it is known that a parametric method is based on the univariate frequency allocation (2), (4), (5), and the rank statistics is related to (3), (6). The acceptance of the null-hypothesis (H_0) by the homogeneity of variance of deviation and the normalcy of distribution is discussed in [5; 7]. To decrease labor content of the statistical calculations the software *Statistica* 6.1.478.0. is used in the research work.

The implementation of the fuzzy logic modeling for (2)-(6) is carried out in the *MATLAB* using a special bump pack of *Fuzzy Logic Toolbox*. The latter has a simple and well-engineered interface making it possible to easily design and diagnose fuzzy models [8-12]. For the assessment of the parts surface roughness quality Harrington's desirability function d is used [13]. In its construction there is an idea of natural value conversion of individual response into a dimentionless desirability or preference scale being related to psychophysical ones. Its purpose is to set up compatibility between physical and psychological parameters. The physical parameters are thought to be some possible responses characterizing the functioning of the object under study. Experimentally assigning the desirability scale is the identifying of the compliance between the received surface quality values and their desirability assessment.

3. Research results and their discussion

Searching for HPW in the *Matlab* the stress has been made on the reduction of the position measures and the scattering of the quality parameters excepting the reference value *HV*. The microhardness should be maintained at the top level. As for microasperity it is a priori known that the most high level parameters are positioned in the cross-feed direction (R_{a1}, R_{max1} and etc.) and the step-by-step ones such as S_2 and especially S_{m2} are located along the vector s_l. As far as reference lengths of the profile t_p are concerned, it is known that, if a profile section level is constant, they do not depend on the grinding conditions [1]. Hence, they are excluded from the multicriteria optimization of grinding. Therefore, all of the above taken into account the process optimization according to the surface microgeometry is made with three parameters R_{a1}, R_{max1} and S_{m2} being the most important ensuring reliability and durability of machines.

The testing (1) on the homogeneity of variance of deviation is conducted according to three groups of criteria: Hartley, Cochren, Barlett (in the software they are united into one package); Levene; Brown-Forsythe. It showed that H_0 concerning the homogeneity (1) at 5% level of importance is accepted at all parameters. Much worse are the results of the normalcy of distribution testing according to Shapiro-Wilk criterion (Table 1), for which H_0 is accepted carrying out the inequality: α_l>0.5, where α_l is acceptance reliability H_0 for the grinding parameters: $(R_a, R_{max}, S_{m2}, EFE_{max}, EFE_a, EFE_q, HV)_l$, $l = \overline{1;4}$. As is clear from Table 1, the normal distributions are found

Table 1. Normalcy of distribution testing according to Shapiro-Wilk criterion

Parameter	Calculated significance level α_l with variables $l = \overline{1;4}$			
	1	2	3	4
R_{al}	0.0100	0.0175	0.2897	0.0184
R_{max1}	0.1324	0.0239	0.7746	0.9949
S_{m2l}	0.00005	0.0827	0.0551	0.0129
EFE_{max}	0.0584	0.2144	0.0028	0.0412

EFE_a	0.0635	0.0864	0.0112	0.0686
EFE_q	0.0562	0.1359	0.0115	0.0451
HV	0.0270	0.0077	0.0462	0.4736

only in the two underlined cases out of 28 analyzed. Under the conditions of the normalcy of distribution violation (1) it is reasonable to use the rank statistics method with its one-dimensional distribution of frequencies (3) and (6). Modeling in the MATLAB the input data are assumed to be the parameters given in Table 2.

Table 2. Part surface quality input data for fuzzy logic modeling

l	Quality parameters													
	R_{a1l}, μm		R_{max1l}, μm		S_{m2l}, μm		EFE_{maxl}, μm		EFE_{al}, μm		EFE_{ql}, μm		HV_l, MPa	
	$\tilde{y}$	QL	$\tilde{y}$	QL	$\tilde{y}$	QL	$\tilde{y}$	QL	$\tilde{y}$	QL	$\tilde{y}$	QL	$\tilde{y}$	QL
1	0.61	0.07	3.72	0.54	83.36	28.39	12.00	11.00	5.88	4.92	6.96	6.01	4763.30	601.80
2	0.58	0.12	3.32	0.44	96.75	57.24	25.00	18.00	12.92	8.67	15.28	10.6	3958.80	1164.30
3	0.60	0.27	3.58	1.27	84.91	48.24	21.50	10.00	11.75	5.83	13.55	6.68	3376.20	506.00
4	0.56	0.11	3.37	0.79	80.52	36.88	14.50	12.00	7.46	5.17	9.17	6.22	3871.00	560.76

Note. Wheels CBN30 100 OVK27-KF40 l: 1- B76, 2-B107, 3-B126, 4-B151

The methods of statistics do not permit to fully evaluate the quality of the ground surface, while the microgeometry and precision of the parts form can be assessed by three characteristics for every parameter. Meanwhile, every variable is additionally presented by position measures and scattering. The above mentioned is even related to the parameter of HV_l. The statistical methods can only separately predict according to microhardness which HPW is better by reference values or which one is better under the precision of its forming. Having regard to the above the method of fuzzy logic is used making it possible to create an expert system of wheel classification in accordance with all the researched parameters. At the model input the attributes $(R_{a1}, R_{max1}, S_{m2}, EFE_{max}, EFE_a, EFE_q, HV)_l$, $l = \overline{1;4}$ simultaneously satisfying (3) and (6) are sent. The output of the model is the assessment of the surface quality for every HPW.

Table 3. Fuzzy logic results according to the choice of HPW grain for every researched parameter

Wheel l	Desirability function d_l according to parameters						
	Surface roughness			Flatness deviation			Microhardness
	R_{a1l}	R_{max1l}	S_{m2l}	EFE_{maxl}	EFE_{al}	EFE_{ql}	HV_l
1	0.500	0.499	0.823	0.827	0.870	0.870	0.825
2	0.656	0.827	0.173	0.173	0.130	0.132	0.173
3	0.175	0.174	0.541	0.620	0.367	0.473	0.500
4	0.809	0.821	0.821	0.797	0.827	0.820	0.787

Note. Grain l: 1- *B76*, 2-*B107*, 3-*B126*, 4-*B151*

The received input attributes are given in Table 3 and estimated with desirability function d [13].

According to the results shown in Table 3 a general model for HPW grain choice, $l = \overline{1;4}$, is developed in the conditions of synchronous position measures and scattering decreasing by all parameters of the surface roughness and flatness deviation. The model consists of four variables: three of them are input and one is output (Figure 1). Every input variable value is represented by a numerical range [0; 1] and its layout into three classes is shown in Table 4. From the analysis of the input parameters the microhardness is excluded, as it is represented by one attribute.

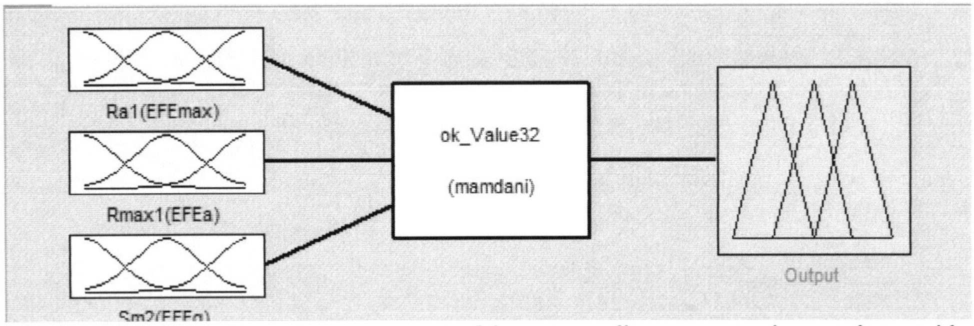

Figure. 1. The system of complex assessment of the parts quality parameters in accordance with all parameters of the surface roughness and flatness deviation

Table 4. Input parameters researching for HPW grain according to complex quality assessment of ground steel parts 06Cr14Ni6Cu2MoVaTi-SH

Type of assessment	Input parameters		
Linguistic	bad	middle	good
Numerical	[0.1; 0.5]	[0.1; 0.5; 0.5; 0.9]	[0.5; 0.9]

The membership function of «Output» input variable is presented by five classes of ground parts quality as in Table 5 and Figure 2.

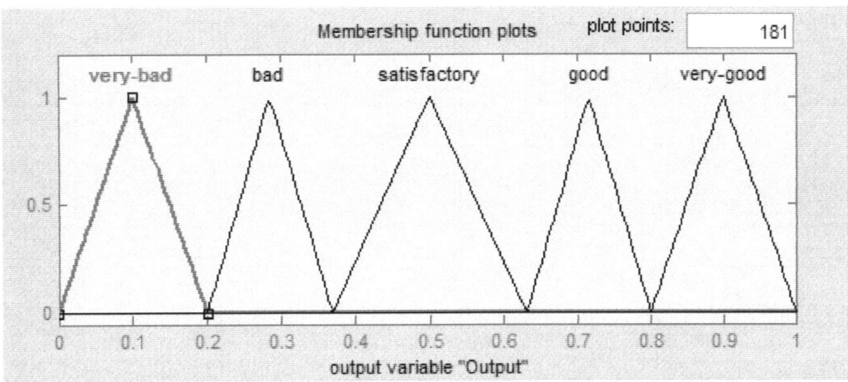

Figure 2. The membership function for the variable «Output» solving the problem of the search for the HPW optimum grain according to the complex assessment of the surface roughness and flatness deviation

Table 5. Parameter of «Output» searching for HPW optimum grain according to complex quality assessment of ground parts

Type of assessment	Input parameters				
Linguistic	Very bad	bad	satisfactory	good	very good
Numeric *d*	[0; 0.2)	[0.2; 0.37)	[0.37; 0.63)	[0.63; 0.8)	[0.8; 1]

The rules of fuzzy reasoning for the developed system include $N = 3\times3\times3 = 27$ of possible combinations of output parameters. In Table 6 there are the fuzzy rules for the system of parts surface quality classification. In the batch of *Fuzzy Logic Toolbox* there is a graphic interface simplifying the user's view of the rules of reasoning (*Rule Viewer*). Using the graphic format all possible reductions are shown, that makes it possible to forecast an output variable - Output. Every change of parts feature is displayed in the view of the output rules changing it accordingly.

Table 6. Fuzzy rules of parts classification system

Variant	$R_{a1}(EFE_{max})$	$R_{max1}(EFE_a)$	$S_{m2}(EFE_q)$	Output
1	good	good	good	very good
2	good	good	middle	very good
3	good	good	bad	good
4	good	middle	good	very good
5	good	middle	middle	good
6	good	middle	bad	satisfactory
7	good	bad	good	good
8	good	bad	middle	satisfactory
9	good	bad	bad	bad
10	middle	good	good	good
11	middle	good	middle	good
12	middle	good	bad	satisfactory
13	middle	middle	good	good
14	middle	middle	middle	satisfactory
15	middle	middle	bad	bad
16	middle	bad	good	satisfactory
17	middle	bad	middle	bad
18	middle	bad	bad	very bad
19	bad	good	good	good
20	bad	good	middle	satisfactory
21	bad	good	bad	bad
22	bad	middle	good	satisfactory
23	bad	middle	middle	bad
24	bad	middle	bad	very bad
25	bad	bad	good	bad

Variant	$R_{a1}(EFE_{max})$	$R_{max1}(EFE_a)$	$S_{m2}(EFE_q)$	Output
26	bad	bad	middle	very bad
27	bad	bad	bad	very bad

The received results of the HPW grain influence on the complex assessment of the parts surface microrelief of corrosion-resistant steel 06Cr14Ni6Cu2MoVaTi-SH are presented in Table 7 .

The results in Table 7 show that the HPW with the maximum grain of B151 increasingly satisfy reducing the surface roughness, according to flatness deviation and microhardness the best HPW is the one with the lowest grain of B76.

Table 7. HPW grain influence on complex assessment of surface roughness, flatness deviation and microhardness

Grain ($l = \overline{1;4}$)	By surface roughness		By flatness deviation		By microhardness	
	d	Output	d	Output	d	Output
B76 (1)	0,676	Good	0,872	Very good	0,825	Very good
B107 (2)	0,566	Satisfactory	0,114	Very bad	0,173	Very bad
B126 (3)	0,184	Very bad	0,425	Satisfactory	0,500	Satisfactory
B151 (4)	0,820	Very good	0,822	Very good	0,787	Good

The finite problem of the research is the look for the optimum grain between HPW $l = \overline{1;4}$, that would minimize the position measures for the surface roughness and flatness deviation at the same time, the microhardness would reach its maximum. In this respect the precision of all quality parameters should be the largest, i.e. their measures of dispersion should be minimized. With this in mind, a model of four variables is generated: three of them are input and one is output (Figure. 3). The values of every input variable fit the output ones shown in Table 7.

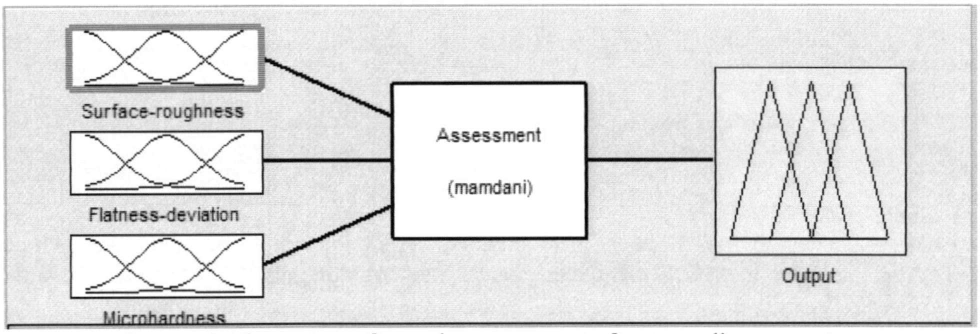

Figure. 3. The system of complex assessment of parts quality parameters

The results of HPW grain influence on the complex assessment of parts surface quality of corrosion-resistant steel 06Cr14Ni6Cu2MoVaTi-SH are given in Table 8.

Table 8. HPW grain influence on complex assessment of parts surface quality

Grain ($l=\overline{1;4}$)	d	Output
B76 (1)	0.7959	good
B107 (2)	0.1613	very bad
B126 (3)	0.3034	bad
B151 (4)	0.8234	very good

The fuzzy logic modeling helped us determine that the HPW with the grain of B151 to the fullest extent ensures the multicriterion quality optimization of part surfaces. The received data have been similar to those ones foreseen at the first stage of the investigation. The multicriterion optimization of HPW features is very important during robust design of grinding operations, that should be carried out with the help of the best basic tooling. This will optimize all target process functions with much more effectiveness.

4. Conclusion

1. The surface roughness, microhardness and flatness deviations are important factors estimating the microrelief of part surfaces. They are characterized by a great number of standardized parameters assessed by the position and scattering measures. This complicates the choice of optimum characteristics of the wheels using the statistical methods.
2. In conditions where homoscedasticity and normalcy of distribution of the experimental data are violated, to realize the fuzzy logic it is reasonable to consider nonparametric assessment of the position and scattering measures, to which medians $\tilde{y}_l$ and quartile latitude QL_l relate.
3. As a result of the fuzzy logic modeling in the MATLAB it is found that HPW CBN30 B151 100 OVK27-KF40 during grinding parts from 06Cr14Ni6Cu2MoVaTi-SH steel ensure the best quality of the surface according to the complex quality assessment of the ground parts. The use of the wheels increase the efficiency of the robust design of grinding operations, as the search for the basic models of the multiple variance analysis is fulfilled with the tools of the optimum features.
4. The modeling results can be used in local grinding problem solving. For example, to choose the optimal wheel grain by surface roughness, flatness deviation or microhardness. It has been established that the HPW with the greatest grain of B151 reduces all surface roughness and the HPW with the lowest grain of B76 has advantages of flatness deviation and microhardness.

References

[1] Suslov A G, BezyazichnyV F, Panfilov Y V et al. 2008 Surface Engineering details Mechanical Engineering (Mashinostroyenie) (Moscow) 320.
[2] Zadeh L A 1988 Fuzzy logic IEEE Transactions on Computers 21 4 83–93.
[3] Hollander M and Wolfe D A 1999 Nonparametric statistical methods, Second Edition Wiley-Interscience 787.
[4] Soler Ya I and Nguyen V L 2015 Selection of synthesis corundum Crain in Grinding flat parts from hardened steel 30ChGSA the macrogeometry criterion Applied mechanics and material 788 95-101.
[5] Zaks L 1976 Statistical estimation (Statisticheskoye otsenivanie) // Transl. from German M.: Statistika. 598.
[6] Willer D and Chambers D 2009 Statistical process management (Statisticheskoye upravlenie protsessami) // Transl. from English M.: Alpina Business Books 469.
[7] Soler Ya I and Kazimirov D Yu 2010 Selecting abrasive wheels for the plane grinding of airplane parts of the basis surface roughness Russian engineering research.30 3 251-261.

[8] Ali Y M and Zhang L C 1999 Surface roughness prediction of ground components using a fuzzy logic approach Journal of Materials Processing Technology 561–568.
[9] Ali Y M and Zhang L C 2004 fuzzy model for predicting burns in surface grinding of steel Int J Mach Tool Manu 44 563 .
[10] Vyatchenin D A 2004 Fuzzy methods of automatic classification UP Technoprint (Minsk) 219.
[11] Kofman A 1982 Introduction to the theory of fuzzy sets M. Radio and communication (Radio i svyaz) 432.
[12] Leonenkov A V 2005 Fuzzy modeling in MATLAB and FuzzyTech BXV-Petersburg (St. Petersburg) 736.
[13] Harrington E C 1965 The desirability function // Industrial Quality Control Industrial Quality Control 21 494-498.

IOP Conf. Series: Materials Science and Engineering **125** (2016) 012018 doi:10.1088/1757-899X/125/1/012018

Size Influence of High Pores on Cutting Ability of Cubic Nitride Boron Wheels for Pendulum Grinding of High-Speed Steel W9Mo4Co8 Plates by Surface Roughness Criteria

Ya I Soler [1] and A I Shustov[2]

[1] Associate Professor, Irkutsk national research technical university, Irkutsk, RF
[2] Postgraduate Student, Irkutsk national research technical university, Irkutsk, RF

e-mail: solera@istu.irk.ru

Abstract. The grinding wheel CBN30 B126 100 L V K27 KF40 is chosen as preferable according to the statistical interpretation of cases of surface roughness of high-speed steel plates W9Mo4Co8.

1. Introduction

Grinding is a major process that determines endurance and workability of machine tools, therefore grinding is often a completion phase of an engineering tool production. In mechanical engineering, high-speed steels (HSS) are referred to hard-to-machine materials due to the high content of carbide-forming elements. In order to ensure optimal grinding conditions, it is essential to take into account not only a reserve strength coefficient of an abrasive grains but high heating temperatures as well [1]. In real-time grinding, more than 80% of power transforms into heat. The heat sources are abrasive grains which can be classified into cutting, pressing and non-cutting ones by their different radial orientation. The most projecting grains first cause an elastic deformation of metals, then – a plastic one, and achieving the ultimate strength - chipping. The instantaneous heating of a workpiece surface reaches a temperature of 700-800 °C, in case of intensive grinding - 1200-1500 °C. Heat rate reaches 10000 °C/s [2], and temperature decreases in a split second as well, as most heat passes to the underlying metal. Traditional grinding is often used for machining of HSS, but the high grinding forces and micro cracks on a workpiece surface are major problems [3, 4]. Use of cubic nitride-boron (CBN) abrasive wheels might solve the problem. CBN has a high hardness that exceeds the hardness of traditional abrasives several times. Moreover, CBN has a high thermal resistance: grain oxidation begins only at temperature of 1100-1200 °C. As HSS is high alloyed steel, another advantage of CBN is their chemical resistance that makes it possible to grind avoiding adhesive and diffusion grain abrasive wear [1;8].

In the pores of a grinding wheel, vacuum is created under the action of centrifugal forces. As a result, the lubricant and coolant fluid is sucked through the pores to the ends of the wheel, with subsequent release into the grinding zone. The pore size has the primary influence on the permeability (corresponds to the ability of the wheel to pass through the material; under the action of the applied pressure gradient), as established in [5]. We have estimated cutting abilities of CBN grinding wheels with different pore-forming agents (PA) to provide the lowest roughness and highest grinding stability.

2. Statistical analysis methods of experimental data

Surface microhardness parameter Ra_d is estimated statistically using software Statistica 6.1.478.0 (where d is a direction of roughness). Applied statistics makes it possible to analyze the sequences being the sets $l = 1,\ldots k$ where $v = 1,\ldots n$.

$$\{y_{dlv}\} \tag{1}$$

There are two statistical methods - parametric and non-parametric ones. Each of these methods has more stability zero-hypothesis (H_o) «on the home court» [6]. To present and analyze output data, homogeneous distribution parameters [6] have been used. Among them, one can mention:

position measures (State Standard GOST R ISO 5721-1-2002):

averages $\overline{y}_{dl} = \overline{y}_{dl\bullet}$, (2)

medians $\tilde{y}_{dl}$, (3)

which defines the cutting ability of wheels;

dispersion (precision) measures:

deviation standards SD_{dl}, (4)

ranges $R_{dl} = \left(y_{\max} - y_{\min}\right)_{dl}$, (5)

quartile latitude $QL_{dl} = \left(y_{0,75} - y_{0,25}\right)_{dl}$, (6)

which defines the process stability.

We use (2), (4), (5) and (3), (6) for parametric and non-parametric methods respectively. Asymmetry (skewness) based on the following equation:

$$As_{dl} = \left[3\left(\overline{y} - \tilde{y}\right)/SD\right]_{dl}, \tag{7}$$

where A_{Sdl} is a positive value when $\tilde{y}_{dl} < \overline{y}_{dl}$, otherwise, skewness is negative.

For a parametric statistical method «on the home court» we assumed the conditions when (1) is a normal distribution with homogeneous variances of deviations. In this case, their criteria are robust to an insignificant random variables (RV) distribution violation. However, it cannot be applied to the homoscedasticity of distributions (1). The violation of distribution laws can cause accepting a wrong hypothesis, because maximum permissible deviation values are not stipulated anywhere. In these cases, a non-parametric method which does not depend on the properties of any distribution family is the most suitable.

Kolmogorov-Smirnov (D) and Shapiro-Wilk (W) test statistics can be used to prove H_o for the normal distribution in software. D-test is widely used in technical applications. However, in [8], the authors have revealed the weakness of the criterion. Therefore, in our paper H_o for (1) has been tested according to Shapiro-Wilk test statistics which satisfies the inequality:

$$\alpha_{dl} > 0.5 \tag{8}$$

There are three tests for homogeneity variance (q=1,...3): 1 – Hartley, Cohran and Bartlett's tests; 2 – Levene's tests; 3 – Brown-Forsythe's tests. In all cases, they satisfies the inequality (α – acceptance reliability for H_o when d,l – *const.*):

$$\alpha_{dl} > 0.05. \tag{9}$$

Stochastic nature of hypothesizes assumes that decisions f, based on criteria q=1,...3, may differ. Thus, final results on homogeneity (H_o) and heterogeneity (H_l) deviations variances have been accepted for following reasons:

$$H_o \text{ at } f_o \in [2;3]; \quad (10)$$

$$H_l \text{ at } f_l \in [2;3]. \quad (11)$$

Predicted averages and medians have been obtained by multiple comparison [6,7] and designated as $\hat{y}_{dl\bullet}$ and $m\hat{y}_{dl}$ respectively.

3. Methodology

The experiments have been carried out under the following conditions: plan grinding-machine 3E711V; form and sizes of wheels (State Standard GOST R 53923-2010): 1A1 200×20×76×3; HSS plates made from steel W9Mo4Co8 (63-64 HRC); grinding scheme – by the side of a wheel without nursing courses; cutting parameters: cutting speed v_w=28 m/s, traverse motion s_{tr} = 7 m/min, cross-feed motion $s_{c\text{-}f}$ =4 mm/double pass, cutting depth t=0.01 mm, stock allowance z=0.10 mm. The lubricant is a 5 % Akvol–6 emulsion (Technical Specifications TU 0258-024-00148845-98) supplied by watering to the workpiece at a rate of 7-10 l/min. The wheel is being lowered at depth t when the bench with HSS plate is being displaced fully to the left relative to the operator. Its movement from left to right is a working stroke. Otherwise, it is a nursing stroke with a final formation of a HSS plate surface micro-relief according to the wind grinding scheme as the spindle rotates clockwise. HSS plates are samples sized D×H = 40×50 mm. They are fixed on the magnetic bench and undergone face grinding. The number of repetitions of experiments is n=30. Roughness parameter Ra_{de} is measured by profilograph-profilometer 252 “Kalibr”. Output process data are presented as y_{dev}, where d = 1,2 is a code of direction of roughness: 1 – corresponds to cross-feed motion $s_{c\text{-}f}$, 2 – corresponds to traverse motion s_{tr}; e = 1,2 is a code of high-porous CBN wheels (HPCBNW): 1 - for CBN30 B126 100 L V K27 KF25 (the basic wheel), 2 - for CBN30 B126 100 L V K27 KF40. The direction of roughness d = 1 has chosen as general to regulate quality of HSS plate surface.

For a quantitative analysis of cutting ability of HPCBNW according to predicted analogues of medians (3), we use

$$K_{1e} = \left(m\hat{y}_2 / m\hat{y}_1\right)_{1e}. \quad (12)$$

Grinding is a stochastic process, which often don't satisfies the normal distribution with homogeneous variances of deviations. Therefore we have analysed its stability according to the deviation standards (SD_e), ranges (R_e) and quartile latitude (QL_e) of experimental cases (1), and obtained:

$$\left(K_{SD}\right)_{1e} = \left(SD\right)_1 / \left(SD\right)_2, \quad (12)$$

$$\left(K_R\right)_{1e} = R_1 / R_2, \quad (13)$$

$$\left(K_{QL}\right)_{1e} = \left(QL\right)_1 / \left(QL\right)_2. \quad (14)$$

The value of (12)-(14) < 1 corresponds to the higher process stability for the basic wheel (e=1), otherwise – for wheels e = 2.

4. Result and Discussion

The results in table 1 prove the hypothesis H_l for the homogeneity of variances for all wheels e = 1,2, and the decision (11) is made for all three tests. Moreover the distribution normality condition is not fulfilled only for both wheels. Therefore, a non-parametric method based on Kruskal-Wallis range criteria is the most suitable.

Table 1. Results of tests for homoscedasticity and normality of distributions for a Ra_1 parameter.

Inequality	CBN30 KF25 ($e = 1$)	CBN30 KF40 ($e = 2$)
(8)	H_1	
(9)		

Figure 1 illustrates case results of (1) based on parametric (figure 1A) and nonparametric (figure 1B) statistics methods for Ra_1 parameter at grinding with e = 1,2 wheels where "square" means position measure for $\overline{Ra}_{1e}$ (figure 1A) and $\tilde{Ra}_{1e}$ (figure 1B). The other marks have different meanings. In figure 1A, "rectangle" means dispersion borders for error standards ($\pm SDE_{1e}$) and "moustache" means dispersion of deviation standards ($\pm SD_{1e}$). In figure 1B, the quartile latitude $|y_{0.75} - y_{0.25}|_{1e}$ and ranges $R = |y_{\max} - y_{\min}|_{1e}$ are shown. Median and ranges in figure 1B show a form of a distribution curve (7). Categorical values (CV) (State Standard GOST 2789-73) has shown in the baskets.

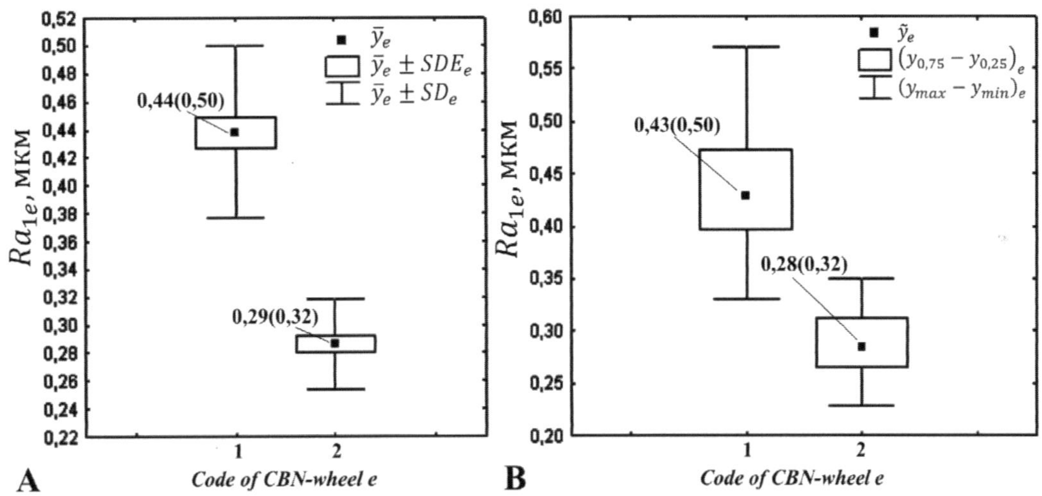

Figure 1. Parametric (A) and nonparametric (B) statistics of wheel characteristic influence to Ra_1 parameter

First, let us analyze the results for roughness position measures. As seen from table 2, the wheels e = 2 showed the best W9Mo4Co8 steel grinding results for expected medians of Ra_1 parameter. At the same time, the wheel e = 1 showed the worst result greater than 2 CV than e = 2: $m\hat{R}a_{11}0,43(0,50)$ for KF25 PA and $m\hat{R}a_{12}0,29(0,32)$ for KF40 PA with higher permeability. Thus, increasing the permeability changes the thermal conditions in the grinding zone, as indicated by contact-temperature data for the grinding of W6Mo5 steel. Thus, the grinding of AEROBOR wheels with A40 filler (correspond to KF40) reduces the temperature greater then grinding of wheel with smaller pores (with A25 filler) and correspondingly with smaller permeability [5].

There is a positive asymmetry of distributions for both HPCBNW (e=1,2). It has a positive impact on process reliability as it reduces position measures. However, the best is when cases (1) are located below the medians. For example, while CBN30 KF25 wheel grinding (e = 1), only 1 case from n = 30 turned out to be smaller than an experiment median. With the same n = 30, the majority of cases turned out to be smaller than experiment medians - 29 with $\tilde{R}a0,28(0,32)$ for CBN30 B126 100 L V K27 KF40 (figure 1, table 2).

Table 2. Prediction results and wheel influence on position measures for Ra_1 parameter

Grinding wheels	$Ra_{1e\bullet}$ [μm]	$\tilde{R}a_{1e}$ [μm]	$\hat{R}a_{1e\bullet}$ [μm]	$m\hat{R}a_{1e}$ [μm]	$As_{1e} > 0$ (7)	K_{1e} (12)
CBN30 KF25 (e = 1)	0.44(0.50)	0.43(0.50)	0.44(0.50)	0.43(0.50)	+	1.00
CBN30 KF40 (e = 2)	0.29(0.32)	0.28(0.32)	0.29(0.32)	0.28(0.32)	+	0.66

Wheel grinding stability analysis e = 1,2 showed that reproducibility ratios (12)-(14) coincided (table 3). Wheel CBN30 KF40 (e = 2) demonstrated the highest level of stability of deviation standards, ranges and quartile latitude.

Table 3. Prediction results and wheel influence on dispersion measures for Ra_1 parameter

Grinding wheels	SD_{1e} [μm]	R_{1e} [μm]	QL_{1e} [μm]	$(K_{SD})_{1e}$	$(K_R)_{1e}$	$(K_{QL})_{1e}$
CBN30 KF25 (e = 1)	0.06	0.24	0.08	1.00	1.00	1.00
CBN30 KF40 (e = 2)	0.03	0.12	0.05	1.88	1.99	1.64

According to our experiment, grinding wheel CBN30 B126 100 L V K27 KF40 (e = 2) is preferable.

5. Conclusion

1. The rationale of grinding high-speed steels using HPCBNW instead of traditional abrasives has been proved.
2. It has been proved that the cutting ability of abrasive wheels in assumed technological conditions should be estimated using non-parametric statistical method «On the home court».
3. PA of HPCBNW influence on the process stability has been established. KF25 PA has lower reproducibility coefficients than KF40 PA.
4. CBN30 B126 100 L V K27 KF40 has shown the best results according to position measures and precision of roughness when grinding HSS plate W9Mo4Co8.

References

[1] Kremen' Z I and Baboshkin A F 2007 Grinding technology in mechanical engeneering (St. Petersburg: Politehnika)
[2] Ravindra N Y, Vinod Y and Singh G K 2014 Application of non-dominated sorting algorithm for multi-objective optimization of electrical discharge diamond face grinding process Journal of Mechanical Science and Technology 28 6 2299-2306

[3] Kopac J and Krajnic P 2006 High-perfomance grinding-A review J. Mater. Process. Technol. 175 278-284
[4] Hageman J B J W, Hosson J T M G and With G 2001 Grinding of WC-Co hard metals Wear 248 187-196
[5] Kremen' Z I and Lebedev A I 2011 High-Porisity Wheels Based on Cubic Nitride for Defect-Free Grinding Russian Engineering Research 9 31 867-869
[6] Hollander M and Wolfe D A 1999 Nonparametric statistical methods, Second Edition (New-York: Wiley-Interscience)
[7] Sachs L 1976 Statistical estimation (Moscow: Statistica)
[8] Soler Ya I and Prokop'eva A V 2009 Information base for regulating the grinding of high-speed cutting plates of variable rigidity by a cubic-boron-nitride tool Russian Engineering Research 29 9 955-956

IOP Conf. Series: Materials Science and Engineering **125** (2016) 012019 doi:10.1088/1757-899X/125/1/012019

Mechanical Activation of Construction Binder Materials by Various Mills

R S Fediuk

Far Eastern Federal University, Vladivostok, Russia

e-mail: roman44@yandex.ru

Abstract. The paper deals with the mechanical grinding down to the nano powder of construction materials. During mechanical activation a composite binder active molecules cement minerals occur in the destruction of the molecular defects in the areas of packaging and breaking metastable phase decompensation intermolecular forces. The process is accompanied by a change in the kinetics of hardening of portland cement. Mechanical processes during grinding mineral materials cause, along with the increase in their surface energy, increase the Gibbs energy of powders and, respectively, their chemical activity, which also contributes to the high adhesion strength when contacting them with binders. Thus, the set of measures for mechanical activation makes better use of the weight of components filled with cement systems and adjust their properties. At relatively low cost is possible to provide a spectacular and, importantly, easily repeatable results in a production environment.

1.Introduction

Modern building materials technology tends to a constant reduction of the dimension of the materials. The increase in the specific surface binding powders can increase their surface area, respectively, increasing the area exposed to chemical hydration reaction when mixed with water. Milling binders - an essential process step consisting in mechanical grinding of raw mix. With increasing fineness increased activity binders, mechanical strength hardening. At the same time, it should be noted that for each binder set its own optimal fineness at which most fully use its properties. Needless high fineness is not always desirable because with increasing fineness of powder require more water to make a dough, which, in turn, leads to increased shrinkage and cracking during hardening.

Mineral binders - is finely crushed mineral powders, which form when mixed with water, plastic mass, which over time under the influence of physical and chemical processes passes in stone condition. This property is used for binders for producing the artificial stone materials (concrete, etc.). In this case, the mechanical process of natural raw materials is increasingly replaced by chemical - a simple, efficient and cost-effective.

The aim of this study sought to determine the effect of various industrial milling machines on the properties of composite cementations materials.

2. Results and Discussion

To achieve the goal have been developed composite binders obtained by co-grinding cement, hyperplastisizer, fly ash and limestone.

The problem of increasing the density and strength of the concrete in a large number of works of both Russian and foreign scientists [1-3]. It is known that one of the ways to improve the performance of concrete, reduced permeability parameters is the use of highly active additives of different composition and origin, both micro- and nanosized level, which contribute to process optimization of structure by initiating the formation of hydrated compounds. So in the work carried out previously, it has been proven effective use as an active mineral additives of nanostructured silica-modifier composition. In addition, the parameters studied the possibility of reducing the permeability of concrete by mechanical grinding of the components of the composite binder.

The properties of composite materials largely depend on the structure of disperse systems on which they are obtained [4]. The structural strength of the dispersed system, its stability and the behavior in the flow, the rate of destruction and restoration of the structure are interconnected. Conglomerate building material - concrete refers to a class of composites. As a concrete matrix substance, successively ambitious outline levels are the cement-sand stone, cement stone (microconcrete), cementing substance, neoplasms cementing substance, solid phase tumors, essence single structural element of tumors, which corresponds to a large-scale order from the macro- to the nanoscale structure [5]. Ordered structure of the composites due to the proportionality of scale levels of the structure - the corresponding properties of the composite at each scale level [6]. The achievement of high strength concrete contributes to a combination of factors: increasing density of systems by optimizing the composition of the grain; reducing the number of long cement matrix by reducing the water-cement ratio; filling the pores between the cement particles and to improve the rheology due to the effect of lubrication; formation of secondary products of hydration in the pozzolanic reaction with Ca $(OH)_2$ for the introduction of concrete additives microfill effect [7].

To select the optimal method of grinding tests were carried out in a ball, vibration and vario-planetary mill.

Ball mill is a hollow rotating drum around its axis, which is about half-filled with crushing balls (percussion elements). As a result of the rotation, the balls rise in the upper part of the drum, and then under the influence of gravity fall down [8-10]. After one of the pivots is constantly fed material to be ground, and going through another bowl discharge (figure 1).

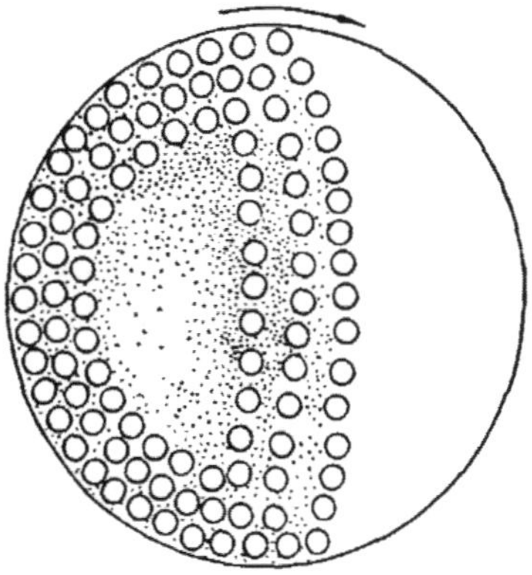

Figure 1. Operating principle of the ball mill

One of the features is a crushing blow to free the fact that the destruction of the material occurs at the weakest bond, structural defects in the joints of crystal grains, layers, etc. In the production of graded gravel or artificial sand is an undoubted advantage because the product is represented by impact crushing grains of isometric shape with no internal defects with a small amount overgrinding product. At the same time, to obtain greater fineness, particle hardening that occurs with decreasing their size, it creates additional difficulties.

At some point, when the structural strength of each individual particle reaches its maximum, and its weight is negligible, free kick almost completely replaced by abrasion. Rotor centrifugal mill accelerator ceases to function as soon as the running and the swirler material airflows. Is interested in to the walls of the grinding chamber large particles are squeezed smaller, which, moving from the periphery to the center, are crushed only by mutual abrasion of turbulent flows.

Judging by the consumption of energy on the surface of the formation of a new unit of solid materials is one of the most inefficient ways of grinding.

The operating principle of a vibration mill (figure 2) based on the prompting vigorous grinding bodies, when instead of gravitational forces causing the drop of balls is used the inertia, centrifugal force, etc. [11-13]

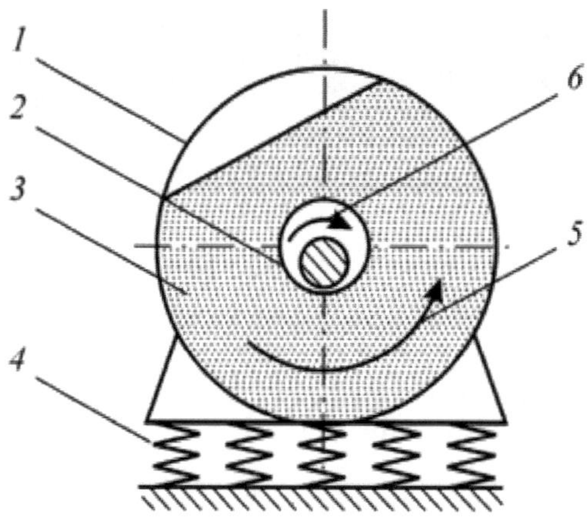

Figure 2. Operating principle of the vibration mill

1 - body of the mill; 2 - the eccentric (or unbalanced) shaft; 3 - ball boot from the material being ground; 4 - spring supports; 5 - the direction of the circulation load; 6 - the direction of rotation

Rotation of the shaft of the vibrator, and behind them of the mill housing causes the grinding media to be moved according to the magnitude of the eccentricity or the radius of the carrier. Energy transfer grinding load through the mill housing. Under the influence of inertia, centrifugal force, alternating loads balls inside the body move along a complex trajectory, pressed against the wall of the drum, hit against each other, as well as the particles of the crushed material, breaking, crushing and rubbing them.

For the production of fine materials, vibrating mills are more efficient than the ball. An impact grinding effect on the material in this case, a small but intense abrading, to achieve fineness

The vario-planetary mill speed grinding bowls and supporting disc can be set completely independently of each other. By varying the ratio, it can affect the movement and trajectory of grinding balls so that the balls hit horizontally on the inner wall of the grinding bowl (high impact energy), approach each other tangentially (high friction) or just roll along the inner wall of the grinding bowl (centrifugal mills) [14-16]. All intermediate stages and a play between the friction and blow pressure can be freely set (figure 3).

Accordingly, the grinding in vario-planetary mill is more energy-efficient compared to the ball and vibration mills. In addition due to the combined action of the drum and centrifugal drum and abrasive efforts, it is possible to achieve a highly dispersed powders.

Through comparison fineness of the claimed composite binder on different mills, revealed that the ball mill (1.1kW, 90 rpm) can be milled to 400 m^2/kg, a further unit operation is uneconomic. Vibration Mill (2.2 kW, 1500 rpm) showed the ability to effectively grinding composite binding to a specific surface of 430 m^2/kg (figure 4).

Due to the fact that according to the declared method of experiment planning interval varying specific surface 500-900 m^2/kg, a ball mill and vibration is not able to meet these conditions.

Vario-planetary mill "Pulverisette 4" (9 kW) capable of providing grinding to a predetermined specific surface area (900 m^2/kg).

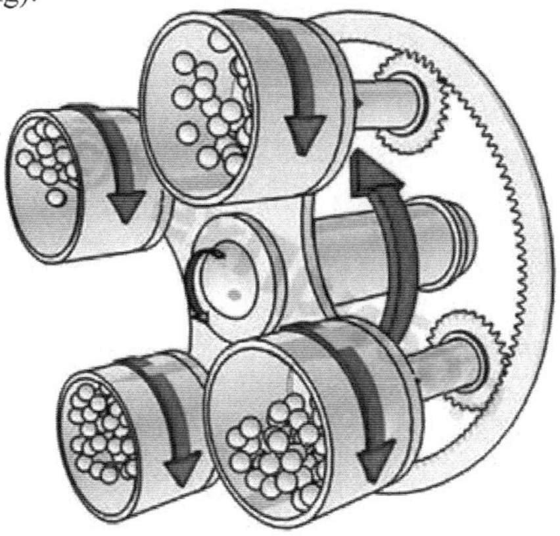

Figure 3. Operating principle of the vario-planetary mill

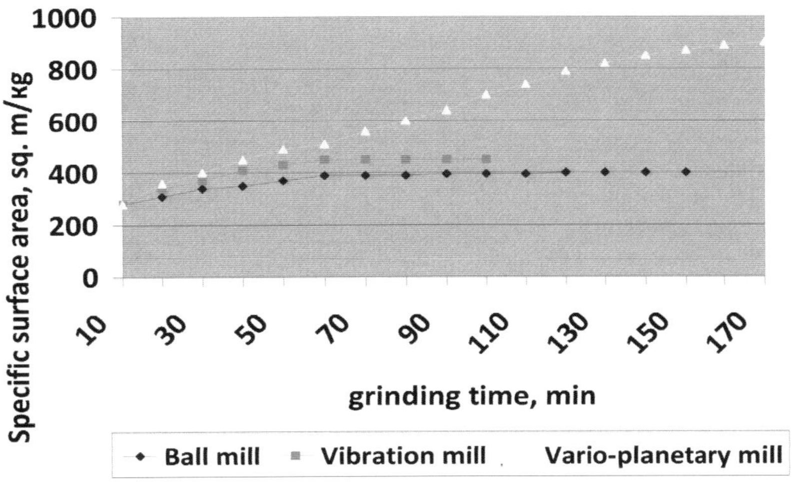

Figure 4. The dependence of the specific surface area of the composite binder of the time for any grinding mill

In order to determine the optimum particle size was produced intergrinding cement hyperplastisizer, ash and limestone to varying the specific surface area: 500, 550, 600, 700, 800, 900 m^2/kg (table 1).

Table 1. The results of determination of compressive strength (MPa) of cement samples

Age of the sample, d.	The specific surface area of the composite binder, m^2/кg					
	500	550	600	700	800	900
3	46.1	47.4	47.2	46.0	45.6	45.5
7	50.3	54.2	54.1	49.1	48.6	48.4
28	68.1	77.3	70.2	65.8	55.0	65.0

It was found that the optimum surface area binder 550-600 m^2/kg. Increased activity in excess of these values has a negative effect on the structure. Using a binder with increased activity significantly speeds up the process of setting setting the end of the mixture ends after 35-40 min., while the developing temperature 95-97°C. Fast setting raw prevents the formation of spherical particles uniformly distributed in the macro-structure of cement stone.

During mechanical activation a composite binder active molecules of cement minerals occur in the destruction of the molecular defects in the areas of packaging and breaking metastable phase decompensation intermolecular forces. The process is accompanied by a change in the kinetics of hardening of portland cement.

Mechanical processes during grinding mineral materials cause, along with the increase in their surface energy, increase the Gibbs energy of powders and, respectively, their chemical activity, which also contributes to the high adhesion strength when contacting them with binders.

Thus, the set of measures for a mechanochemical activation makes better use of the weight of components filled with cement systems and adjust their properties. At relatively low cost is possible to provide a spectacular and, importantly, easily repeatable results in a production environment.

Increased strength binders when administered in their composition finely additives besides hydraulic activity could also be explained by the formation of fine grains more nucleating additives in the contact zone of cement. "Microfiller effect" can not be explained only by the formation of additional nucleation, as their direct effect is to accelerate the initial stage of the chemical hardening. The basis of "micro filler effect" are both chemical processes of interaction of cement hydration products and the physico-chemical phenomena, such as the effect of the surface energy of the particles of fine additives.

In the presence of mill ground additives in concrete is hardened the contact zone between the cement stone and aggregate. In portland cement concrete without nanoadditives contact area is usually less dense than the cement paste, and includes a large number of plate-like crystals of calcium hydroxide, whose longitudinal axis is perpendicular to the surface of the aggregate. Consequently, it is more susceptible to the formation of microcracks in the tensile forces arising from the changes in the normal conditions of temperature and humidity. Thus, the contact area due to its structure is weakest in the concrete and therefore has a great influence on its toughness. Introduction of fine additives greatly reduces capillary porosity contact zone due to a sharp reduction in the total content of $Ca(OH)_2$.

The positive effect on the microstructure of the contact zone is achieved by introducing a relatively small amount of fine ground active mineral additives, such as fly ash. The cementitious systems containing the hydraulically active mineral additives, the formation of hardening an additional quantity CSH of by the interaction $Ca(OH)_2$ with active silica or aluminosilicate filler. The consequence of these processes is the formation of additional phase contacts (contacts between fusion crystal hydrates) and the increase in the density of the cement stone, which determines the high strength cement system.

In addition, we should not forget that in addition to the introduction of cement mill ground supplements increase the specific surface of the cement also allows you to adjust the activity of the binder. It is known that by grinding the same Portland cement clinker and thus altering the proportion of particles 5-20 microns in the total mass of the cement powder can be prepared portland cement grades 600, 700 and 800 and very rapidly hardening cement.

It was found that the combined effect of mechanical and chemical activation (the presence of limestone particles) increases the pozzolanic activity of acidic angry. Analysis of the microstructure showed that the particles of ash and limestone formations completely surrounded by the gel. Individual particles are bonded together to form clusters. This cement matrix characterized by the presence of needle-like lesions, length of 2 m and a diameter of about 0.2 microns (figure 5).

Thus, from fine limestone additive is a chemical way to improve the activity of ash and sand. It has a catalytic effect on the reaction activity of the surface of ash and sand during machining in vario-planetary mill. Furthermore, the introduction of limestone increases the alkalinity of the concrete, resulting in a greater formation of cement hydration products in a unit time.

The possibility of increasing the impermeability of the concrete due to the varying amount and type of additives, fineness, and synthesis parameters, allowing you to create materials for multilayer walling with a compressive strength of up to 100.9 MPa, with low permeability under real operating conditions. Intergrinding cement hyperplastisizer, acidic fly ash, and limestone in an amount of 55%, 40% and 5% respectively accelerates the synthesis of neoplasms, linking allocated at alite hydration of CaO in calcium hydrosilicates of different basicity. Thus optimizing the microstructure of cement paste, you can get an astringent activity of 77.3 MPa.

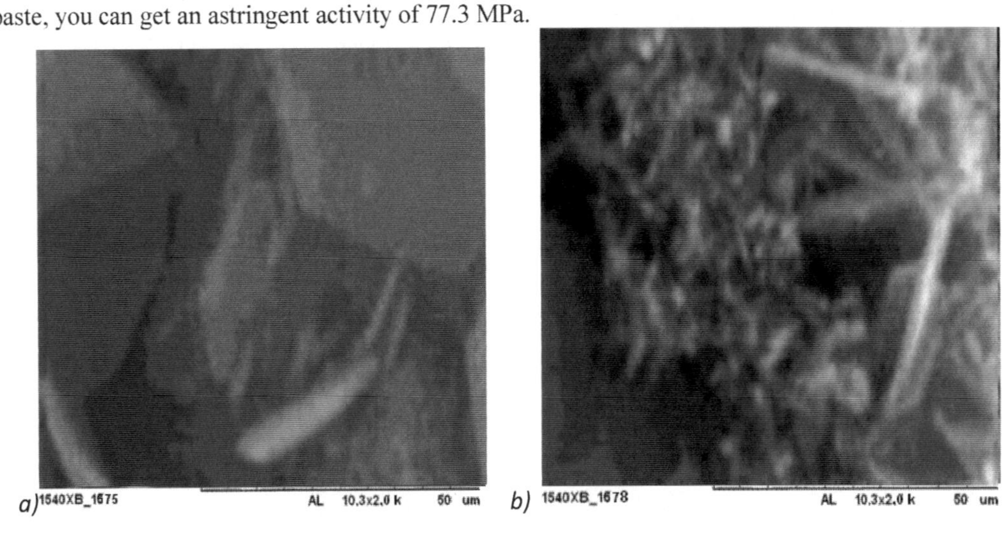

Figure 5. The micrographs neoplasms: *a* - cement stone without additives; *b* - cement stone on the basis of the composite binder

3. Conclusion

It was proposed the mechanism explaining the effect of the use of cement-ash binder in conjunction with hyperplastisizer and limestone in the a mechanochemical activation in the vario-planetary mill: by mechanical grinding is an increase in the reactivity of the cement-ash binder, hyperplastisizer combined with limestone increases the density of tumors, and the adhesion of the binder with filler to provide increased impermeability of concrete.

It was designed the principles of optimizing the structure of fine-grained concrete on the nanoscale by using nanosized composite binder; at the micro level through the creation of high-density packaging filler; at the macro level through the introduction of steel and basalt fiber. It is possible to develop a wide range of fine-fiber-reinforced concrete with a water vapor permeability to 0.021 mg /

(m-h-Pa), permeability to W14, breathability to 0.0253 cm^3/s, the water absorption by weight to 2.5%, gas permeability, effective diffusion coefficient of $1.34 \cdot 10^{-4}$ cm^2/s and high strength characteristics.

References

[1] Lesovik V S, Zagorodnyuk L K, Mestnikov A E, Kudinova A I, Sumskoi D A 2015 International Journal of Applied Engineering Research 10(5) 12383-12390
[2] Ichiko Misumi, Jun-ichiro Kitta, Hiroyuki Fujimoto, Satoshi Gonda, Yasushi Azuma, Keisuke Maeda, Tomizo Kurosawa, Yoshiyasu Ito, Kazuhiko Omote, Yoshinori Nakayama and Hiroki Kawada 2012 Meas. Sci. Technol. 23 015002
[3] del Cerro J, Manchado J, Romero F J and Gallardo M C 2012 Meas. Sci. Technol. 23 035003
[4] Camila Barreneche, Aran Solé, Laia Miró, Ingrid Martorell, A Inés Fernández and Luisa F Cabeza 2012 Meas. Sci. Technol. 23 085606
[5] Md Jashim Uddin, N H Md Yusoff, O Anwar Bég and Ahamd Izani Ismail 2013 Phys. Scr. 87 025401
[6] Cuevas J M, Rubio R, Laza J M, Vilas J L, Rodriguez M and León L M 2012 Smart Mater. Struct. 21 035004
[7] Zhi Chen, Dai Mochizuki, Masato M Maitani and Yuji Wada 2013 Nanotechnology 24 265602
[8] Chadwick A V, Düvel A, Heitjans P, Pickup D M, Ramos S, Sayle D C and Sayle T X T 2015 IOP Conference Series: Materials Science and Engineering 80
[9] Polat R, Demirbofa R and Khushefati W H 2015 Construction and Building Materials 81 pp 268-275.
[10] Raheek I Ibrahim Wong Z H and Mohammad A W 2015 IOP Conference Series: Materials Science and Engineering 80
[11] Baeza F J, Galao O, Zornoza E and Garcés P 2013 Materials 6 (3) pp 841-855.
[12] Al-Rahman L A and Raja R I 2012 Attenuation of Noise By Using Absorption Materials and Barriers: A Review 2 (7) pp 1207-1217.
[13] Murtazaiev S-AY, Lesovik V S, Bataiev D K-S., Chernysheva N V, Saidumov M S 2015 Modern Applied Science 9 (4) pp 233-245.
[14] Suleymanova L A, Lesovik V S, Lukuttsova N P, Kondrash E V, Suleymanov K A 2015 International Journal of Applied Engineering Research 10 (5) pp 12399-12406
[15] Lesovik R V, Klyuyev S V, Klyuyev A V, Netrebenko A V, Kalashnikov N V 2014 World Applied Sciences Journal 30 (8) pp 964-969.

Experimental Study of Sliding Friction for PET Track Membranes

E O Filippova[1], A V Filippov [1,2,*] , I A Shulepov[1]

[1] Tomsk polytechnic university, Tomsk, Lenina av. 30, Russia
[2] Institute of strenght physics and materials sciencs SB RAS, Tomsk, per. Academicheskii 2/4, Russia

e-mail: andrey.v.filippov@yandex.ru

Abstract. The article is presented results of a study of the process for a dry friction metal-polymer couple on scheme disc- finger. Track membrane from polyethylene terephthalate was a research material. Membrane had pores with 0.4 and 0.8 μm diameters. The effect of the sliding velocity for membranes with pores of 0.8 microns was determined. Research was shown that increasing pore's diameter caused a reduction of the friction coefficient and downturn its magnitude vibrations. The study showed that track membrane have adequate resistance to wear and can be successfully used in surgical procedures in the layers of the cornea.

1. Introduction

Polyethylene terephthalate (PET) is the one of the most common materials used for the manufacture of track membranes well-established in such areas as electronics, food, chemical and automotive industries, cryogenics and medicine [1-5].

PET has hydrophobic, non-hygroscopic, biologically passive, non-radioactivity properties, resistance to most acids and organic compounds, a high gear transmission of the light flux.

There is the need for further processing of PET materials due to expansion of its application. Further processing of PET materials includes chemical etching, ionic raying, and plasma process. PET physical-chemical and mechanical material's properties change after reworking. These changes remain understudied.

The study of friction and wear thin films of PET was performed in scratching single crystal natural diamond. Film thickness was 14.2 μm. Also the study of friction and wear PET films was made with reciprocating sliding friction regime of flat finger along the surface of the 80.0 μm thickness film and reciprocating sliding friction regime of steel ball along the surface of the 125 μm thickness film.

We can identify the following features of PET thin films friction based on works of authors [6-8].

1) Plastic deformation of the film is done by scratching. Changing the coefficient of friction depends on the unevenness of the film surface [6].

2) Authors [7] have shown that the change of surface topography and the creation of projections and depressions result in a change of the wear and friction coefficient. The friction coefficient of film with projections is higher than that of films with depressions.

* Corresponding author: A V Filippov - andrey.v.filippov@yandex.ru.

3) Authors [8] have shown that the friction coefficient is 0.2-0.5. Wear and destruction of films increase in the center of the track wear after increasing numbers of sliding circles. Microcutting occurs as a result of delamination of the film surface under the influence of rider. Wear particles are removed from the path of friction under the influence of the hard counter body. Friction process is accompanied by the transfer of the polymer to the surface of rider.

Application of PET track membranes is of the utmost interest in the medical field such as ophthalmology [9]. According to the authors [10] implantation PET track membrane into corneal layers can cause mechanical damage of film. Therefore, the PET membrane should be resistant mechanical properties.

Purpose of the work is conducting tribological research of PET thin track membranes in sliding friction in contact with a metal rider.

2. Research method

The research was made with Tribometer THT-S-AX0000. Sliding friction was done according to the scheme finger-disc with a normal load 1 N. The diameter of friction surface was 10.0 mm. The metal rider was steel ball with 3.0 mm diameter. Tests were done under dry friction at room temperature 27±1°C and a relative humidity of 50%. Sliding speed for track membrane with 0.4 μm pore's diameter was 4.2 m/min. Sliding speeds for track membrane with 0.8 μm pore's diameter were 3 m/min and 4.2 m/min.

Samples of track membrane were fixed on metal substrate (figure 1). The steel ball was attached to cylindrical finger motionlessly.

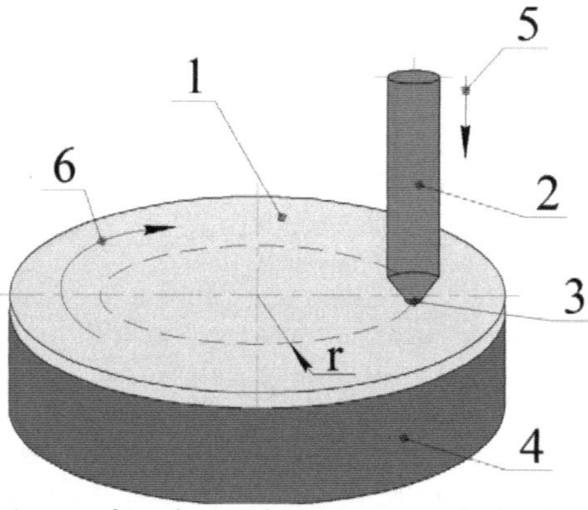

Figure 1. Tribological scheme of track membrane's research: 1 – track membrane, 2 – finger, 3 – steel ball, 4 – disc, 5 – normal load, 6 – rotating direction

Track membranes (TM) were made from polyethylene terephthalate. Technique of producing porous track structures is based on irradiated different polymers by high-energy heavy ions. These ions induce latent narrow tracks through the entire thickness of the polymeric material. Latent tracks are defective area with a diameter of 5-12 nm. Subsequent selective etching can remove these defective areas and receive the porous material with nano pores density from 10^6 cm^{-2} to 10^9 cm^{-2}. Oriented polymer films are irradiated by Ar ion beam with a maximum energy of 41 MeV in a specially designed vacuum chamber with a tape transport mechanism. Accelerated beam of argon ions was extracted from the chamber with an electrostatic deflector. Argon ion beam was directed into the channel, in which there were a uniform system for scanning the ion beam and the camera exposure film. The original scheme with electrostatic scanning beam in the horizontal direction was designed and used to scan the beam in the horizontal direction. A beam of argon admitted to the vacuum chamber, where the irradiation of the film was done after passing through the scanning system. Energy of the accelerated ions was measured before irradiation of the film with high accuracy using the method of recording backscattered ions (POP). It allowed to identify the acceleration ions by mass and to determine the energy of the incident ions . The intensity distribution of the beam on the film was controlled by measuring the current on the lamellae arranged in the film over its entire width. Ions passing through the film created an area of high density ionization. It was the track of the ion. Selective etching of the alkaline material in the region of track allowed to get a porous film in the original system with cylindrical holes with a typical symmetric structure The film was irradiated with UV light for an additional sensitization before etching. The etching was carried out in an aqueous solution of NaOH 1.5 N concentration at temperatures in the range 72- 82C .

The density and pore sizes were controlled with the electron microscope Hitachi TM – 1000. Pore diameters were 0.4 and 0.8 μm. Density was $5 \cdot 10^6$ pores/cm^2. Electron micrograph TM with pore's diameter of 0.4 μm is shown in figure 2a. Research of surface properties was made on Scanning AFM/Confocal/Raman/Fluorescence system «Centaur HR». 3D-surface TM profile with a pore's diameter of 0.4 μm is shown in figure 2b.

Membrane thickness measure was performed using an electronic thickness gauge «Tesa Unit». Measurement accuracy was ± 0.1 μm. Thickness of TM was 7.0 μm.

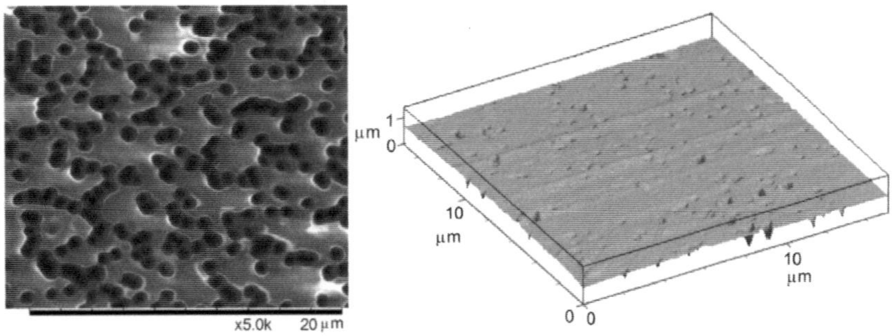

a) b)

Figure 2. Electron micrograph TM fragment (a) and its profile surface (b)

3. Results and discussions

Results of track membranes friction coefficient are shown on the figure 3 received 6000 laps at a constant speed drive slip of 4.2 m/min. Curve 1 shows the friction coefficient of the TM with 0.4 μm pores. Curve 2 shows the friction coefficient of the track membrane with 0.8 μm pores.

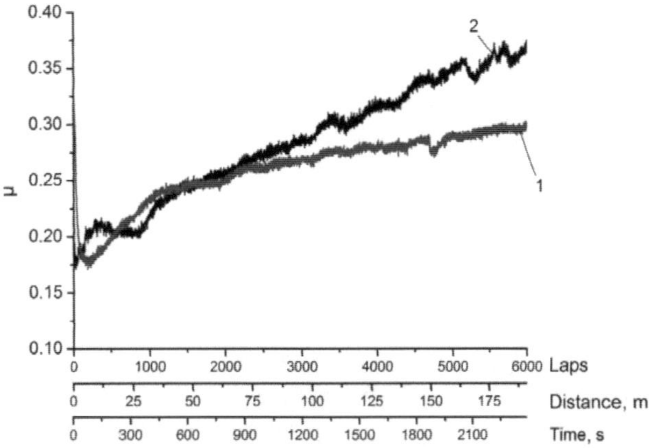

Figure 3. Friction coefficient of track membranes with 0.8 μm (1) and 0.4 μm (2) pore's diameters. Sliding speed is 4.2 m/min

Friction coefficient (μ) is 0.3 for the TM with 0.4 μm and 0.8 μm pore's diameters at the initial stage μ of 0.4 pore's diameters TM drops to 0.17 in the first 30 laps. Friction coefficient increases to 0.2 from 30 to 175 laps, and after coefficient of friction is stabilized to 850 laps. In the figure we can observe μ increasing to 0.29 after 850 laps and sudden upsurge to 0.37 before the end of the test.

Friction coefficient of 0.8 pore's diameters TM decrease to 0.2 in the first 50 laps. The next 150 laps μ continues smooth decreasing to 0.17. We can observe an increase of the friction coefficient from 0.17 to 0.23 in the range 200-850 laps and smooth growth period of μ to 2000 laps. In the range of 2000-6000 laps μ increases skippingly and reaches a value of 0.3.

Results of friction coefficient of TM with 0.8 μm pore's diameters are shown in the figure 4 received 6000 laps at a constant speed drive slip of 4.2 m/min and 3.0 m/min. Curve 1 shows 3.0 m/min sliding speed. Curve 2 shows 4.2 m/min sliding speed.

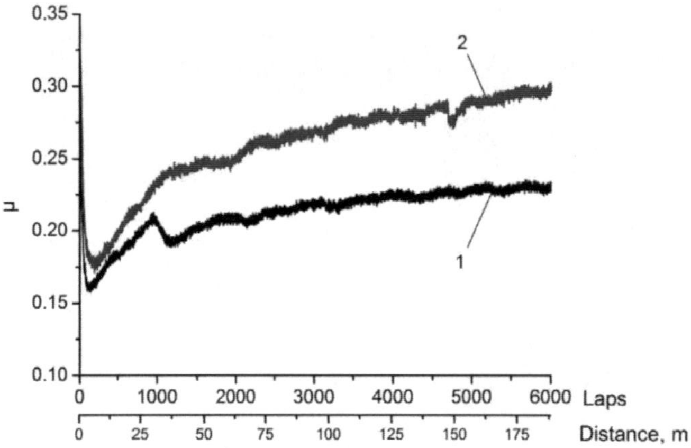

Figure 4. Friction coefficient of track membranes with 0.8 μm pore's diameters. Sliding speeds are 3.0 m/min (1) and 4.2 m/min (2)

The initial value of friction coefficient is 0.347 (3.0 m/min sliding speed). After 35 laps μ gradually decreases to 0.16. Coefficient of friction increases to 0.21 in the period of 100-950 laps. Decline to 0.19 μ magnitude happens in between 950-1200 laps, and after this the friction coefficient increases to 0.233. Also it should be noted that the reduction sliding speed from 4.2 to 3 m/min leads to a decrease the coefficient of friction.

Reduction of the friction coefficient at the initial stage dues to the formation of the friction transfer film. It is typical for a pair of metal-on-polymer [12] and also for a smoothing of surface roughness TM counterbody. The coefficient of friction decreases with decreasing sliding speed.

Figures 5-7 are shown the surface profiles of the track membranes, and cross-sectional view of wear track. Surface profile of wear track of track membrane with 0.4 μm pore's diameters has cross-undulation, which is a result of the plastic edging. Using software Gwydion we determined that fluctuation of depths and holes is 2 μm. Wear track has flat surface for a TM with 0.8 μm pore's diameters. Therefore, plastic edging of TM doesn't happen. Surface profile of wear track of track membrane with 0.8 μm pore's diameters doesn't change longitudinally with decreasing sliding speed of 4.2 m/min to 3 m/min.

Sections of track wear have a wavy profile with heights of 0.5-1 mm in all cases. The maximum depth of wear of track membrane with 0.8 μm pore's diameters is 4.41 μm (sliding speed of 4.2 m/min). Wear depth is reduced to 2.34 m with reducing the sliding speed of 3 m/min. The maximum depth of wear of track membrane with 0.4 μm pore's diameters is 4.2 μm (sliding speed of 4.2 m/min).

IOP Conf. Series: Materials Science and Engineering **125** (2016) 012020 doi:10.1088/1757-899X/125/1/012020

Wear depth increases from 0.38 to 0.63 mm with increasing pore's diameter from 0.4 to 0.8 μm. It appears that it is the result of pushing material TM counter body on the side surfaces of the track wear. With decreasing sliding speed the wear depth reduces from 0.63 to 0.52 μm.

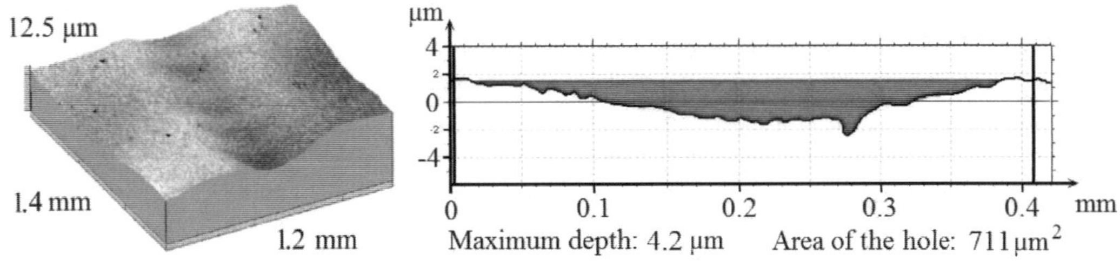

Figure 5. Profile of track membrane surface and sections of wear track of track membrane with 0.4 μm pore's diameters. Sliding speed is 4.2 m/min

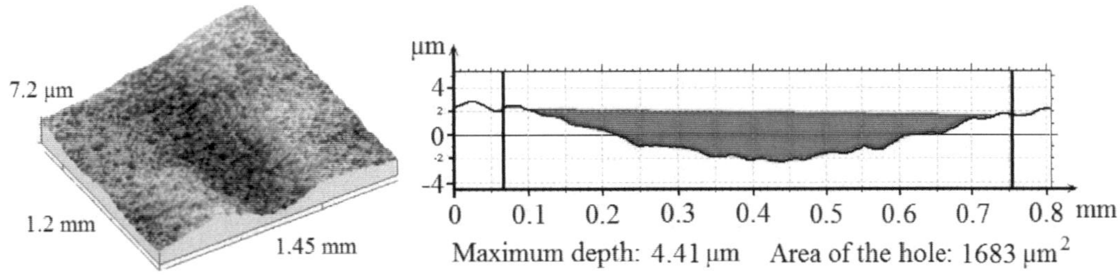

Figure 6. Profile of track membrane surface and sections of wear track of track membrane with 0.8 μm pore's diameters. Sliding speed is 4.2 m/min

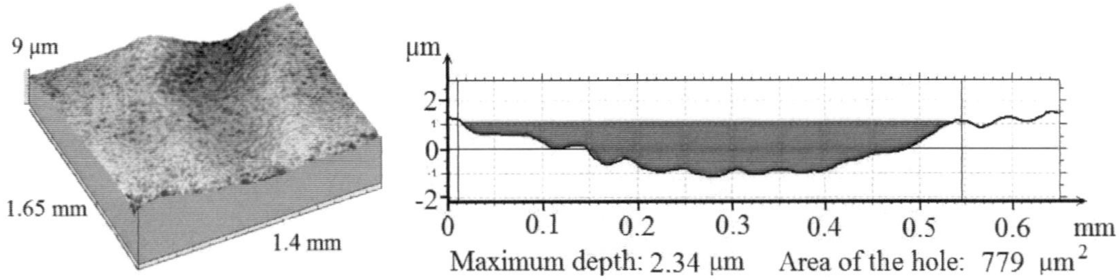

Figure 7. Profile of track membrane surface and sections of wear track of track membrane with 0.8 μm pore's diameters. Sliding speed is 3.0 m/min

4. Conclusion

The process of unlubricated friction in contact "metal-polymer" has next stages. These are a sharp decline of the coefficient of friction (1 stage), a rapid increasing of friction coefficient (2 stage), gradual increasing of friction coefficient (3 stage), stepped variation (4 stage).

The coefficient of friction decreases with increasing pore diameter TM.

Based on the results of determining the depth of wear track we can say that increasing the amount of wear of the pore diameter does not change significantly.

Implantation of the track membrane takes about 10-15 minutes during surgery. According to research active period corresponds to considerable variations of the friction coefficient, but it does not destroy the TM. PET track membrane have adequate resistance to wear and can be successfully used in surgical procedures in ophthalmology.

Acknowledgments
Work performed under the program of basic scientific research of the state academies of sciences for 2013-2020.

References

[1] Oganesyan U Ts et. all A Method of Manufacturing Track Membranes Russian Patent № 2077938, 02.12.94
[2] Kravets L I et. all 2011 Surface and Coatings Technology 205 455-461
[3] Zapuskalov I V, Bosykh E O and Sohoreva V V 2013 Proceedings of the Higher Educational Institutions. Physics 56 303-305
[4] Jing Zhao et. all 2014 Progress in Polymer Science 139 1668-1720
[5] Ramakrishna S et. all 2001 Composite Science and Technology 61 1189-1224
[6] Bharat B and Vilas N 1995 Tribology Transactions 38 119-127
[7] Hiroshi T et. all 2014 Tribology Letters 54 221-227
[8] Sierros K A and Kukureka S N 2007 Wear 263 992-999
[9] Bosykh E O, Sohoreva V V and Pichugin V F 2014 Petroleum Chemistry 54 669-672
[10] Filippova E O and Sokhoreva V V 2014 Second International Conference on Radiation and Dosimetry Fields of Research (Nis) 119-122
[11] Apel P U and Kuznetsov V I 1985 Colloid Journal 48 3-8
[12] Belii VA et. all 1976 Friction and wear of the materials based on polymers (Minsk: Science and Technology)

Model of Layered Weld Formation Under Narrow Gap Pulse Welding

A G Krampit, Doctor of Technical Sciences
Yurga Institute of Technology, TPU Affiliate, Russia, Kemerovo Region, Yurga

e-mail: akrampit@mail.ru

Abstract. The model parameters of narrow gap pulse welding can be divided into input, internal and output ones. The breadth of gap, that is, clearance breadth between upright edges is one of key parameters securing high quality of a weld joint. The paper presents theoretical outcomes for the model of layered weld formation under narrow gap pulse welding. Based on these studies is developed model of processes, which occur in the weld pool under pulse grove welding. It comprises the scheme of liquid metal motion in the weld pool, scheme of fusion with the side edge and in the bottom part, and the scheme of welding current impulse effect on the structure of a weld joint.

1. Introduction

Welding arc is to provide spatial stability, controlled transfer of electrode metal, required geometry of arc column, and necessary force and thermal action to produce a defect-free weld joint.

The model parameters of narrow gap pulse welding can be divided into input, internal and output ones. Input parameters include values of welding current and voltage at various points of pulse welding, as well as their rates of change; shielding medium and electrode composition; rate of electrode supply; diameter of electrode; and stick-out distance. Internal parameters comprise load pressure of welding arc; geometry of arc column; parameters of electrode metal transfer (flight rate and mass); light emission of the arc column; and heat emission. Output parameters are geometry and structure of a weld.

Currently, remains relevant the question of obtaining the quality welding when welding in narrow gap [1-2].

2. Methods of Research

The breadth of gap, that is, clearance breadth between upright edges is one of key parameters securing high quality of a weld joint. Power supply to the electrode wire (electrode) is necessary for stable arcing as stick-out distance usually amounts to 10 – 15 diameters of electrode when applying thin wires.

If the thickness of metal to be welded exceeds the stick-out distance electric current is to be supplied into the gap, requiring for supplementary current insulation of a contact tip. Actually, standard current-carrying tips are over 6 mm, so welding gets complicated; if the gap is less than 7 mm. Arcing gets unstable, if the gap breadth is scaled down due to the effect of ferromagnetic masses. The arc can be thrown over an edge of a weld joint, as the consequence, incomplete melting of a bead to the second edge is possible. On the assumption of the afore-mentioned facts, the author has specified conditions to restrict gap dimensions.

The first condition is a gap breadth. Minimal gap breadth is dependent on the amplitude of welding current, and exceeds the diameter of arc column in pulse (1), that is, being conditional on the peak value of welding current amounts to 4-8 mm (experimental values obtained on the basis of film shots) under conditions of controlled transfer in shielding gases. Minimal gap breadth can be the reason for the arc "throwing over" onto the edges of a weld joint (Fig. 1).

$$b_{min} \geq D = \sqrt{\frac{I \cdot 4}{i \cdot \pi}}, \tag{1}$$

of where
bmin – minimal gap breadth, mm;
D - diameter arc column in the pulse, mm;
I – current strength, A;
i – current density, A/mm^2.

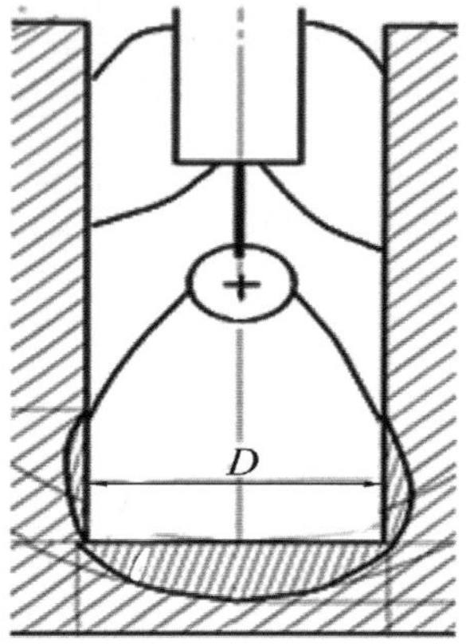

Figure 1. Influence of minimal gap breadth on arcing
(*D* – diameter of arc column)

Substantial gap narrowing necessitates precise electrode running along the joint, electrode wire is to be supplied exclusively parallel to the edges of a weld joint, therefore, the requirements of edge preparation, assembly accuracy and welding equipment are getting stricter. Needed tolerances and parameters of a gap are easily-secured in laboratory conditions, although they are rarely possible in the manufacturing environment.

The second condition is a slope angle of the electrode. The process of impulse power supply to the arc is distinguished by longer applied impulses (fully or partially overlapping melting and transfer) if the amplitude of welding current is greater than that under pulse welding (melting and transfer of electrode metal are separated). As impulse effect is not ceased after bead detachment (as recommended for welding in inert gases or their mixtures), melting occurs on the end of the electrode and the next bead emerges.

Under pulse welding bead detachment is in the range of pulse rise provided that peak values of welding current are substantial. Liquid metal of the weld pool is completely displaced from the head at the point of bead detachment and arc, deviating from the electrode coaxial position, starts burning on the front edge. Out-of-line position of the electrode makes electro-dynamic forces enormously distort shape of the arc and even blow it out. This fact was noted in paper [3].

The paper presents theoretical outcomes for the model of layered weld formation under narrow gap pulse welding.

The basic part. Modeling process of narrow gap welding is depicted in Figure 2. At the initial point of time welding current and voltage are not too big in the pause; therefore, arc column has a small diameter (d), specified in a dash-line. Deepening under the arc is not big as the pressure on the side of the arc (P) is low. Current rise causes arc column widening up to the diameter (D) proportionally to the current strength. The change in diameter, which is relevant for welding in carbon dioxide, can be described by the dependence:

$$D = k \cdot I^{0.5} \tag{2}$$

where
D – diameter of arc column, mm;
k – proportionality coefficient mm/A;
I – current strength, A.

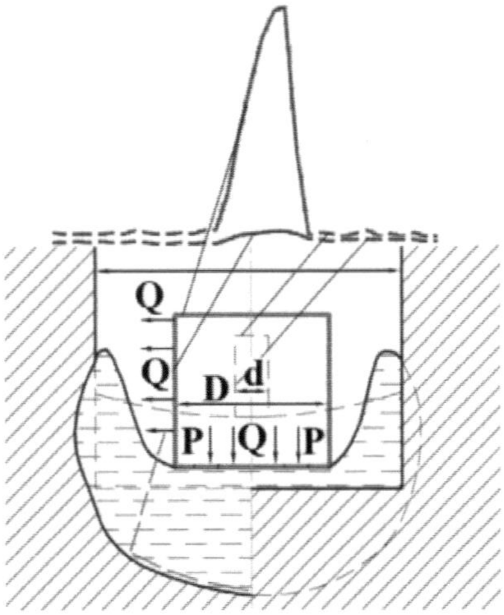

Figure 2. Modeling gap welding process:
D – arc column diameter when impulse piling up; d – arc column diameter in pause;
P – force pressure of welding arc; Q – heat flow from the arc.

Arc length gets extended mainly due to liquid metal displacement from the weld pool, which is conditional on arc pressure rise (quadratic dependence). As far as current strength, voltage and arc length, consequently, go up, the gap between arc spot and weld pool bottom decreases. This fact conditions increase in heat flow and heat input area, as the result, arc transfers more heat into the base metal; that is the reason for rising rate of melting and obtaining the required penetration depth of the product. As far as a part of arc column is submerged into the weld pool, liquid metal is displaced onto the edges of a weld joint due to growing overall level of the weld pool, and because of kinetic energy of liquid metal flow, displaced from under the arc, and rolling of liquid metal under action of Lorenz forces. Liquid metal of a weld pool will bring overheated masses of liquid metal from high temperature zone under the arc column and additional heat flow will be generated as emission energy is absorbed by arc column.

The model of layered weld formation is to meet the following conditions and admissions.

The general condition consists in a stable transfer of electrode metal (controlled range of modes: pulse duration 3 ms to 12 ms; pulse frequency 30 Hz to 120 Hz; pulse amplitude 300 A to 1200 A; electrode wire diameter 0.8 mm to 1.6 mm).

Admissions:

- unfilled part of opening contains no metal, so any change in heat spreading direction towards the upper surface, that is along the axis OZ, can be stipulated;

- on the object – ambient medium boundary, that is, on the plane Z=O`, heat exchange with the ambient medium is equal to zero (adiabatic boundary – second-order boundary condition);
- heat source is an instantaneous point source, acting at the point of current impulse;
- the main share of heat is transferred into edges of the product from the arc through the layer of liquid metal, located under the arc, and the layer of liquid metal between the arc and the side edge;
- the depth of displaced metal depends on the aggregate of forces, acting on the phase boundary;
- heat transfer from the arc into the bottom part does not depend on the own vibration frequency of a weld pool due to prevailing effect of welding current impulses.

Requirements to each layer:

1. root layer is to provide penetration in the corners of opening (energy transferred to the front edge over the current impulse is to be balanced with the energy necessary for penetration of a particular area Sp over the period: ${}^{Q_i}/_{Q_p} \geq 1$);
2. filling layers are to keep up the edge under liquid metal layer at the temperature of melting and secure required fusion depth – angle of fusion φ is to be below the critical value;
3. outer layer is necessary to take into account the reflected heat flow (a fictitious heat source is introduced into the computation scheme) and determine the conditions of welding, which are to keep up the surface temperature below the temperature of melting.

Let us apply the model to each layer.

Root layer welding. Peculiarities of specifying root layer parameters, and their influence on the weld geometry are presented in papers [4, 5].

Weld bed fusion of a root layer in the bottom part depends on the correlation of energy required for melting a specified area over a definite period to the energy transferred to the front edge. This energy can be calculated via summing up heat flow in the current impulse range through a variable of liquid metal layer under the arc as the balance of forces changes near the arc spot on the weld pool surface.

Filling layer welding. The author supposes the breadth of a gap (minimum b_{min} and maximum b_{max}) and penetration depth are the key parameters of narrow gap welding.

Conditions, restricting gap dimensions:

1. minimal breadth (b_{min}): significant gap narrowing conditions additional difficulties, that is, electrode is to be moved precisely along the joint, electrode wire is to be supplied coaxially to the weld joint, as the result, requirements to the welding equipment, accuracy of edge preparation and assembly as well, get exacter (necessary allowances and grove parameters are rather obtainable in laboratory conditions as against to the industrial ones);
2. maximal breadth (b_{max}) results from the sum of arc column diameter and doubled critical value of liquid layer on the gap wall in impulse (Fig. 3).

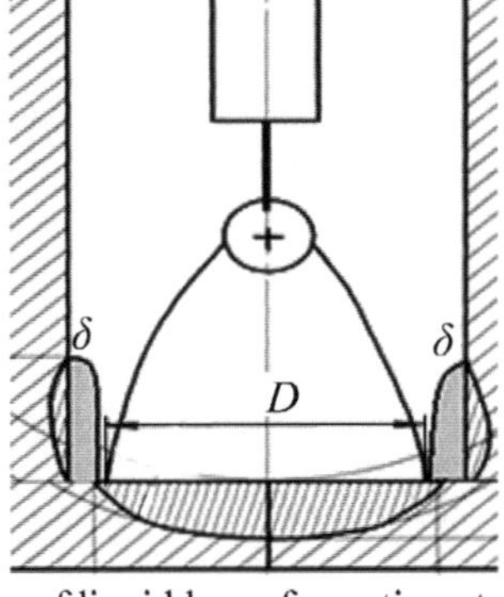

Figure 3. The scheme of liquid layer formation at maximal gap breadth: D – arc column diameter, δ – dimension of liquid layer

IOP Conf. Series: Materials Science and Engineering **125** (2016) 012021 doi:10.1088/1757-899X/125/1/012021

The maximal gap breadth b_{max} is specified by the value of heat flow put in from the arc into the weld joint edges through the liquid layer between the arc column and upright gap edges. This heat is quite sufficient, so the temperature of weld joint edge can provide secure fusion with the weld bed.

$$b_{max} = D + 2 \cdot \delta_{cr}, \quad (3)$$

where:
b_{max} – maximal gap breadth, mm;
D – arc column diameter, mm;
Δ_{cr} – critical value of the liquid layer, mm.

The scheme [6] of weld pool liquid metal flow into the gap assumes a liquid layer between the arc and gap edge. The dimension of this layer is a criterion for weld bed fusion with the base metal, as the consequence, defect-free weld formation.

If the liquid layer is thin, introduced through it energy is sufficient for melting a hard edge and wetting an edge with liquid metal (Fig. 4).

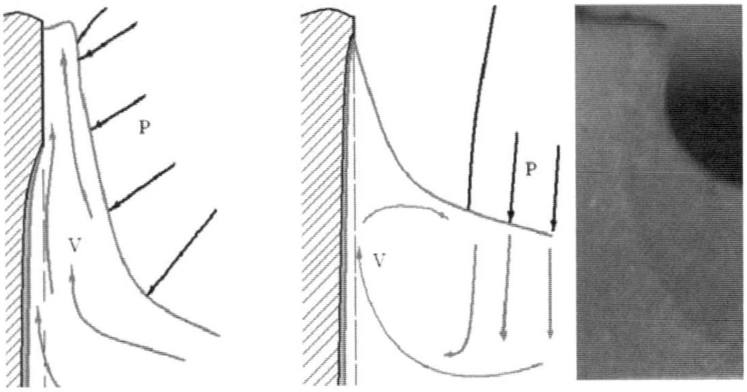

Figure 4 Influence of layer dimensions on arc heat input into the edges (layer dimensions are below the critical value): P – arc pressure strength; V – rate of liquid metal flow

If the layer exceeds the critical dimension energy introduced through it by the arc is not sufficient for melting a hard edge. Under significant heat removal into the base metal cooling liquid metal of the layer crystallizes from the edge surfaces (Fig. 5), because the heat conductivity of liquid metal is lower than that of metal in a solid state.

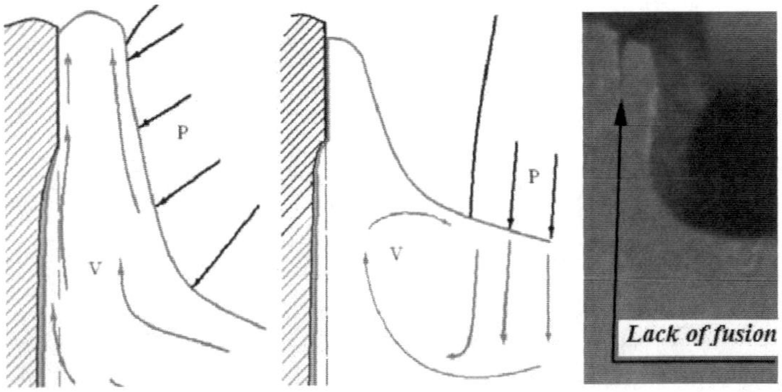

Figure 5. Influence of layer dimensions on arc heat input into the edges (layer dimensions exceed the critical value): P – arc pressure strength; V – rate of liquid metal flow

Calculation of a critical value for the liquid layer is required for defect-free weld formation. Maximal acceptable gap breadth can be specified on the base of layer dimensions.

Layer dimensions reduced enormously to improve weld and edge fusion under narrow gap welding let the arc go outside after covering the gap edges. This "tough" heat input into the solid phase causes significant structural consolidation of the superheated zone and deterioration of mechanical properties of metal in the superheated zone and weld joint as the whole.

Heat transfer into the solid metal through the liquid layer can be depicted in the diagram (Fig. 6).

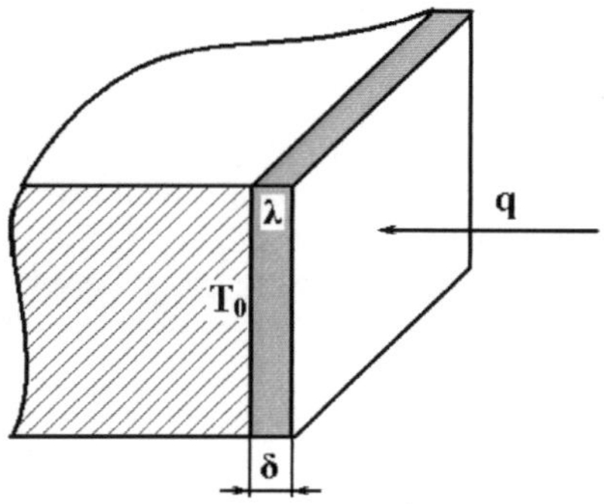

Figure 6. Heat transfer diagram through liquid layer in form of flat one-layer wall:
q – surface density of heat flow; δ – layer dimensions;
λ – thermal conductivity coefficient; T_0 – temperature

The heat influence from the arc column on the product edge can be viewed as the diagram of a point heat source on the surface of a semiinfinitive solid. Temperature increment of the side edge in the depth R from the surface can be described in the following equation [7]:

$$\Delta T = \frac{2Q}{c\rho(4at)^{3/2}} e^{-\frac{R^2}{4at}} = \frac{2qt}{c\rho(4at)^{3/2}} e^{-\frac{R^2}{4at}}, \qquad (4)$$

where
Q – heat amount, introduced per an impulse, V·s
q – surface density of heat flow, V;
v –rate of welding, cm/s;
λ – coefficient of thermal conductivity, V/cm^3 K;
cρ – heat capacity, J/cm^3 K;
t – time, s;
a – coefficient of temperature conductivity, cm^2/s.

However, one should take into consideration energy from the arc is input through liquid metal layer, which is placed out by the arc onto the side edge.

Heat transfer through liquid metal layer. Heat transfer from one medium (welding arc) to the other (product edge) through the liquid layer comprises heat output from the arc column to the outer part of the layer, heat conductivity in liquid metal and heat output from the inner surface (fusion boundary) to the side edge. The process on the wall boundaries under heat transfer is characterized by the third-order boundary conditions determined by liquid metal temperatures (boiling and melting) on both sides of the layer, as well as respective coefficients of thermal conductivity. Thermal conductivity lets heat flow through the layer, whereas heat output transfers it from the second surface of the layer to the cool medium [8]:

$$q = \frac{\lambda}{\delta}(T_1 - T_2), \qquad (5)$$

where
λ – coefficient of layer heat conductivity, V/(cm·K);
δ – layer thickness, cm;
T_1 – temperature of more heated surface of the layer, K;
T_2 – temperature of less heated surface of the layer, K.

Substituting q in equation (4) by expression (5) and designating thermal conductivity coefficient λж and the difference of temperatures ΔTl, ΔTs for liquid and solid phase, respectively, we obtain:

$$\Delta T_{\text{т}} = \frac{2\frac{\lambda_l}{\delta}(\Delta T_l)t}{cp(4at)^{3/2}} e^{-\frac{R^2}{4at}} \quad (6)$$

Expressing δ from equation (6), we obtain:

$$\delta = \frac{2\lambda_l \cdot (\Delta T_l)t}{cp(4at)^{3/2}} e^{-\frac{R^2}{4at}}. \quad (7)$$

We assume: a=0,08cm^2/c (for low-carbon steels); λ_l =0,26 V/cm·K (for liquid layer) [6, 9, 10].

Varying time and rate of welding a dependence of liquid layer critical value on impulse rate and time has been deduced from the equation (7) (Fig. 7).

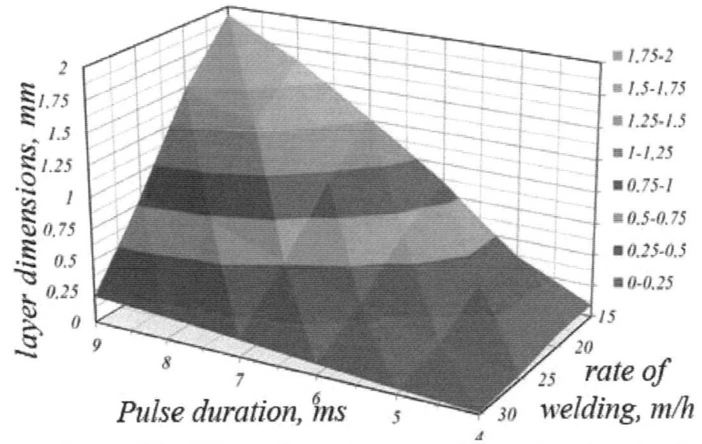

Figure 7. Dependence of liquid layer dimensions on welding rate and peak value of welding current in the impulse

3 Let us give an example

Maximal breadth of clearance under gap welding is specified as doubled sum of arc column radius and critical value of the layer. The data in paper [11] is used to calculate arc column radius. Arc column radius is 5 mm for peak value of current 650-680 A. As the consequence, maximal gap breadth is 12÷13 mm if welding rate is 15 m/h and impulse time 6÷8 ms. For further calculations we confine ourselves by maximal gap breadth b_{max} = 12 mm.

The dimensions of liquid layer (δ) on the gap edge at the time of welding current impulse is relevant for security of bead and upright wall fusion. Maximal dimensions of the layer providing secure fusion when welding filling layers, depends on heat input and thermophysical properties of metal.

Therefore, the area of reliable fusion with product edges, as well as defect-free formation of a weld can be obtained on the base of gap breadth (minimal and maximal) calculation.

Outer layer welding. Application of modes providing secure fusion when welding filling layer to the outer one is the reason for the defect called "breakup" of edges (Fig. 8).

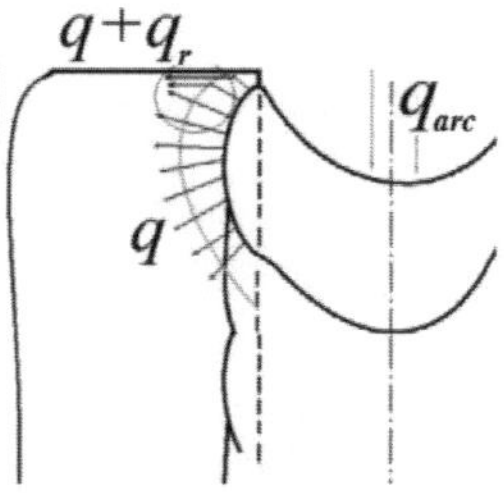

Figure 8. Scheme of arc heat input into the edges at the point of impulse when outer layer welding

It is conditioned by changing heat removal into the edges of a weld joint. Supplementary surface distinguished by insignificant heat removal into the surroundings over the impulse pile-up causes the edge overheating and its significant melting. Overheating is possible due to reflection of a heat flow share from the internal edge, that is why, reflected flow qref adds to heat flow q. Heat flow gets concentrated near the edge and its direction is distorted. As the result, the edge of a weld joint melts strongly (breakup of edges) when outer layer welding. To eliminate the "breakup" of edges heat removal is to be compensated by reducing heat input q_{arc}/v on upper layers. Reducing heat input is possible through increasing rate v or scaling up the arc power. Welding techniques were developed to provide low heat input under the stable arc [12].

Defect-free outer layer (eliminated "breakup" of edges) can be produced according to the following methods and in the following ways:

- power input redistribution in impulse and in the pause (pulse welding);
- current mean value decrease due to cyclic variations of electrode wire supply rate (welding with double modulation);
- combined control over melting and electrode metal transfer (welding with impulse feed of electrode wire under pulse power supply of welding arc).

Therefore, a layered model of processes, occurring in a weld pool when narrow gap welding, is developed and demonstrated below (Fig. 9).

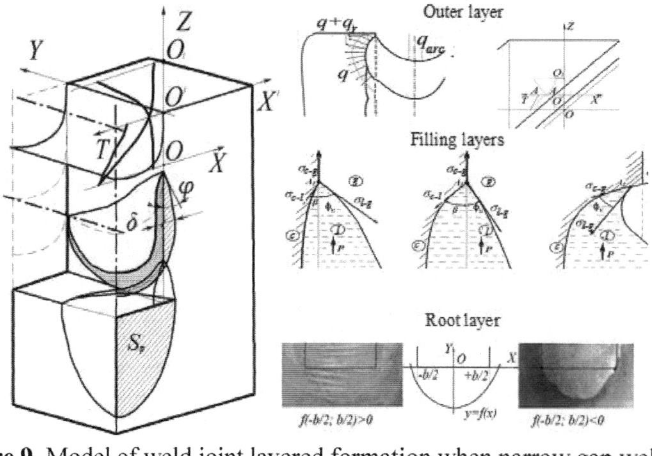

Figure 9. Model of weld joint layered formation when narrow gap welding

Root layer:

- kinetic parameters of liquid metal motion in the weld pool – liquid metal is displaced from under the arc when pulse piling up. The depth of displaced metal depends on the forces on the inter-phase boundary;
- the change in heat flow, transferred into the bottom part through changing dimensions of the layer

under the arc when pulse piling up;

- thermal conditions of melting in the bottom part and providing necessary penetration depth.

Filling layers:

- liquid metal motion in the weld pool – displacement of liquid metal onto the edges of a weld joint when pulse piling up;
- thermal conditions of fusion with the side edge and support of required penetration depth with the side edge;

Outer layer:

- heat transfer from the arc close to the surface.

Therefore, presented model of layered weld joint formation under pulse welding makes control over weld geometry possible.

4 Conclusions

1. The conditions of stable arcing have been deduced:
- minimal gap breadth depends on peak value of welding current and current density in the arc column;
- welding is to be carried out with electrode inclined under acute angle to avoid bead detachment from the weld pool by the reflected anode flow.

2. Some liquid metal of the weld pool is displaced into its tail part under narrow gap pulse welding, while the rest of it is transferred onto unfused gap edges. The dimensions of liquid layer between the arc and weld joint edge when pulse piling up is a key parameter supporting secure fusion under filling layer welding. It depends on heat input and thermophysical properties of metal (maximal layer dimension is 0,5÷1,7 mm for rate of welding 15÷30 m/h);

3. Elimination of edge "breakup" under outer layer welding is possible through compensation of heat removal by reduced welding arc power or via pulse welding, distinguished by lower power.

4. A model of processes is developed, which occur in the weld pool under pulse grove welding. It comprises the scheme of liquid metal motion in the weld pool, scheme of fusion with the side edge and in the bottom part, and the scheme of welding current impulse effect on the structure of a weld joint.

References

[1] Chinakhov, D.A. Gas dynamic control of properties of welded joints from high strength alloyed steels China Welding (English Edition) 2014 23(3), 26-30
[2] Chinakhov, D.A. Regression models of mechanical properties of the multi-layered welded joints of 30KhGSA steel Svarochnoe Proizvodstvo, 2002 №5. – P. 3-5.
[3] Finkelnburg V., Mekker G. Electric arcs and thermal plasma. Moscow. 1961.
[4] Krampit A. G. Calculation of pulse welding parameters of a root layer with edge preparation // Welding Production. – 2013. – №7. – P. 3-5.
[5] Krampit A.G., Krampit N.Yu. A method for the determination of the geometrical dimensions and area of the welded joint // Welding International. – 2013. – Volume 27. – Issue 10. – P. 834-836.
[6] Krampit A.G., Krampit N.Yu. Control over weld joint formation under welding in carbon dioxide with pulse arc power supply // Tomsk: TPU Publishing. – 2009. – 109p.
[7] Theory of welding processes // Edited by Frolov V.V. – M.: Higher School, 1988. 559 p.
[8] A.M. Arkharov, S.I. Isaev, I.A. Kozhinov et al. Edited by V.I. Krutov. Heat engineering / M.: Machinebuilding, – 1986. – 432 p.
[9] Chernyshov G.G. Rybachuk A.M. Determination of liquid metal thickness on the front of the weld pool // Welding Production. – 1979. – №10. – P. 9-10.
[10] Arsentyev P. P., Koledov L. A. Metal melts and their properties. M.: Metallurgy. 1976. 375 p.
[11] Krampit A.G., Krampit N.Yu. Control over welding process in carbon dioxide by long arc // LAP LAMBERT Academic Publishing GmbH & Co. KG. – 2012. – 255p.
[12] Invention patent №2429111, from 20.09.2011. Hybrid mode of impulse control over consumable electrode welding. Knyaz`kov A.G., et al.

Nano-Disperse Borides and Carbides: Plasma Technology Production, Specific Properties, Economic Evaluation

G V Galevskii, Doctor of Engineering, Professor, V V Rudneva, Doctor of Engineering, Professor, S G Galevskii*, PhD in Economic Sciences, Associate Professor, K I Tomas, PhD in Engineering, Associate Professor, M S Zubkov**, Student, Group 10A42**

Siberian State Industrial University,
654007, Novokuznetsk, 42, Kirov str., tel. (3843)-74-89-13
*National Mineral Resources University (University of Mines)
199106, Saint-Petersburg, Vasilyeski Island, 2, 21st Line, tel. (812) 328-89-16
** Yurga Institute of Technology, TPU Affiliate, Yurga
652055, Kemerovo Region, Yurga, 26, Leningradskaya str., tel. (38451) 7-77-66

e-mail: kafcmet@sibsiu.ru, feanor1985@mail.ru*, tki@tpu.ru**

Abstract. The experience of production and study on properties of nano-disperse chromium and titanium borides and carbides, and silicon carbide has been generalized. The structure and special service aspects of utilized plasma-metallurgical complex equipped with a three-jet direct-flow reactor with a capacity of 150 kW have been outlined. Processing, heat engineering and service life characteristics of the reactor are specified. The synthesis parameters of borides and carbides, as well as their basic characteristics in nano-disperse condition and their production flow diagram are outlined. Engineering and economic performance of synthesizing borides in laboratory and industrial conditions is assessed, and the respective segment of the international market as well.

The work is performed at State Siberian Industrial University as a project part of the State Order of Ministry of Science and Education of the Russian Federation № 11.1531/2014/K.

1. Introduction

Over 1.5 milliard tons of steel – the most common structural material were manufactured for the first time in 2011 all over the world. In the pattern of structural materials the share of alloys based on ferrum is 95%, that of non-ferrous metals and alloys – 4%, all the others – below 1%. However, the last group includes materials of special purposes, has a wide range and is extraordinary important for present day civilization. It consists of materials meeting the criteria "refractoriness", "super-hardness", "heat resistance" and "high-temperature strength", which are necessary to solve numerous innovative academic and technological, and engineering problems. That is why, materials based on carbides, borides, nitrides, silicides and their composites are of key importance in this group. In development of their national technological base the following stages can be singled out. In the 50-60s the Institute for Problems of Materials Science and Superhard Materials Institute of National Science Academy of

Ukraine promoted the study on their properties and developed technologies of their production and use. In the 70-80s some regional scientific centers – branches of academic Institutes in Moscow, Kiev, Riga and Novosibirsk made efforts to develop production and support the use of these materials in fine dispersion. The 2000s were distinguished by engineering progress, which is still in process and involves development of nano-technologies and nano-materials at a new level.

At present Russia is among the leaders in terms of carried out research into nano-technology but is rather inferior in respect of manufacturing nano-products and their export [1-5]. In line with the President's initiative, dated 24th April 2007 «The strategy of nano-industry development» the system of state support for this fundamental and applied research was established, which provides integration and concentration of state and private business resources, development of national nano-technological infrastructure, enhancement of efficiency of nano-products commercialization and their transfer, and focusing the efforts on the most commercially advantageous ones. Among these products there are functional, structural and composite nano-materials.

In the Siberian region of the Russian Federation the foundation of plasma-metallurgical nano-technologies of high temperature super-hard materials was laid by the school of academician of the Russian Academy of Sciences M.F. Zhukov, who joined efforts of research teams of academic, industrial and educational institutions in Western and Eastern in the early 1970s Siberia in order to produce and apply various nano-powders in laboratory conditions. The results of this research were presented in multi-volume sets published in Russia and abroad «Low temperature plasma» (editor-in-chef academician RAS M.F. Zhukov), «Thermal plasma in the technology of new materials» (science editor academician RAS M.F. Zhukov), «Nano-materials and nano-technologies in silicon carbide production» (science editor professor G.V. Galevskii) etc. In the course of investigations the functionality of plasma-technological equipment was proved, as well as the possibility to synthesize and apply nano-scaled carbides, borides, nitrides in new fields. However, social and economic transformations in Russia at the turn of the 1990s were the reason for cancelled investments in this research domain, made impossible and uncalled the transition to the industrial scale and commercialization of the products.

In conditions of the evolving in the 21st century nano-technological development we suppose further research into organization of industrial plasma-metallurgical production of high temperature chemical compounds and identification of foreground spheres of their application in ceramic, galvanic, metallurgical and other technologies is the important scientific and economic problem, which is relevant for the State Program «Science and technologies development for the period through 2020».

The aim of this paper is to generalize the long-term experience and results obtained by Siberian State Industrial University when developing main equipment and adopting production technologies of nano-disperse chromium, titanium and silicon borides and carbides. Production technologies of borides and carbides were selected as the object of research due to the favourable combination of their consumer properties (hardness, refractoriness, wear resistance and high temperature strength); availability of raw materials; relatively simple synthesizing in the furnace as the basic technology; stable demand of consumers, particularly for coating materials, materials of functional coatings and modifying complexes, which provide three(four)fold increase in the life cycle of products and tools; and real possibility of attaining new effects when using them in nano-condition.

2. Plasma-technological equipment: structure, operation, characteristics.

For experimental study on the synthesis and production of borides and carbides Siberian State Industrial University in collaboration with experimental and production company "Polymet" developed a plasma-technological complex (Figure 1), excelling previous experimental facilities in the key characteristics. The structures of plasmatrones, mixing chambers, dosing units of burden charge, and bag collector are protected by patents of the Russian Federation № 66877, 107440, 108319, 184916. The main characteristics of the reactor are listed in Table 1. In papers [6, 7] heat engineering, performance and processing characteristics are described in more details. The plasma-technological complex consists of a three-jet reactor and the systems of power, gas, and water supply and ventilation, supervisory instruments and automatic devices, burden charge dosing units, and devices of nano-disperse products entrapping and neutralization of emitted process gases.

To generate the plasma flux three arc heaters of gas (plasmatrones) EDP-104AM with the capacity 50kW per a facility are used, they are mounted in the mixing chamber at an angle 30° to the reactor axis. Plasmatrones EDP-104AM run on direct current and require for the following parameters of the electric arc: arc voltage to 250V, current to 200A. The stabilization of the electric arc is gas-vortex due to the tangential input of plasma-generating gas through a special tightening ring. The water-cooled copper anodes of plasmatrones with the inner diameter of 0.008 m have practically unrestricted life time provided that they are cooled and run in the mixing chamber with plasma jets inclined at an angle 30°. The cathodes of plasmatrones consist of water-cooled copper bodies and cathode inserts made of thoriated tungsten (to reduce the work function of electrons) with the diameter of 0.003 m and life time 100 – 120 hours. Plasmatrones are switched on by oscillators. Plasmatrones EDP-104AM differ from the basic model as technically pure nitrogen with 1.5 – 2.0 volume % oxygen meeting the actual standard of currently supplied technical nitrogen can be used as a plasma-generating gas. Plasmatrones are supplied with electric power from the thyristor transformer AT4–750/600, distinguished by the steeply-falling volt-ampere characteristic and the following operating parameters: power, kW – 450; rectified voltage, V – 600; rectified current, A – 750; efficiency factor in nominal conditions, % – 96; voltage in the supply main, kW – 6.

Fine-grained raw materials are efficiently input into the reactor, mixed with a plasma flux due to the structure of the mixing chamber, moreover, the life time of plasmatron anodes is quite unrestricted. The mixing chamber is joined to the divided water-cooled channel with the inner diameter of 0.064 m. The mixing chamber and reactor sections are made of stainless steel. Fine-grained raw materials are fed into the reactor by a water-cooled tuyere. The tuyere is used also to supply gas hydrocarbon into the reactor. The reactor channel is lined inside with high-temperature heat-insulating material in order to reduce the radial gradient of the temperature near its walls.

Figure 1. Industrial plasma-technological complex

Table 1. Basic characteristics of the reactor

Characteristics	Values
Capacity, kW	150
Type of the reactor	Three-jet Direct flow Vertical
Type of the plasmatron, power, kW	EDP-104A, 50
Plasma-generating gas	Nitrogen
Mass of the heated gas, kg/h	32.5
Inner diameter, m	0.054
Volume of the reactor, m^3	0.001
Lining of the reactor channel	Zirconium dioxide
Temperature of plasma flux, K	5400 (L*=0) – 2200 (L=12)
Temperature of lining, K	1549 (L=0) – 770 (L=12)
Specific electric power, MW/m^3	2142
Life time: - anode - cathode	3125 112
Contamination of borides and carbides with erosion, % - anode - cathode	 Cu – 0.0000954% W – 0.0000002%
L* - relative length of the reactor	

Powdered raw materials are measured out in doses by the batch type electromechanical and gas vortex dosing unit with the removable cylinder – receiver of powdered raw materials, intended for poorly flowing fine-dispersed raw materials.

The entrapping system consists of a settling chamber, providing the temperature decrease of process gases to 2800 – 2000 K and trapping of 10 % nano-powder, and two bag filters working by turns (trapping up to 85 % nano-powder). The reagent is fed by the water-cooled probe into the settling chamber, which passivates and coagulates nano-powders. The filters are made with a water-cooled body, and a filtering bag is refreshed by compressed air (nitrogen) blowing. Filter cloth is twill weave chromium-nickel steel wire mesh.

The plasma-technological complex excels available laboratory and experimental industrial facilities in power four-five folds, life time three-four folds, efficiency 2.5 – 3.5 folds.

3. Production and properties of boride and carbide nano-powders

The results of research into plasma-technological production of borides and carbides are presented in details in a great number of papers published at different times; inter alia [8-15].

The development and adoption of plasma-technological production of chromium, titanium and silicon borides and carbides comprises two phases: 1) the study on formation processes of borides and carbides and physical-chemical properties of nano borides and carbides; 2) industrial adoption of the studies technological samples.

The first phase focuses on the simulation of reaction of material flow with the plasma flux, experimental study, discussing the mechanism of boride and carbide formation, and physical and chemical assessment of nano-dispersed products. The equations describing the relation of borides and carbides concentration to the basic process factors (1.1 – 1.5), tolerances of the change in formation of borides and carbides in industrial reactor with the capacity of 150 kW, main characteristics of borides and carbides (Table 2) and their photomicrographs (Figure 2) are given below.

$[CrB_2]$ = -413.53 + 0.09695 T_0 + 2.283 [B] + 0.1736 $\{H_2\}$ – 0.00058 T_0 [B]; (1.1)
$[Cr_3(C_{0.8}N_{0.2})_2]$ = -66.12 + 0.03 T_0 – 0.42 $\{H_2\}$ – 0.14 {N} – 0.00002 T_0 {N}; (1.2)
$[TiB_2]$ = -412.41 + 0.09489 T_0 + 2.196 [B] + 0.1597 $\{H_2\}$ – 0.00061 T_0 [B]; (1.3)
[TiC] = 17.3211 + 0.0105 T_0 – 0.0156 $T_з$ + 0.1859 $\{CH_4\}$ – 3.432 $\{H_2\}$ – 0.4078{N}; (1.4)
[SiC] = 86.50 + 0.00273 T_0 – 0.0064 $T_з$ – 0.144 $\{CH_4\}$ + 0.00007 T3 $\{CH_4\}$, (1.5)

where T_0 – initial temperature of the plasma flux, K; $T_з$ – hardening temperature of products of boride and carbide formation; [B] – concentration of boron in the furnace burden (% of stoichiometrically required); $\{CH_4\}$ – hydrocarbon content (%of stoichiometrically required); $\{H_2\}$ – hydrogen concentration in a plasma-generating gas, volume %; {N} – content of elemental nitrogen in a plasma-generating gas (% of stoichiometrically required for hydrogen cyanide).

The second phase was concentrated on development of reference documents (technical conditions and technological processes) and production flow diagrams of technologies to synthesize borides and carbides proposed by experimental and production company "Polymet", main engineering and economic factors are calculated. Figure 3 outlines an example of a production flow diagram of chromium carbonitride (I) and carbide (II) synthesizing.

Table 2. Tolerances of the change in the parameters of borides and carbides synthesizing in the industrial reactor with the capacity 150 kW and their main characteristics

Parameters of synthesis and characteristics	CrB_2	$Cr_3(C_{0.8}N_{0.2})_2$	TiB_2	TiC	SiC
Chemical composition of the exchange gas, volume %: - nitrogen / hydrogen / methane	74/25/1	99/–/1	74/25/1	99/–/1	99/–/1
Technological variant of synthesis	Cr+B+H_2	Cr+CH_4	Ti+B+H_2	Ti+CH_4	Si+CH_4
Efficient use of raw materials, kg/h	3.6	3.1	3.6	3.2	3.0
Boron content in the furnace burden, % of stoichiometrically required	100-120	–	100-120	–	–
Content of carbonization agent, % of stoichiometrically required	–	120-140	–	120-140	120-140
Initial temperature of the plasma flux, K	nm 5400	nm 5400	nm 5400	nm 5400	nm 5400
Hardening temperature, K	2600-2800	2000-2200	2600-2800	2600-2800	2800-3000
Phase composition	CrB_2	$Cr_3(C_{0.8}N_{0.2})_2$	TiB_2	TiC	β-SiC
Main phase content, %	92-93	92-93.5	92-93	93-93.5	91-92
Main phase output, %	91-92	90.5-91.5	91.5-92.5	92-92.5	87-90
Efficiency, kg/h	3.0	3.4	3.4	3.7	4.05
Intensity, kg/h·$м^3$	1364	2010	1980	2105	2200
Specific surface, m^2/kg	33000-35000	31000-35000	46000-48000	33000-35000	40000-44000
Size* of particles, nm	42.0	34.0	36.0	35.0	55.0
Form of particles	Spherical	Spherical	Spherical	Faceted cube	Faceted
Oxidation** of nanopowder x10^7, kg O_2/m^2	9.0-9.7	8.0-10.0	5.8-7.6	8.5-9.5	6.5-8.0

* - specified according to the specific surface;
** - specified after air storage for 24 hours

IOP Conf. Series: Materials Science and Engineering **125** (2016) 012022 doi:10.1088/1757-899X/125/1/012022

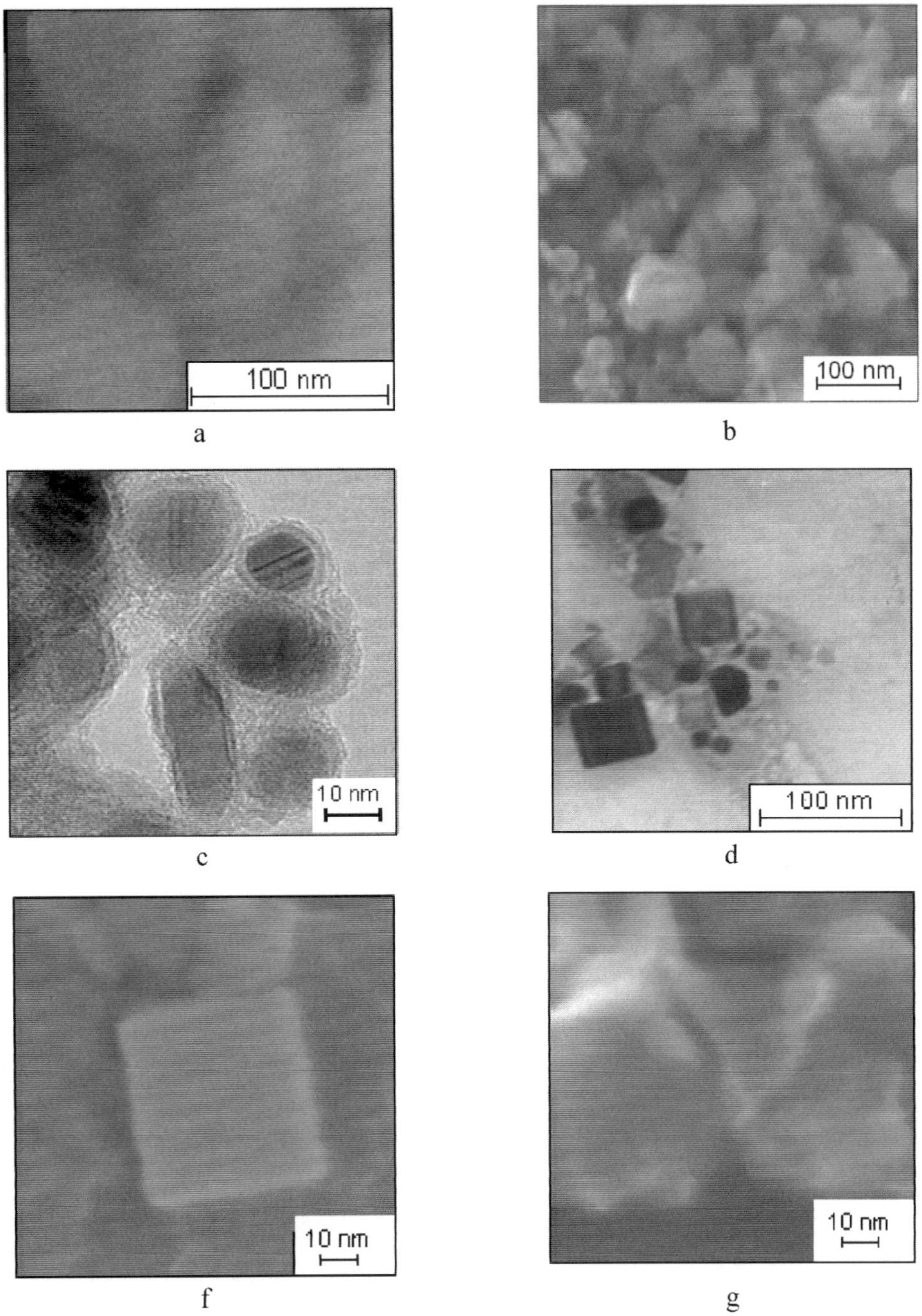

Figure 2. Photomicrographs of chromium boride and carbonitride nanopowders (a, b), titanium boride and carbide (c, d), silicon carbide (f, g)

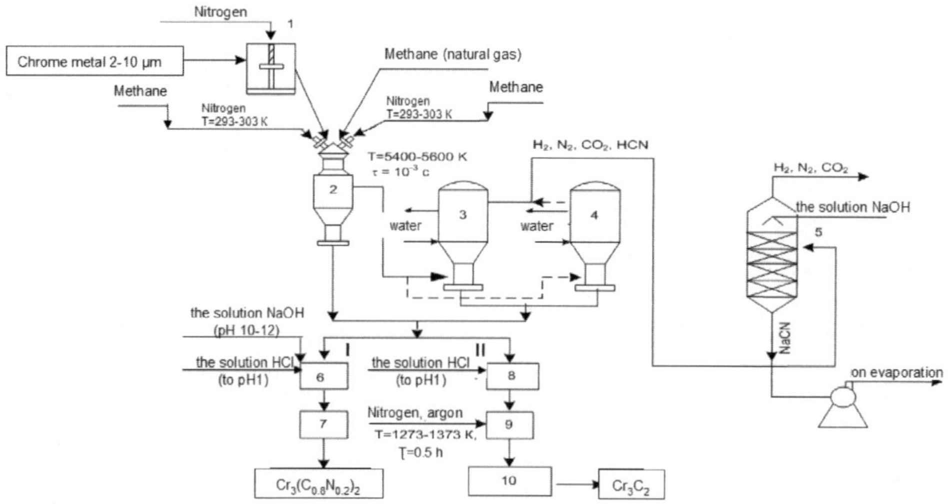

1 – furnace burden measuring out in doses; 2 – synthesis; 3, 4 – separation of the end product; 5 – absorption process of decontaminating emitted gases; 6, 7 – refining chromium carbonitride and control of its characteristics, 8, 9 – refining chromium carbonitride and its additional carbonation, 10 – control of chromium carbide characteristics

Figure 3. Production flow diagram of chromium carbonitride (I) and carbide (II) synthesis

4. Economic assessment

Introduction of boride and carbide nanopowders into industrial production create the necessary prerequisites for their wide application, at least on the homeland market of nano materials. Table 3 outlines the results of comparing main technical and economic factors of synthesized chromium boride and carbide, and silicon carbide in laboratory and industrial conditions.

Table 3. Comparison of technical and economic factors of nanopowders synthesized in laboratory and industrial conditions

Technical and economic factors	CrB_2	Cr_3C_2	SiC
Maximal capacity of the reactor, kW	150/50	150/50	150/50
Main phase content, %	95.0*/81,0**	96.0/81.0	99.0/94.0
Oxidation x10^7, kg oxygen/m^2	8-10/18-20	12-14/9-11	0.8/6.7
Efficiency, t/year per a reactor	3.2/0.5	3.6/0.6	3.1/1.8
Intensity, kg/h·m^3	1365/265	1360/245	1210/605
Specific electric energy consumption, thousands kW·h/t	75/145	69/140	74/115
Technical and economic factors	CrB_2	Cr_3C_2	SiC
Cost price, thousand rubles/kg	6.2/15.8	6.6/14.5	6.0/11.0
*/** - laboratory and industrial conditions			

World key developers and producers of nano materials based on borides and carbides are experimental and production companies «Nanostructured & Amorphous Materials, Inc.» (the USA), «Tokyo Tekko Co» (Japan), «Hefei Kaier Nanotechnology & Development Ltd. Co» (China), «NEOMAT Co» (Latvia), «PlasmaChem GmbH» (Germany), selling them at the price about 1200 $/kg; quality and dispersion are comparable [2-5].

Conclusions

The carried out analysis proves the possibility of industrial plasma-technological synthesizing high-temperature superhard titanium, chromium and silicon borides and carbides provided that technological complex equipped by three-jet direct flow reactor with the capacity of 150 kW is used as a base model. Industrial synthesis supports achieving quality, technical and economic standards similar to the world counterparts, and competitive ability of scientific and technical products.

References

[1] Galevskii G V Nanomaterials and nanotechnologies: assessment, tendencies, and forecasts/ G.V. Galevskii, V.V. Rudneva, E.K. Yurkova // Russian Journal of Non-Ferrous Metals – 2007. V. 48. –№2. –P. 157-160.

[2] Plasma Chem [Electronic resource] /Nano Powders; Webmaster PIXXL.WEBDESIGN.– Electronic data. – Berlin: Plasma Chem Gmbh, [2008]. – URL: www.plasmachem.com

[3] NEOMAT NANO POWDERS [Electronic resource]/ Products. – Electronic data. – Salaspils: Neomat Co., [2008]. – URL: www.neomat.lv

[4] Nanoceramics Powders [Electronic resource]/ Hefei Kaier Nanotechnology & Development Ltd. Co. – Electronic data. – Hefei Kaier Nanotechnology & Development htd. Co., [2008]. – URL: www.hfkiln.com

[5] NanoAmor. Nanostructured & Amorphous Materials. Inc. [Electronic resource]/ Products.– Electronic data. – Huston: Nanostructured &Amorphous Materials. Inc., [2008]. – URL: www.nanoamor.com

[6] Rudneva V V Thermal characteristics of three – jet plasma reactor / V.V. Rudneva, G.V. Galevskii, E.K. Yurkova // Steel in Translation. – 2007. Vol. 37. – No. 2. – P.115–118.

[7] Nozdrin I V Reactor for plasmometallurgical production of refractory borides and carbides / I.V. Nozdrin, L. S. Shiryaeva, G. V. Galevskii, V. V. Rudneva// Steel in Translation. – 2011. Vol. 41. – No. 8. – P. 644–648.

[8] Nozdrin I V Particle size of vanadium and chromium borides and carbides in a plasma flux / I.V. Nozdrin, G. V. Galevskii, L. S. Shiryaeva, M.A. Terenteva// Steel in Translation. – 2011. Vol. 41. – No. 10. – P. 799-801.

[9] Nozdrin I V Plasma synthesis and physicochemical certification of chromium nanoboride / I.V. Nozdrin, L. S. Shiryaeva, V. V. Rudneva// Steel in Translation. – 2012. Vol. 42. – No. 12. – P. 797-801

[10] Efimova K A Synthesis and properties of nano-scale titanium boride // K.A. Efimova, G. V. Galevskii, V. V. Rudneva//IOP Conf. Series: Materials Science and Engineering – 2015. Vol.91

[11] Garbuzova A K Titanium Carbide: Nanotechnology, Properties, Application / IOP Conf. Series: Materials Science and Engineering – 2015. Vol.91

[12] Rudneva V V Investigation of the morphology and particle size of silicon carbide nanopowders using electron microscopy/ V.V. Rudneva , G.V. Galevskii, E.K. Yurkova // Russian Journal of Non-Ferrous Metals – 2010. T. 51. –№2. –P. 158-164.

[13] Kuznetsov M A, Zernin E A, Danilov V I, Kolmogorov D E, Zoubenko L N Corrosion and mechanical properties of austenic steel weld joints. // IOP Conference Series: Materials Science and Engineering. – 2015. – №91 – http://iopscience.iop.org/article/10.1088/1757-899X/91/1/012010/pdf.

[14] Kuznetsov M A, Barannikova S A, Zernin E A, Filonov A V, Kartsev D S Methods for defining the concentration of nanostructured powders in protective gas and its effect on the micro-

structure of deposit metal // Applied Mechanics and Materials. - 2015 - Vol. 770. - p. 28-33.
[15] Kuznetsov M A, Zernin E A, Lukashov A S, Zhuravkov S P Influence of Molybdenum Nanopowder Concentration on Welded Joint Microstructure // Applied Mechanics and Materials. - 2014 - Vol. 682. - p. 438-442.

Specifics of Mode Parameters Choice Under Twin Arc Welding of Fillet Welds

A U Melnikov[1], A M Fiveyskiy[1] and M A Sholokhov[2]

[1]Ural Federal University named after the first President of Russia B N Yeltsin, Yekaterinburg, Russia
[2] Ltd SHTORM, Yekaterinburg, Russia

e-mail: au_melnikov@mail.ru

Abstract. The present article covers the specifics of mode parameters choice under twin arc welding of fillet welds. The necessity of mode parameters adjustment at the second arc due to heated metal of the first arc was proven. The obtained correction indexes allow us to determine with satisfactory accuracy the mode parameters under given dimensions of weld joint.

1. Introduction

Application of twin arc welding allows us to significantly increase the performance of the process and widen technological capability of welding [1-2]. The article [3] covers the technology of twin submerged arc welding in separate melting pools with inclined electrode of H-beam circumferential seams, this technique allows us to eliminate all main weak points of underhand welding, i.e.: low productivity, unallowable deformations, difficulties of through penetration of a beam web.

The application of the proposed method requires strict choice of welding modes and the electrode shall be placed in a right position against the joint [4]. At the present time there is no data presented regarding the choice of weld modes and parameters under twin welding in separate melting pools with inclined electrode.

At the present moment there are only several known calculation methods of modes for twin arc welding [5-7]. But they are designed only for calculations of twin arc welding in one melting pool and do not consider inclined electrode.

2. Specifics of mode parameters choice for first arc

The present work covers distinctive points of welding mode parameters choice under twin arc welding of fillet welds in separate melting pools with inclined electrode.

At the first step we calculate mode parameters for the first arc regarding needed depth of penetration. For doing so:

2.1. Make a draught of weld joint (Figure 1). Based on the given algorithm we define the offset distance from a beam flange.

Geometrical construction of the offset distance shall be carried out according to the following algorithm:

1) Draw a fillet weld legs on the beam flange and beam web. Draw a line AB that forms the surface of a flat joint.

2) Draw a line DG mid-perpendicular to AB. But line AD shall be equal to line DB.

3) From A draw a line AOM with an angle α to a beam web.

4) Through the cross of line DG and line AOM we draw an interval OF which is perpendicular to AOM. The resulting interval is the axis of weld.

5) The cross point of interval OF and a beam web defines the distance a, to which we shall move an electrode to obtain needed weld size.

6) To get a guaranteed overlap of joints, depth of penetration for the flange needed to be taken as 0,7δc.

7) Accepting the form for depth of penetration– as a semi-ellipse, we can draw melding line a. So that axis of a semi-ellipse will be equal to intervals AO and OF.

2.2. According to Figure 2 and Equation 1 we identify the area of a metal under extension and we measure the total height of a joint and its width.

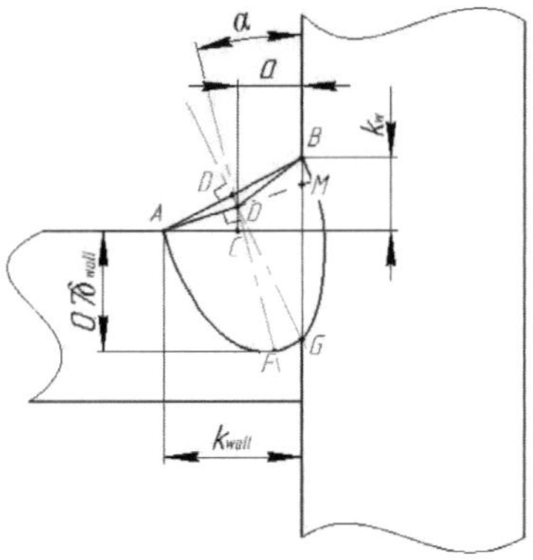

Figure 1 – Diagram to identify the distance «a»

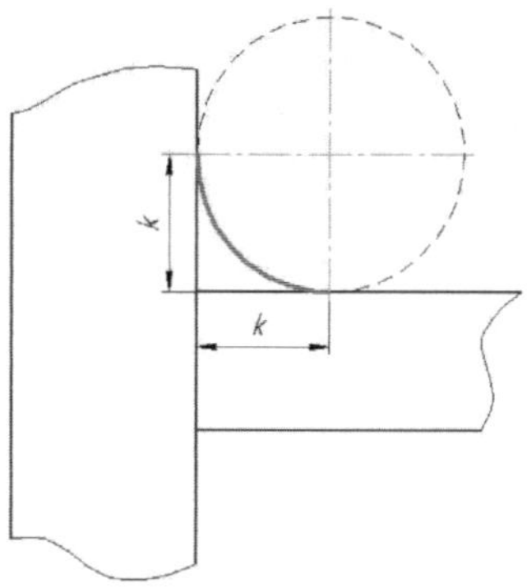

Figure 2 – Draught to identify the critical area of a metal under extension

The area of a metal under extension for a weld seam with not identical legs might be found according the equation:

$$F_y^{cr} = k_1 \cdot k_2 \cdot (1 - \frac{\pi}{4}), \quad (1)$$

where k_1 and k_2 – fillet weld legs, mm

2.3. We identify a calculated depth of penetration according to the following equation:

$$h_p = H - \frac{F_H}{0.73 \cdot e}, \quad (2)$$

where $F_н$ – the area of a welded layer, mm^2;

H – the total height of a weld seam, mm;

e – width of a weld seam, mm.

Under calculation of the joints with edge preparation or root opening – the area of a welded layer will include not only metal under extension but also the area of edge preparation or the area of a root opening.

2.4. Identify the range of a weld current

$$I_w = (80...100)h_p \quad (3)$$

From obtained range we choose approximate current rate.

2.5. Under chosen current rate we identify according to the following equation a welding arc voltage, needed for acceptable weld seam formation

$$U_a = 20 + \frac{0.05 \cdot I_w}{\sqrt{d_e}}, \tag{4}$$

where d_e –an electrode diameter, mm

2.6. Calculate the welding speed based according to the following equation:

$$V_w = \frac{A}{I_w}, \tag{5}$$

where A – a coefficient, that depends on the electrode diameter, $A \cdot m/h$

Values of a coefficient A are given in Table 1.

Table 1 – Values of coefficient A

Electrode Diameter d_e, mm	Values of Coefficient A, $A \cdot m/h$
1.2	от $2*10^3$ до $5*10^3$
1.6	от $5*10^3$ до $8*10^3$
2.0	от $8*10^3$ до $12*10^3$
3.0	от $12*10^3$ до $16*10^3$
4.0	от $16*10^3$ до $20*10^3$
5.0	от $20*10^3$ до $25*10^3$
6.0	от $25*10^3$ до $30*10^3$

2.7. Identify the value of a heat input:

$$q_p = \frac{36 \cdot I_w \cdot U_a \cdot \eta_e}{V_w}, \tag{6}$$

where η_e - effective efficiency of a heat treatment of a surface by arc ($\eta_e = 0.8...0.85$);

2.8. Calculate a coefficient of a penetration form:

$$\varphi_p = k' \cdot (19 - 0.01 \cdot I_w) \cdot \frac{d_e \cdot U_a}{I_w}, \tag{7}$$

where k' - a coefficient that depend on the type of a current, its polarity and its density in electrode;

φ_p - a coefficient of a penetration form.

Under current density $j < 120 A/mm^2$ and DC welding on reverse polarity a coefficient k' is defined according to the equation:

$$k' = 0.367 \cdot j^{0.1925} \tag{8}$$

Under current density $j < 120 A/mm^2$ and DC welding on direct polarity a coefficient k' is defined according to the equation:

$$k' = 2.82 / j^{0.1925} \tag{9}$$

Under AC welding within all range of current density a coefficient k' equals 1.

Under current density $j > 120 A/mm^2$ and DC welding on reverse polarity a coefficient $k' = 0.92$, on direct polarity - $k' = 1.12$.

Current density can be identified according to the equation:

$$j = \frac{4 \cdot I_w}{\pi \cdot d_e^2} \quad (10)$$

2.9. The actual depth of penetration under welding on chosen mode parameters is defined according to the equation:

$$h_p' = 0.076 \cdot \sqrt{\frac{q_p}{\varphi_p}} \quad (11)$$

2.10. Define the width of a seam according to the equation:

$$e_p' = \varphi_p \cdot h_p' \quad (12)$$

The value of an actual depth of penetration and an actual width of a seam should not differ from set values more than 5%. When the difference exceeds this 5% the mode parameters shall be adjusted and the calculation shall be carried out once more.

2.11. We define the wire feed speed based on the condition of an arc temperature balance according to the following dependency:

$$V_f = \frac{(I_w \cdot (U_{an} + \varphi) + I_w^2 \cdot \rho \cdot \frac{l_e}{F_e}) \cdot 36}{F_e \cdot \gamma_e \cdot \Delta h_e}, \quad (13)$$

where U_{an} –anode drop, B;

φ - electronic work function, B;

ρ - electrical resistivity of a metal at 300 °C, $microohm \cdot cm^2 / mm$;

F_e – cross-section area of an electrode, cm^2;

γ_e - density of an electrode metal, g/cm^3;

Δh_e - enthalpy change of electrode metal at 300 °C, J/g.

2.12. Calculate an area of a deposited metal

$$F_d' = \frac{F_e \cdot V_f}{V_w}, \quad (14)$$

Obtained actual value of the deposited metal area shall not differ from set values more than 10%. When the difference exceeds this 10% the mode parameters shall be adjusted and the calculation shall be carried out once more.

2.13. Define actual height of reinforcement

$$g' = \frac{1.375 \cdot (F_d' - F_\Delta)}{e_p'}, \quad (15)$$

where F_Δ - area of penetration, mm^2

2.14. Define given depth of penetration according to the following equation:

$$h_p'' = h_p' + \frac{F_\Delta}{0.735 \cdot e_p'} \quad (16)$$

2.15. Calculate the height of a seam

$$H = h_p'' + g' \quad (17)$$

2.16. Draw a melting line. Form of penetration will be as semi-ellipse. Major semiaxis equals the height of a seam, and minor semiaxis equals half of seam width.

3. Specifics of mode parameters choice for second arc

To go to the next step we have to determine heating tepmereture of a metal under second arc. So that the distance between arcs will be in a range from 100 to 120 mm.

The initial temperature of a metal under the second arc we define according to the equations proposed by определяем по уравнениям Rykalin NN.

To calculate the temperature after the first arc passing we accept the following model of a heated field (Figure 3):

a) we consider a heat input as a combination of two processes:

1) heating of an edge of web with focused point source with wattage qст;

2) heating a a flange with with focused point source with wattage qп.

Values qw and qweb depend on the angle of inclination of an electrode and ration of thickness between web and flange. The fraction of a heat input for the flange according to the recommendations presented in work [8] shall be calculated based on the following equation:

$$k_{web} = 0.374 \cdot \sqrt{\frac{\delta_w}{\delta_{web}}}, \tag{18}$$

Then the fraction of a heat input for the web might be calculated according to the following equation:

$$k_w = 1 - k_{web}, \tag{19}$$

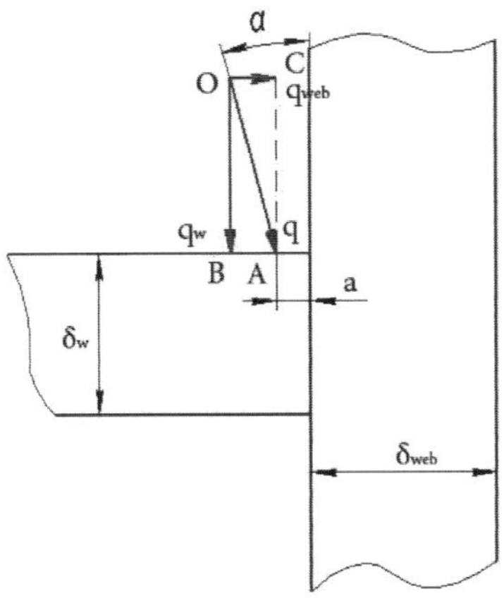

Figure 3 - Scheme of a heat input from the first arc

After calculation of a temperature of a metal under the second arc we define mode parameters for the second arc.

The second arc shall provide needed legs of a seam and its formation with graded junction to the main metal.

To calculate mode parameters for the second arc we need to make a draught of a weld joint with indication of a seam from the first arc so that we can draw a leg to a flange and following the scheme, given above, define the offset distance of an electrode to the web for the second arc.

The calculation of weld modes and parameters of a seam shall be carried out according to the same scheme as we used for the first arc with allowances made for heating of a metal by the first arc. For doing so we shall find current load according to the following equation:

$$I_w = B \cdot (80...100) h_p \tag{20}$$

where B – a coefficient that represents the influence of a heating by the first arc - $B = \sqrt{\frac{\Delta T_2}{\Delta T_1}}$

To identify the depth of penetration we need to use the following equation:

$$h = \sqrt{\frac{\Delta T_1}{\Delta T_2}} \cdot 0.076 \cdot \sqrt{\frac{q_p}{\varphi_p}} \quad (21)$$

4. Conclusion

Carried experimental accuracy check of mode parameters with the given recommendations showed acceptable repeatability (within 10%) of calculated results and parameters of actual weld seams. As can be seen from the above the recommendations given in the present article might be used for calculation of mode parameters under twin welding of fillet welds in separate melting pools with inclined electrode.

References

[1] Kuanfang He: Ethernet solutions for communication of twin-arc high speed submerged arc welding equipments / Kuanfang He, ZhuoJie Zhang, Jun Chen, Qi Li // Journal of computers 2012 (12) 3052-3059

[2] Cho D.-W. Analysis of the flux consumption and metal transfer for tandem submerged arc welding process under iso-heat input condition / Cho D.-W., D.V. Kiran, S.-J. Na // Welding Journal 2015 (94) 396-401

[3] Chechulin U.B.: Совершенствование технологии сварки поясных швов двутавровых балок [*Improvement of welding technologies under welding of circumferential seams of H-beams*] / Chechulin U.B, Melnokov A.U., Sjukaev G.M., Kozhuhkar A.F., Shumyakov V.I. // Welding and Diagnostics: Abstracts from reports presented at scientific and technical conference. – Ekaterinburg, 2011 145-150.

[4] Buzorina D.S.: Расчет параметров режима многопроходной механизированной сварки в защитных газах для получения качественных сварных соединений [Calculation of mode parameters of mechanized multi-pass welding in protective gases for quality of welded joints]. Welding and Diagnostics, 2014 (3) 13-17.

[5] Popkov A.M.: Расчет параметров режима двухдуговой сварки угловых и стыковых швов [*Mode parameters calculation under twin arc welding of fillet and butt-joint seams*]. Welding International, 1998 (7) 3-5.

[6] Melnik S.S.: Многодуговая автоматическая сварка стальных корпусных конструкций под флюсом АН-66 [*Multiarc automatic submerged arc welding of steel housings of AN-66*] / Melnik S.S., Ksenz N.D., Kukharenko V.V.: Welding International, 1984 (8) 37-38.

[7] Popkov A.M. Методика расчета параметров режима сварки тавровых соединений с неравнокатетными угловыми швами [*Calculation method of mode parameters for T-joints with not identical legs for the fillet joints*]. Welding International, 2002 (12) 3-4.

[8] Belchuck G.A. Распределение тепла между элементами таврового соединения при дуговой сварке [*Distribution of heat within elements of T-joint under twin arc welding*]. Welding International, 1959 (5) 13-15.

IOP Conf. Series: Materials Science and Engineering **125** (2016) 012024 doi:10.1088/1757-899X/125/1/012024

Pulsed Long Arc Welding

N Yu Krampit, PhD in Technical Sciences

Yurga Institute of Technology, TPU Affiliate, Russia, Kemerovo Region, Yurga

e-mail: nkrampit@mail.ru

Abstract. The paper presents a method and an appliance for pulsed arc welding. The method supports dosage of energy required for melting each bead of electrode metal starting from the detachment of a bead. The appliance including a sensor to register bead detachment shows this moment due to the voltage burst in the arc space. Transferred beads of electrode metal are of similar size because of the dosage of energy used for melting each bead, as the consequence, the process is more stable and starting conditions to transfer electrode metal are similar, as the result, a produced weld is improved.

1. Introduction

Various models of welding equipment are currently available on the market. It should be noted that over the last 5 years nearly all producers have expanded the range of manufactured equipment by appliances for impulse welding. Further development of digital technology and use of invertors support production of new appliances and discovery of new welding techniques: FastMigPulse, DeepARC, ForceArc, SpeedArcTandemControlSystem-KempArcPulse TCS, CMT; Interpulse, SpeedUp, SpeedPulse, SpeedRoot, TwinPulsing, Speed-TwinPulseThermalPulse, SynergicPulseWelding, PulseMigTwinPulse, SynergicPulse MIG welding. However, a great number of processes, their various designations and general insufficient information on their practical application, hardly help to realize the nature of these processes, in particular, modes of welding developed by different manufacturers of welding equipment.

Up to date modes of impulse control over gas-shielded welding can be divided in the following way:

- pulsed shielding gas feed welding [1, 2];
- pulsed electrode wire feed welding [3, 4];
- pulsed current and voltage welding [5-7];
- welding, which involves above mentioned modes (for instance, combining impulse and pulsed electrode wire feed welding [8]).

The author supposes pulsed current and voltage welding can be further classified according to the type of electrode metal transfer:

- with short circuits ("short arc", "forced short circuits", "stimulated short circuits");
- without short circuits ("long arc");
- alternate mode involving periods with and without short circuits.

2. Methods of Research

At present, mechanisms of melting and electrode metal transfer are in focus of researchers [9-11]. It is a vital task to develop efficient welding modes and appliances for their implementation. Impulse active gas welding is distinguished by combination of melting process and electrode metal transfer. The paper deals with the mode of pulsed arc consumable electrode welding [12], an appliance for its implementation is considered [13], and practically-oriented research into this welding mode is outlined.

The point of this mode is as follows: when pulse pilling-up the dosage of energy for melting a bead takes place starting from the instant of bead detachment. Bead detachment is registered due to voltage burst in the arc space. Current and voltage diagrams of pulsed arc welding are shown in Figure 1 [12].

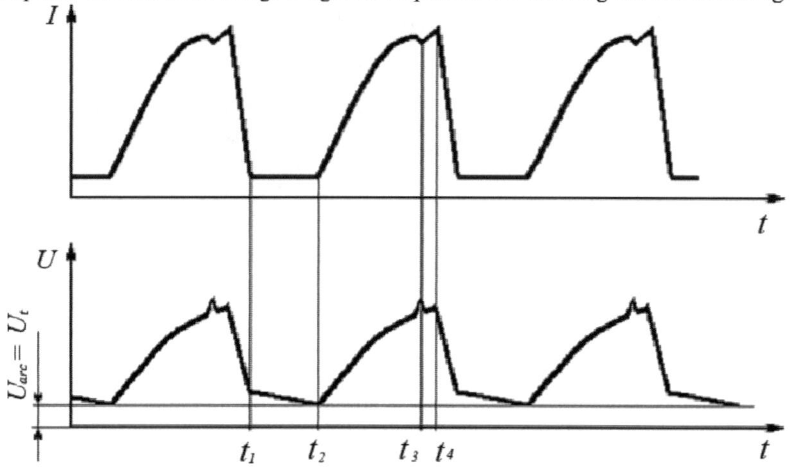

Figure 1. Current and voltage oscillograms of pulsed arc welding:
I – current strength; U – voltage in the arc space;
U_{arc} – arc voltage in the pause; U_t – specified arc voltage in the pause;
t – time; t_1 –pause start time; t_2 – impulse start time;
t_3 – instant of bead detachment from electrode tip; t_3 -t_4 – time for energy dosage

Pilot arc burns in the time interval t_1-t_2 at the minimal current (15-40) A. At the same time a bead tends to take a coaxial position relative to the electrode. The length of the arc space gets shortened because of sharp current decrease. At the time t_2 current impulse is fed. A bead is taken in the arc column by electromagnetic forces and moves towards the weld pool. At the time t_3 a bead is detached and it moves faster towards the pool. The rest of liquid metal is displaced on the side surface. In the time interval t_3-t_4 energy dosage is carried out for melting the forthcoming bead. The dosage start time is determined according to voltage burst of arc space at the time t_3. The impulse energy carried by the detached bead is not taken into account. Energy for melting the last bead alone is considered. That is why, conditions are provided for exact dosage of energy and dimensional stability of beads.

An appliance to stabilize the length of the arc space is designed to implement the suggested mode of pulsed arc welding [13]. Functional diagram of the appliance is shown in Figure 2.

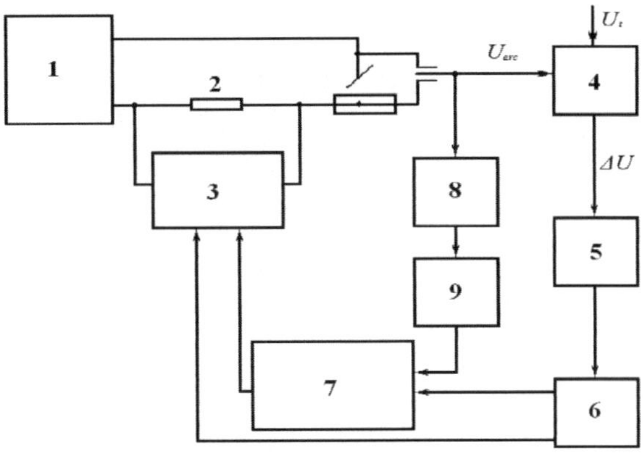

Figure 2. Functional diagram of the appliance for energy dosage:
1 – power source; 2 – ballast rheostat; 3 – switching device; 4 – unit comparing arc voltage with the setting voltage; 5 – unit forming pauses; 6 – relay element; 7 – unit forming pulse times; 8 – sensor registering bead detachment; 9 – amplifying device; U_{arc} – arc space voltage; U_t – specified voltage; ΔU –disparity of arc voltages

A sensor in this appliance registers bead detachment according to voltage burst in the arc space. The sensor is connected in series with the delay circuit directly linked to the switching device. The delay circuit functions as a dosage device as soon as detached bead of electrode metal is registered.

This appliance provides energy dosage necessary for melting each bead. Under bead detachment sensor malfunctioning the system switches to the mode of forced switching device shutdown provided that pulse time exceeds the pre-set value.

A chart outlining pulsed arc welding with energy dosage is shown in Figure 3. If we identify the instant of bead detachment and accept it to be the reference point of energy dosage we can support equal energy for melting a bead (highlighted in grey in the Figure). Pulse time is pre-set before power thyristors are switched off. Then current drops, so voltage does during current falling (t_{cn}). Welding current and voltage during the pause are little, so they do not change bead sizes of electrode metal till the next pulse pilling-up.

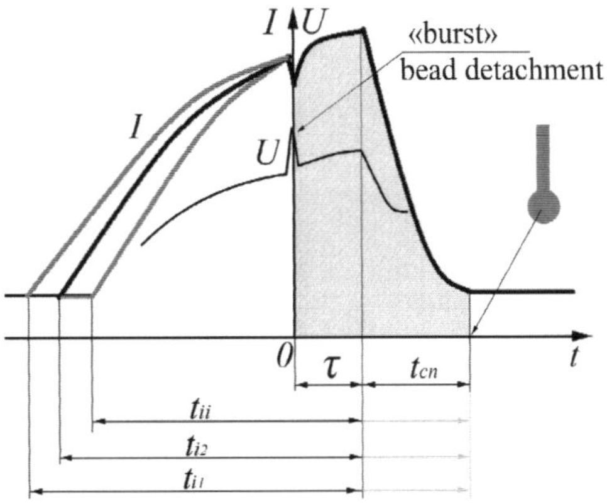

Figure 3. Chart of pulsed arc welding with energy dosage:
U – voltage; I – welding current; t – current pulse duration;
t_{cn} – pulse droop duration;
τ – time for energy dosage in each impulse (τ = const);
t_{i1}, t_{i2}, t_{ii} – duration of the first, second, i-impulse, respectively

Energy dosage supports equality of bead dimensions and provides similar initial conditions for transfer of electrode metal beads.

Investigations into influence of pulse parameters on weld formation. Methods of carrying out experiments. Beading was carried out on 300×150×10 mm plates made of steel St 3 with wire 1,2 mm in diameter. Carbon dioxide consumption, electrode extension and welding velocity were constant: 15 l/min, 16 mm and 15 m/h. To assess each parameter influence welding was carried out at its different values while other parameters were not changed. Melting intensity of the base metal was assessed in view of the following parameters: depth of penetration Hв, weld breadth B, weld reinforcement E, penetration zone F_p.

3. Results and Discussion

The role of welding current pulse amplitude. Welding current pulse amplitude is a key parameter relevant for the geometry of the weld when pulsed arc welding. Figure 4 outlines the dependence of penetration depth, weld breadth and convexity on welding current pulse amplitude.

The arc gets stabilized under action of welding current pulse, a powerful anode stream occurs and furthers the increasing force impact of the arc on the welding pool, as the result, metal is displaced from it. As the consequence, liquid layer gets thinner under the arc improving heat transfer to the base metal. Therefore, there is increase in the penetration depth.

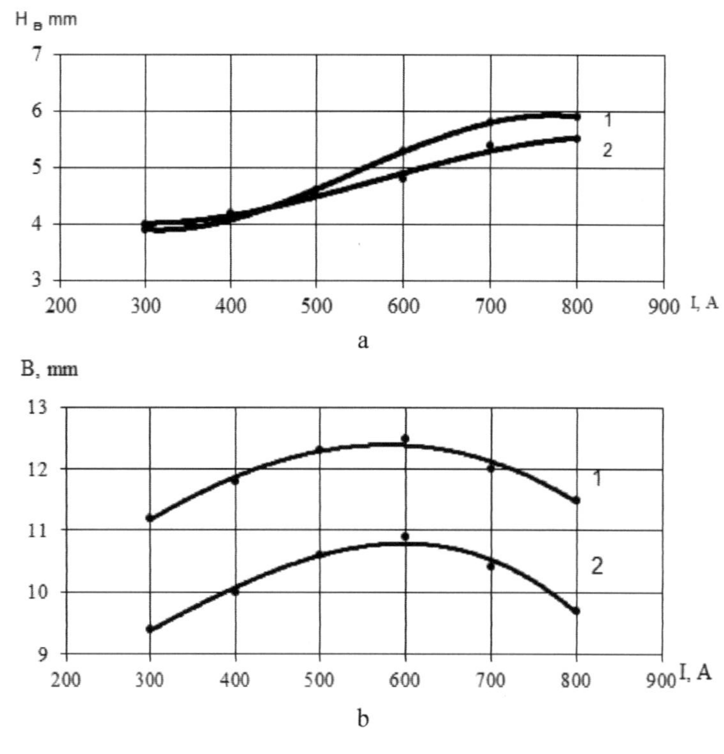

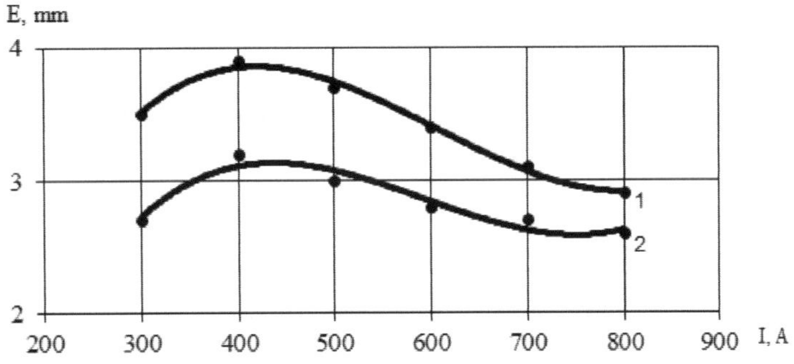

c

Figure 4. The change in penetration depth (a), breadth (b) and weld reinforcement (c) according to current amplitude (t_i = 4.5 mc; f = 90 p/s; 1 – V = 650 m/h; 2 – V = 450 m/h)

The role of welding current pulse. It is pulse time that determines the amount of energy applied to the arc during the pulse under other equal conditions. Growth of the pulse time is to be simultaneous with the decrease in pulse frequency in order to keep the welding current amplitude and arc space length constant. When pulsed arc welding a bead is melted during a pulse mainly, and pulse time is calculated as the sum of bead melting time and time necessary for its detachment. The dependence of weld geometry on pulse time is outlined in Figure 5.

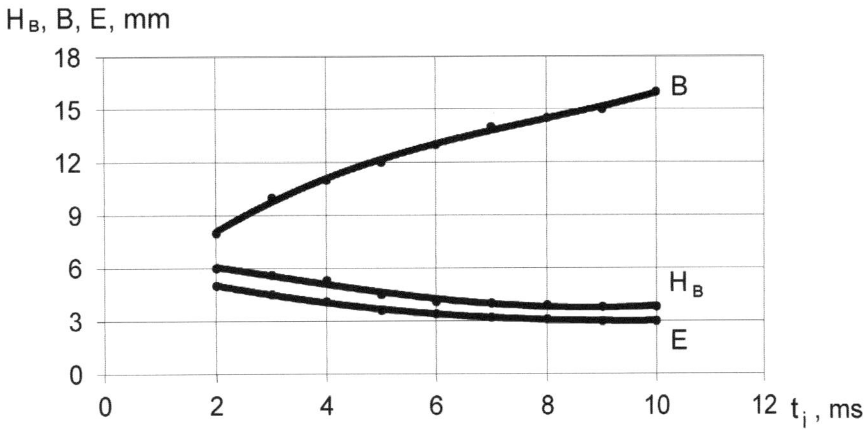

Figure 5. The dependence of penetration depth (Hв), weld breadth (B) and weld reinforcement (E) on pulse time: I_{av}=200A; V=450 m/h; f=60 Hz

Prolonged pulses in conditions of their constant frequency further increase in the weld breadth, whereas penetration depth and weld reinforcement are reduced.

Therefore, pulsed arc welding parameters have a positive effect on the geometry and mechanical properties of a weld joint [14].

4. Conclusions

1. Pulsed arc welding supports energy dosage used for melting each bead of electrode metal, that is, ensures more quasi-stability of welding due to similar dimensions of transferred electrode metal beads.

2. The appliance for implementing this type of welding makes possible the dosage of energy necessary for melting each bead of electrode metal just starting from the previous bead detachment, registered according to voltage burst in the arc space.

3. Arc space gets more stable, so bead transfer does provided that transferred beads are of the same size.

References

[1] Novikov O.M., Radko E.P., Kvon D.Yu. Appliance for pulsed shielding gas feed welding / Patent of invention №2337797, dated 27.07.2007.

[2] Tazetdinov R. G., Novikov O. M., Persidsky A. S. Shielded arc welding with alternate pulsed supply of various gases // Welding production. – 2012. – №1. – P. 38-42.

[3] Mozok V. M., Lebedev V. A. Further peculiarities of mechanized arc and automatic welding with pulsed electrode wire supply / Welding production. – 2010. – №2. – P. 34-38.

[4] Pavlov N.V., Kryukov A.V., Zernin E.A. Distribution of temperature fields in welding in a gas mixture with pulsed electrode wire feed // Welding International. – 2012. – Vol. 26, Issue 6. – P. 483-484.

[5] Dyurgerov N G, Sagirov D H Determination of arc peculiarities under pulsed welding processes // Welding production. – 2004. – №4. – P. 14-18.

[6] Knyazkov A F, Krampit A G et al. Welding mode. / Patent of invention №2191665, dated 27.10.02.

[7] Krampit A Welding with double modulation of the main welding parameters // Welding International. – 2012. – Vol. 26, Issue 11. – P. 867-869.

[8] Krampit A G, Knyazkov A F et al. Integrated approach to pulsed control over consumable electrode welding. / Patent of invention №2429111, dated 20.09.2011.

[9] Kuznetsov M A, Zernin E A, Danilov V I, Karzev D S Application of nanostructured powders to control characteristics of electrode metal transfer and the process of weld structurization // Applied Mechanics and Materials. – 2013. – Vol. 379. – P. 199-203.

[10] Krampit N Yu, Krampit M A, Zubenko L N Definition of peak value of welding current at pulsed arc welding // Applied Mechanics and Materials. 2014. 682 P. 397 – 401.

[11] Chinakhov D A Calculation of Gas-dynamic Impact of the Active Shielding Gas on the Electrode Metal Drop in Gas Jet Shielded Welding. Applied Mechanics and Materials. Vol. 379 (2013). P. 188-194.

[12] Knyazkov A F, Krampit N Yu, Petrikov A V Pulsed arc welding mode. / Patent of invention № 2133660 dated 27.07.99.

[13] Knyazkov A F, Knyazkov S A, Krampit A G, Verevkin A V Welding appliance. / Patent of invention №2185941, dated 27.07.02.

[14] Krampit A G, Krampit N Yu, Krampit M A Mechanical properties of welded joints in welding with pulsed arcs // Applied Mechanics and Materials. 2013. 379. P. 195 – 198.

Influence of Specific Features of Twin Arc Welding on Properties of Weld Joints

M A Sholokhov[1], A U Melnikov[2] and A M Fiveyskiy[2]

[1] Ltd SHTORM, Yekaterinburg, Russia
[2] Ural Federal University named after the first President of Russia B N Yeltsin, Yekaterinburg, Russia

e-mail: au_melnikov@mail.ru

Abstract. The present article covers the influence of standard and narrow gap twin arc welding on properties of weld joints from high-strength steels. While analyzing microsections we established that distribution of microstructure and phase terms, as well the distribution of micro-hardness, were more homogeneous under narrow gap twin arc welding.

1. Introduction

Regarding the manufacturing process of special-purpose machinery the most expensive and time-consuming things to produce are thick-walled housings where welding takes over 50% of the total amount of work [1-2]. Why this process is so time-consuming might be explained by thickness of weld materials and the usage of joints with bevel preparation. Use of high tensile medium alloy steels with derated weldability for manufacturing of housings of special-purpose machinery requires the development of special welding technique that will allow us to get solid weld seams. It is important to note, that the straight welds and fillet or I-shaped welds are pretty extended [3]. All that predetermines worthwhileness of usage of automatic welding. Increase of productivity and decrease of defect formation, improvement of welding process might be achieved by gas-shielded multi arc welding [4-8].

Mechanical, physical-chemical and performance characteristics of weld seam metal and heat-affected zone (HAZ) are determined by whole range of factors, such as structural phase composition of all layers of weld seam because of that a comprehensive research of their formation under different welding techniques appears to be a relevant applied research task.

2. Experimental method

To determine special aspects of the influence of multi arc welding on characteristics of weld joints we carried out a series of experiments. The welding was done as gas-shielded single / multi arc welding (Ar+CO2) with welding wire such as: OK 12.51 - for the first arc, ER307Ti for the second arc. An angle of V-shaped preparation (groove angle) of welded sheets with thickness 20 mm was 60° (area of preparation 231 mm^2) or 12° (area of preparation 122 mm^2). As the material for samples we used high-tensile-strength steel 30CrMnSiN, which is commonly used for manufacture of housings for special-purpose machinery.

After welding of samples, the templates were cut out for micro structural analysis (Figure 1). The analysis was carried out using raster electron microscopy of high resolution Quanta-200 (produced by

FEI Company, Holland, USA) and additionally equipped with Pegasus system, which included a tool for structural and grain flow analyses by Electron Backscatter Diffraction (EBSD) and Energy Dispersive Spectroscopy (EDS).

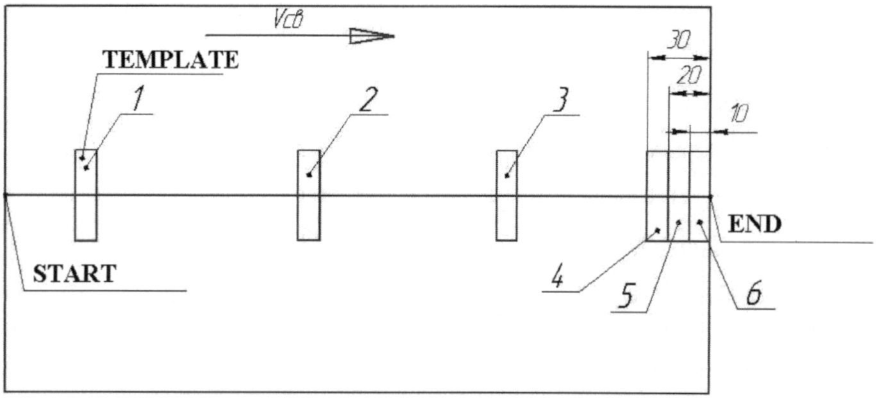

Figure 1. Scheme of templates cutting out from the sample weld seam

By spectrometer, produced by "EDAX", recorded characteristic X-ray excited chemical elements and measurement of their weight (and atomic) ratio in the test material, the elements are detected in a range from Borium to Uranium; energy resolution on line KαMn not less than 160 electrone-volts.

The welding mode is given in Table 1. Regulation of thermocycle under twin arc welding was carried out automatically due to variations of the distance between arcs from 50 to 200 mm. During welding the heating temperature of samples was recorded with 6 thermocouple-sensing elements, 3 of them were set in the middle of the sample in heat-affected zone (HAZ) within different distances from weld seam axis (4, 6, 8 mm) and at the end of the seam within equal distances from weld seam axis, the same as in point 1 (4 mm) in template positions 4, 5, 6. Example of thermal cycles of heat supply and cooling after the second pass in analyzing points, the allocation of thermocouple-sensing elements under twin welding of sample 07 (distance from 100 mm within arcs) are given on Figure 2. Indicative macrosections of weld seams are presented on Figure 3.

Table 1 – Welding mode parameters for samples

Parameter	Single arc welding	Twin arc welding
Current at the first/second arc, A	300	300/300
Voltage at the first/second arc, B	33	31/33
Welding speed, m/h	20	40
Wire feeding speed at the first/second arc, mm	11.0	10,4/11,0
Diameter of welding wire at the first/second arc, mm	1.2	1.2/1.2
Electrode stick-out at the first/second arc, mm	20	20/20

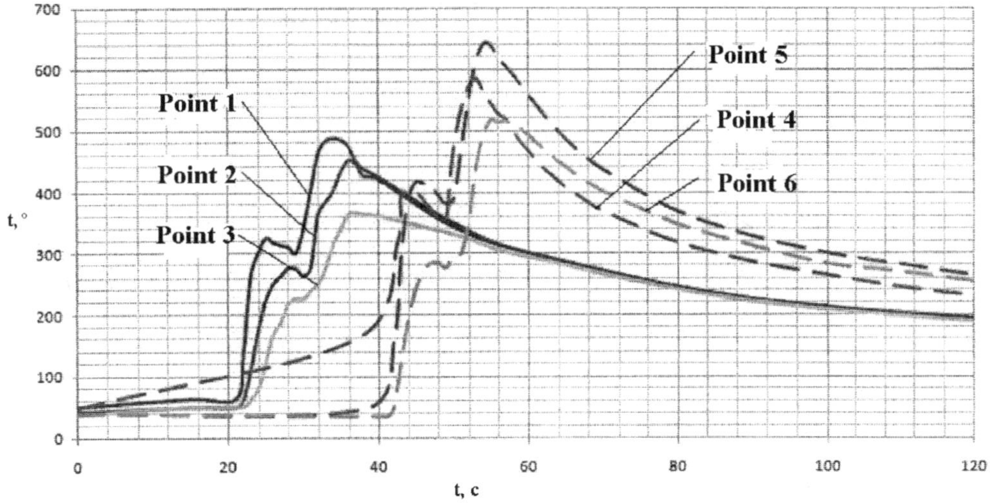

Figure 2. Indicative thermal cycles in analyzing points under twin arc welding (distance between arcs 100 mm), the second pass

a) b)

Figure 3. Macrosections of weld seams under twin arc welding with the distance between arcs 100 mm: a) bevel angle is 30°; b) bevel angle is 6°

3. Results and Discussion

To analyze the influence of twin arc welding on the properties of weld seams we measured micro-hardness at the cross-section. Indicative allocation of micro-hardness is presented on Figure 4.

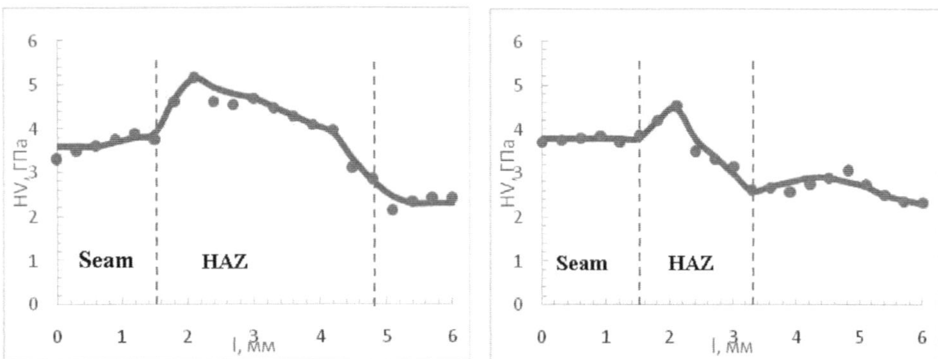

a) b)

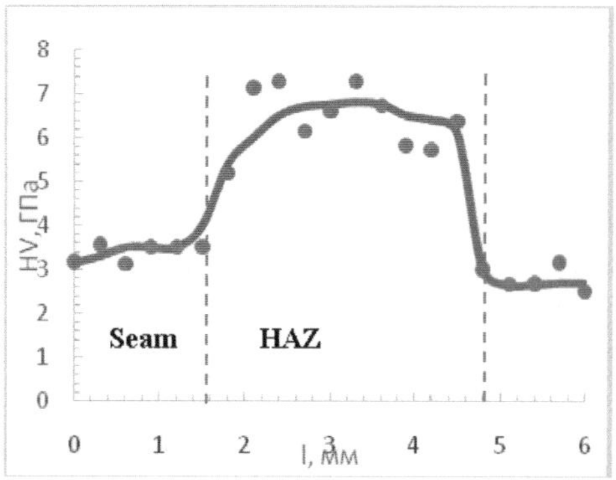

c)

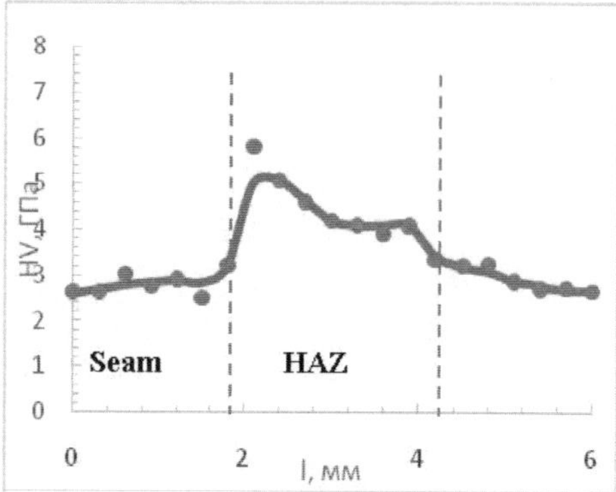

d)

Figure 4. Allocation of micro-hardness at the cross-section under twin arc welding: a) bevel angle 6°, to the surface of a seam; b) bevel angle 6°, to the bottom of a seam; c) bevel angle 30°, to the surface of a seam; d) bevel angle 30°, to the bottom of a seam.

To sum up the obtained results of micro-hardness measurement, we need to indicate the following:

HAZ has greater micro-hardness in comparison to the hardness of weld seam metal, as well as matrix steel plate out of HAZ.

HAZ in all examined cases is more distinctive within the width (reaching approx. 3 mm) to the surface of a seam in comparison with HAZ, that is spread at a seam room, that 1.2 – 1.5 times narrower.

A higher value has micro-hardness of the upper part of HAZ than of the lower part of HAZ that joins the root of a weld seam.

In a way micro-hardness scans are similar under single and twin arc welding.

Less shown as for micro-hardness values near-weld area HAZ of narrow grooves with bevel angle 12°.

Besides measurement of micro-hardness at the cross-sections we carried the analysis of microstructure of seams, heat affected zones and main metal. Indicative microstructure of different parts of cross sections is presented on Figure 5 and Figure 6.

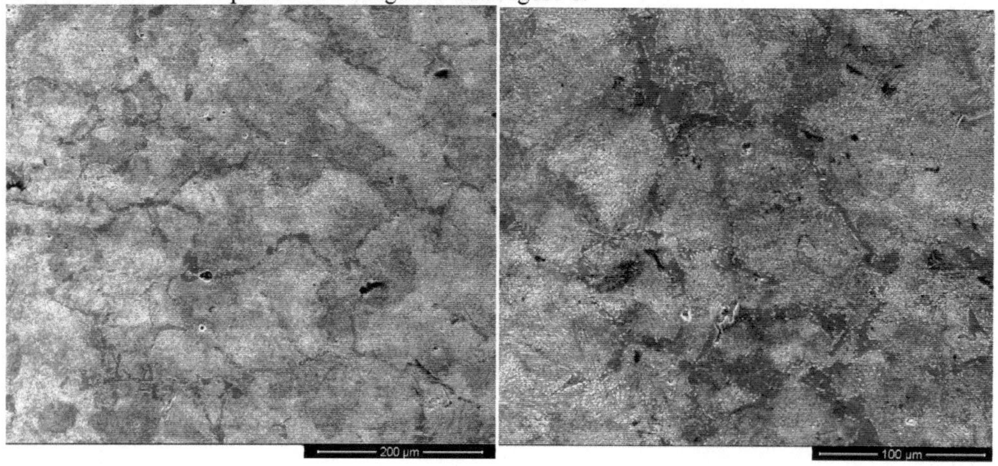

Figure 5. Typical microstructure of main metal of samples within 5 mm from a seam

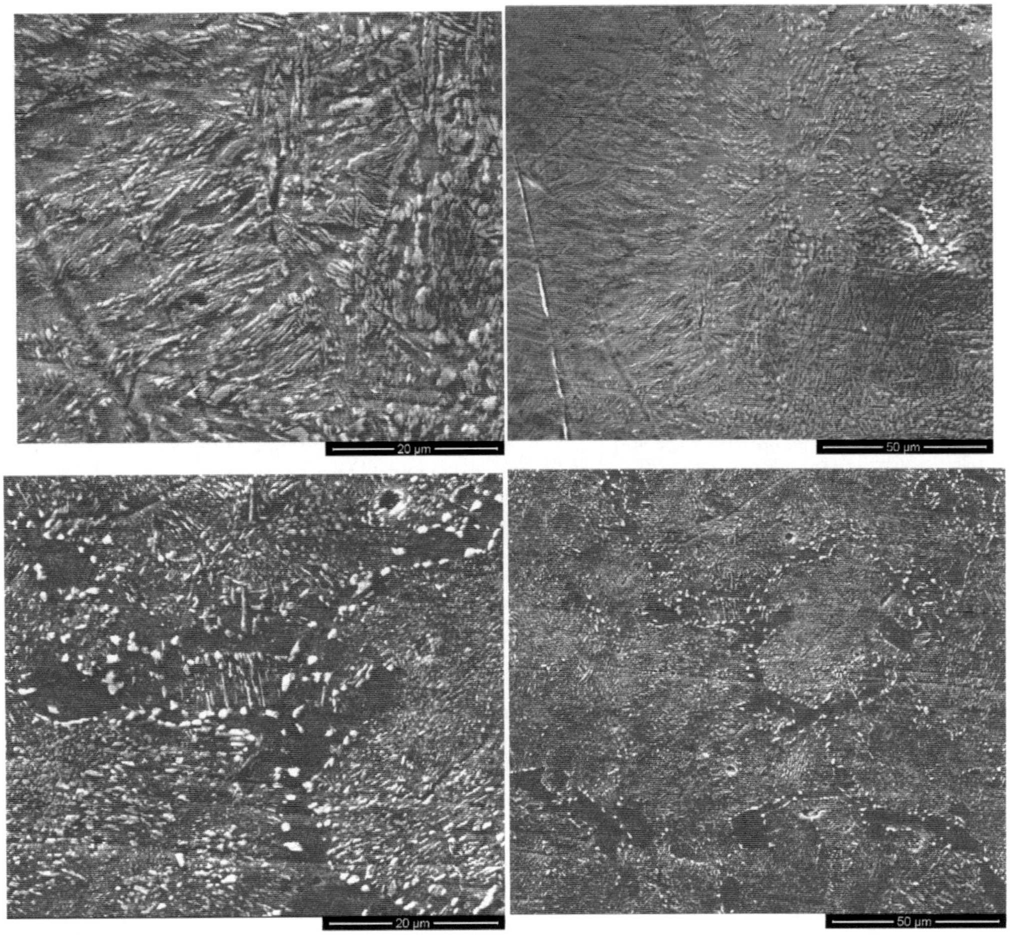

Figure 6. Typical microstructure of a sample in HAZ under twin arc welding (distance between arcs 100 mm)

Based on the analysis of microstructure we can come to the following conclusions:

1) In close to the weld seams areas, HAZ in all cases in prior austenite grain we observed colonies of laminar morphology, but within borders there were no globular crystalline α-phase particles, typical for initial steel structure of a plate. Such structure might be assigned to bainitic phase and determined by intense overheating of steel in a close distance to the weld seam, that assured a reverse change of perlite to austenite and its bainitic hardening.

2) In microstructure of HAZ located between bainitic and main globular perlitic (matrix) that is far from the weld seam we can see domination of thin plate perlite with a certain amount of perlitic areas with globalized finely dispersed cementite. This microstructure set alongside with bainite occurred as a result of dissolution of the initial under heating process and then under cooling in the decay to thin plate perlite.

3) As for the depth of HAZ (from the surface of a cross-section – the first line to the bottom where the root weld is located– the third line) thin structure of bainite and neighboring intermediate perlite changes: the thickness of bainetic paltes increases as well as the thickness of plates and equiaxed particles of cementite in perlite colonies. On the borders of grains we can see areas of α-феррита (on photos wide black layers). In total such morphology is typical more for main fundamental state of steel perlite.

4) Under twin arc welding with bevel angle of 60°, microstructure of bainite and intermediate perlite in HAZ is more organized.
5) In case of single / twin arc welding with bevel angle of 12° in HAZ we can also see beinite layer and only then finely dispersed perlite. With that a perlite area hardly differs from neighboring matrix of steel out of HAZ, but bainite morphology in forms of packs was more structured and contained particles of finely dispersed cementite, in general it represents less distortion and lower initial stresses of microstructure. This correlates with data obtained from measurements of micro-hardness.
6) In general HAZ with narrow gap stood out by larger microstructure and phase homogeneity, that makes her closer to a main metal of steel plate. This results from the measurements of micro-hardness.

4. Conclusion

As can be seen from the above, according to the results of carried examinations we can come to a conclusion that usage of narrow gaps and twin arc welding influence in a positive way microstructure and characteristics of weld seams of high-tensile-strength steel.

References

[1] Rakhmatullin T A: Проблемы внедрения зауженных разделок при сварке корпусных конструкций специальной техники [Problems of implementation of narrow gaps under welding of housing for special-purpose machinery]. Rakhmatullin T.A., Sholokhov M.A., Buzorina D.S. News of Higher Educational Institutions. Mashinostroyeniye, 2012 (4) 64-66.
[2] Sholohov M A Перспективные технологии и оборудование для сварки по узкому зазору корпусных конструкций специальной техники [Advanced technologies and equipment for the narrow gap welding of hull structures of special equipment] Weldihg and Diagnostics 2015 (5) 46-51.
[3] Sterjovski Z Weld-end solidification cracking in pulsed-tandem gas metal arc welding of naval steels / Sterjovski Z., Bayley C., Donato J., Lane N., Lang D. // Welding Journal 2014 (93) 145-152
[4] UEYAMA T Higi-speed welding of steel sheets by the tandem pulsed gas metal arc welding system / Ueyama T., Ohnawa T., Yamazaki K., Tanaka M., Ushio M., Nakata K. // Transactions of JWRI 2005 (34) 11-18
[5] ISO 13847,2000. Petroleum and natural gas industries -- Pipeline transportation systems -- Welding of pipelines.
[6] Industry standard of Gazprom 2.2.3-137-2007. Инструкция по технологиям сварки при строительстве и ремонте промысловых и магистральных газопроводов. Часть II. [Instruction on welding technologies under building and repair of trade and main pipelines]
[7] DDD 153-006-02. Инструкция по технологии сварки при строительстве и капитальном ремонте магистральных нефтепроводов [Instruction on welding technologies under building and repair of main pipelines].
[8] Efimenko L A: Формирование структуры сварных соединений высокопрочных сталей при автоматической сварке в защитных газах [Structure formation of weld joints of high-tensile-strength steel under automatic shielding gas welding]. Efimenko L.A., Kapustin O.E., Kapustin A.A., Shkapenko M.V., Karasev D.N. Rabotinskii D.N. Pipeline transport. Theory and Practice, 2012 (4) 34-42.

IOP Conf. Series: Materials Science and Engineering **125** (2016) 012026 doi:10.1088/1757-899X/125/1/012026

Methods to Predict Stresses in Cutting Inserts Brazed Using Iron-Carbon Brazing Alloy

V V Konovodov[1], A V Valentov[2], O Yu Retuynskiy[2], Sh B Esekuev [2]

[1] Novosibirsk State Agrarian University
147 Nikitin St, 630039 Novosibirsk, Russia; phone: (383) 267-35-07
[2] Yurga Institute of Technology, affiliate of Tomsk Polytechnic University
26 Leningradskaya St, 652050 Yurga, Russia; phone: (38451)-6-05-37

e-mail: retol@list.ru

Abstract. This work describes a method for predicting residual and operating stresses in a flat-form tool insert made of tungsten free carbides brazed using iron-carbon alloy. According to the studies' results it is concluded that the recommendations relating to the limitation of a melting point of tool brazing alloys (950-1100°C according to different data) are connected with a negative impact on tools as a composite made of dissimilar materials rather than on hard alloys as a tool material. Due to the cooling process stresses inevitably occur in the brazed joint of dissimilar materials, and these stresses increase with the higher solidification temperature of the brazing alloy.

1. Introduction

Analytical studies show that about a half of the agricultural machinery failures occur for production reasons. This rate has been maintaining at the same level for the past 15 years. Operation life of complex mechanisms is usually determined by durability of the weakest components and parts.

For example, life of high pressure hoses, composed of high-pressure sleeves and nozzles is determined by durability of sleeve screening braid if careful embedding of nozzles. The service life of internal combustion engines depends on durability of valve springs, specifying the output parameters of engines, namely fuel consumption and capacity.

2. Timeliness of the research

Capabilities of a tool to do its jobs are largely determined by actual defects resulting from deformations caused by manufacturing processes [1]. Pores, any kind of inclusions, structural heterogeneity, grains in different sizes, micro- and macro cracks are the most dangerous defects.

The stresses occurring during the tool making processes and the usage can lead to the following consequences:

reducing in adhesion of brazing alloys to tool materials;

deteriorating physical and mechanical properties of brazing alloys and compositions consisting of tool material – braze – tool body;

forming dangerous internal micro cracks that can cause the destruction of cutting tools while operations.

The known tools, including custom made tools do not meet the requirements in terms of durability, strength and wear resistance. The use of tungsten free hard alloys helps to solve this problem, but the mechanical jointing of such tool inserts due to their low rigidity results in malfunctioning in response to impact and load actions. Brazed cutting tools with tungsten-free carbide inserts are not possible to make using conventional brazing techniques because the contact angle of wetting a brazing alloy on the surface of tungsten-free carbides is low (<300). This leads to the increased tool expenditure, decreased rate of equipment use and higher costs of cutting processes and tool recovery procedures. The demand for quality cutting tools intended for use in repair and reconditioning service for agricultural machinery only in Kemerovo oblast makes 100 thousand units per year, with the tool resistance specified as 45 minutes.

3. Methods of research, authenticity and validity of the results

The simplified and trial techniques and methods as well as the finite element method are used to evaluate residual and operating stresses.

To determine the stresses arising during the tool brazing process we make the following assumptions:

- a brazing alloy is ideally aligned with a tool material, there is no displacement between them;
- brazing alloys and tool materials are anisotropic and obey Hooke's law;
- stresses along the contact length are equal.

Then the total strain can be written as follows:

$$\varepsilon = (\varepsilon_{TC} + \varepsilon_{\Pi}) = \frac{h_{TC} E_{\Pi}}{h_{TC} E_{TC} + h_{\Pi} E_{\Pi}} \tag{1}$$

Where $\varepsilon_{TC} + \varepsilon_{\Pi}$ is the total strain after the brazing process;

$h_{TC,}$ h_{Π} are the thicknesses of a carbide insert and brazing alloy respectively;

E_{TC} и E_{Π} are the values of the Young's modulus relating to the carbide insert and the brazing alloy respectively.

The mean value of residual stress in the carbide insert can be determined by the following expression

$$\sigma_{O_{TC}} = (\varepsilon_{TC} + \varepsilon_{\Pi}) \frac{h_{TC} E_{\Pi}}{h_{TC} E_{TC} + h_{\Pi} E_{\Pi}} E_{TC} \tag{2}$$

Then

$$\varepsilon_{TC} + \varepsilon_{\Pi} = \frac{P_{TC}}{b h_{TC} E_{TC}} + \frac{P_{\Pi}}{b h_{\Pi} E_{\Pi}} \left(1 + \frac{h_{TC} + h_{\Pi}}{2\rho_{TC}} \right) + \frac{h_{TC}}{2\rho_{TC}} + \frac{h_{\Pi}}{2\rho_{TC}} \left(1 + \frac{h_{TC} + h_{\Pi}}{2\rho_{TC}} \right) \tag{3}$$

Where P_{TC}, P_{Π} are the tensile forces in the x-axis direction;

b is the width of supporting area under the carbide insert;

ρ_{TC}, ρ_{Π} are radii of curvature of the carbide insert and the brazing alloy respectively.

And since $P_{TC} = P_{\Pi}$, then we can write the following

$$M_{TC} + M_{\Pi} = \frac{P_{TC}}{2} (h_{TC} + h_{\Pi}) \tag{4}$$

Where M_{TC}, M_{Π} are the bending moments of the carbide insert and the brazing alloy respectively.

$$M_{TC} = \frac{E_{TC} J_{TC}}{\rho_{TC}}; \quad M_{\Pi} = \frac{E_{\Pi} J_{\Pi}}{\rho_{\Pi}}, \tag{5}$$

Where J_{TC}, J_{Π} are the axial moments of inertia of the carbide insert and the brazing alloy respectively ;

$E_{TC}J_{TC}$, $E_{\Pi}J_{\Pi}$ are stiffness when bending the carbide insert and the brazing alloy respectively.

If $\rho_{TC} = \rho_{\Pi}$, formula (3) can be put in the following form

$$\varepsilon_{TC} + \varepsilon_{\Pi} = \frac{h_{TC}^3 E_{TC} + h_{\Pi}^3 E_{\Pi}}{6\rho h_{TC}(h_{TC} + h_{\Pi})} + \frac{h_{TC}^3 E_{TC} + h_{\Pi}^3 E_{\Pi}}{6\rho h_{\Pi} E_{\Pi}(h_{TC} + h_{\Pi})} + \frac{h_{TC}^3 E_{TC} + h_{\Pi}^3 E_{\Pi}}{12\rho^2 h_{\Pi} E_{\Pi}} + \frac{h_{\Pi} + h_{TC}}{2\rho} \quad (6)$$

By substituting expressions (4) and (5) into (6) we get

$$\sigma_{O_{TC}} = \left[\frac{h_{TC}^3 E_{TC} + h_{\Pi} E_{\Pi}}{6\rho h_{TC}(h_{TC} + h_{\Pi})} + \frac{E_{TC}(h_{TC}^3 E_{TC} + h_{\Pi}^3 E_{\Pi})}{12\rho^2 (h_{\Pi} E_{TC} + h_{\Pi} E_{\Pi})} + \frac{E_{TC} E_{\Pi} h_{\Pi}(h_{TC} + h_{\Pi})}{2\rho(h_{TC} E_{TC} + h_{\Pi} E_{\Pi})}\right] \frac{1}{1-\mu^2} \quad (7)$$

Where μ is the Poisson's ratio in carbide – brazing alloy composite.

Therefore, the stress in the carbide cutter insert having a flat form is expressed as follows

$$\sigma_{O_{TC}} = \frac{(\alpha_{TC} + \alpha_{\Pi})\Delta T}{\frac{1-\mu_{TC}}{E_{TC}} + \frac{1-\mu_{\Pi}}{E_{\Pi}} \cdot \frac{h_{TC}}{h_{\Pi}}} \quad (8)$$

Where α_{TC}, α_{Π} are the coefficients of linear expansion of the carbide and brazing materials respectively;

ΔT is the difference between the ambient temperature and the brazing temperature;

μ_{TC} и μ_{Π} are the Poisson's ratios for the carbide and the brazing alloy respectively .

Relative to a two-layer cutter insert a unified theory of stiffness is as follows

$$\begin{cases} \sigma_{o1} \le [\sigma]_p / K_3, & \sigma_{o1} > 0; \\ \sigma_{o2} \le [\sigma]_- / K_3, & \sigma_{o2} < 0, \end{cases} \quad (9)$$

Where $[\sigma]_p$ and $[\sigma]_-$ are the values of tensile strength of material under tension and compression respectively;

K_3 is the strength factor;

σ_{o1} are the residual stresses in the material having a lesser value α_T;

σ_{o2} - the same relating to a greater value α_T.

When the thickness of both layers is equal $\sigma_{o1} = -\sigma_{o2}$ and considering $|[\sigma]_p| < |[\sigma]_-|$, then the limiting condition is the first one in (9). In other words, an interlayer crack will be growing in the tool material with a lower coefficient of thermal linear expansion, which has stretching residual stresses. For different layer thicknesses

$$\sigma_{o1} = -\sigma_{o2} h_2 / h_1, \quad (10)$$

Where h_1 and h_2 are the thicknesses of the first and second layer respectively. To calculate the value of residual stresses at their interface we suggest the following formula

$$\sigma_{o1} = (\alpha_{T1} - \alpha_{T2})\Delta T \Bigg/ \left[\frac{1-\mu_1}{E_1} + \frac{1-\mu_2}{E_2} \cdot \frac{h_1}{h_2}\right], \quad (11)$$

Where α_{T1}, E_1, μ_1 are the coefficient of thermal expansion, the Young's modulus and the Poisson's ratio relating to the first material;

α_{T2}, E_2, μ_2 are the same relative to the second material;

ΔT is the heating temperature when manufacturing or using the two-layer composite.

With regard to a cutting wedge made of two different anisotropic wedges and having of a flat phase boundary, expression (11) can be written as

$$\sigma_{o1} = (\alpha_{T1} - \alpha_{T2})\Delta T \bigg/ \left[\frac{1-\mu_1}{E_1} + \frac{1-\mu_2}{E_2} \cdot \frac{\sin(\nu - \gamma)}{\cos(\alpha + \nu)} \right], \tag{12}$$

Where ν is the angle of wedge interface between phases.

Considering (3.4), conditions for crack resistance (10) are as follows

$$\begin{aligned} &(\alpha_{T1} - \alpha_{T2})\Delta T \bigg/ \left[\frac{1-\mu_1}{E_1} + \frac{1-\mu_2}{E_2} \cdot \frac{\sin(\nu - \gamma)}{\cos(\alpha + \nu)} \right] \le [\sigma]_p K_3; \\ &(\alpha_{T1} - \alpha_{T2})\Delta T \bigg/ \left[\frac{1-\mu_1}{E_1} \cdot \frac{\cos(\alpha + \nu)}{\sin(\nu - \gamma)} + \frac{1-\mu_2}{E_2} \right] \le [\sigma] K_3. \end{aligned} \tag{13}$$

In order to obtain between-group composites, two-component three-layer composites are possible, where the middle layer represents a mixture composed by materials of two other layers. The coefficient of thermal linear expansion α_{TC} and the tensile strength σ_{BC} of the mixture are determined by the rules of mixtures

$$\alpha_{T1} = 100\alpha_{T1}\alpha_{T2} / (P_1\alpha_{T2} + P_2\alpha_{T1}); \tag{14}$$

$$\sigma_{BC} = 100\sigma_{T1}\sigma_{T2} / (P_1\sigma_{T2} + P_2\sigma_{T1}), \tag{15}$$

Where P_1 and P_2 are the material contents in percentage in the mixture.

Based on the analysis of equations (7) and (8) we can make a conclusion that aligning properties of the brazing alloy and the carbide is mainly limited to the maximum reduction of difference in coefficients of linear expansion α_{TC}, α_{Π}. However, making these coefficients equal is impossible in practice, for this reason it is required to follow the condition $\alpha_{TC} < \alpha_{\Pi}$ at least. In this case, compressive stresses appear in the carbide insert, and these stresses are less dangerous in terms of brittle fracture occurrence throughout the carbide material and at the interface between the carbide and the braze.

Decreasing the values of E_{TC}. E_{Π} and μ_{TC}, μ_{Π} parameters results in the reduction of stresses occurring in the carbide and at the interface between the carbide and the brazing alloy, because stiffness of tools and their ability to resist elastic and plastic deformation increase with this.

Especially dangerous are the stresses occurring in carbides when temperatures on tool's contact surfaces sharply change, that is typical for discontinuous cutting processes and cutting where dimensional allowances are varying (in particular along metal deposits). In this case the stresses developing in carbides can be analyzed using the Biot number [6, 7, 8, 10].

$$Bi = \frac{h_{TC}\chi}{2\lambda_{TC}} \tag{16}$$

Where h_{TC} is the thickness of the carbide insert;

χ is the coefficient of heat transfer between the ambient temperature and the tool;

λ_{TC} is the heat conductivity of the carbide.

According to the Biot number the reduction of thermal stresses in a carbide can be achieved by either increasing thermal conductivity of the carbide λ_{TC} the tool body or reducing the carbide insert thickness.

After sintering two-layer composites, compressing stresses occur in materials having a lower coefficient of linear expansion and tension stresses in materials having a higher coefficient. Therefore, the stresses determined by formula (11) must not exceed the compressing stresses for materials with a lesser value α_{TC}. In this case, the tensile strength of materials has been determined by taking into account the safety factor K3=1.5. Putting different values of ratios h_1 and h_2 into the formula we can define the stresses in layers.

4. Results of the research

The test results confirmed the researches performed by authors in [2, 3, 7, 8, 11] indicating that the short-run (≈45 sec.) and high temperature (about 1180C) brazing process do not have a negative effect on cutting and strength properties relating to carbides.

A higher melting temperature results in improving the adhesion and diffusion of materials of brazed components exhibiting good wetting and spread properties that is considered a positive factor relative to the refractory quality of brazing alloys. This results from the energy sufficient to activate brazing materials, as it is believed that the thermodynamic activity of elements becomes substantially higher only when T=2/3Тпл relative to the melting points of main components [4,5,9]. At the same time a brazing alloy with a higher melting point largely maintains its strength qualities at increased operating temperatures.

Laboratory tests of carbide alloys for their wear resistance are reported in Figure 1, showing the high rate of running-in wear of the carbide T15K6 hз up to 0.4 mm in comparison with the carbide KHT16 hз having the running-in wear within 0.2 ... 0.22 mm, depending on a method used to mount carbide inserts. In the second stage the back surface wear of the KHT16 carbide brazed using a FeC alloy remained almost constant; the rest of test samples showed increased values of wear, especially severe wear was detected in the case of mechanically mounted inserts in the T15K6 alloy.

The highly intensive running-in wear is associated with a defective surface layer of up to 0.1 mm, the presence of which is caused by the interaction of hard alloy surface with air and flux during high temperature heating.

The most preferred variants relating to two-layer inserts are composites with the cutting tip and base materials having the same coefficient of linear expansion, because thermal stresses do not occur in such composites. For this reason the base material is required to be a mixture of materials, with one material having a coefficient of thermal expansion less than the cutting tip material coefficient, and the other – higher (a coefficient of thermal expansion of the mixture was calculated using formula (14)). In this connection there is a need for a low cost material having low α_T, whereas in practice the situation is reversed: the cost of tool materials rises with decreasing α_T and increasing the elastic modulus. Therefore, this methodology can be used when composites have a cutting tip material with minimum α_T being (4...5)x10-6K-1.

5 Conclusions

The investigated conditions suggest that the recommendations on limiting a melting temperature of brazing alloys used for tooling (950-1100 ° C according to various sources) are connected with a negative impact on the tool as a composite of dissimilar materials, rather than on the carbide as a tool material. Due to the cooling process the stresses inevitably occur in the brazing joint of dissimilar materials, and these stresses increase with the higher solidification temperature of the brazing alloy.

References

[1] Vereshchaka A S Rabotosposobnost' rezhushhego instrumenta s iznosostojkimi pokrytijami [The efficiency of cutting tools with wear-resistant coatings] Moscow: Mashinostroenie [Engineering], 1993. 336 p.

[2] Loshak M G Prochnost' i dolgovechnost' tverdyh splavov [The strength and durability of hard alloys]. Kiev: Naukova Dumka [Scientific Thought], 1984. 328 p.

[3] Klochko N A, Hersonskij A K Issledovanie vlijanija termicheskoj obrabotki na naprjazhennoe sostojanie i svojstva izdelij iz spechennyh tverdyh splavov, Tverdye splavy i instrument [Investigating the effect of heat treatment on stress condition and properties of sintered hard alloy products, hard alloys and tools] VNIITS Moscow: Metallurgija [Metallurgy]. 1986. pp. 56-60. (in Russian)

[4] Lasukov, A A, Mokhovikov, A A Influence of modified layer of tool on stress - Strain state of cutting wedge, Proceedings - 2012 7th International Forum on Strategic Technology, IFOST 2012
[5] Retyunskiy O Yu, Valentov A V, Solomatin P A Statisticheskaja obrabotka rezul'tatov ehksperimental'nykh issledovanijj tochenija poverkhnostejj detalejj gornykh mashin, vosstanovlennykh naplavkojj. [Statistical analysis of the research results relating to the turning processes on surfaces of mining equipment parts reconditioned using deposit welding]. Gornyjj informacionno-analiticheskijj bjulleten' [Mining informational and analytical bulletin (scientific and technical journal)]. 2012. No. S3. pp. 250-254.
[6] Valentov A V, Retyunskiy O Yu Ispol'zovanie rezcov iz bezvol'framovogo tverdogo splava pri obrabotke detalejj gornykh mashin, vosstanovlennykh naplavkojj [The use of tungsten–free carbide cutting tools in processing mining machinery parts reconditioned using deposit welding]. Gornyjj informacionno-analiticheskijj bjulleten' (nauchno-tekhnicheskijj zhurnal). [Mining informational and analytical bulletin (scientific and technical journal)]. 2012. No. S3. pp. 245-249.
[7] Valentov A V, Konovodov V V, Agafonova E V Prognozirovanie ostatochnykh i ekspluatatsionnykh napryazhenii pri paike reztsov dlya obrabotki naplavlennykh poverkhnostei [Prediction of residual and operational stresses in brazing cutting tools used to machine weld deposits]. Novosibirsk: *Vestnik Novosibirskogo gosudarstvennogo agrarnogo universiteta* [Novosibirsk State Agrarian University Bulletin]. 2013. No. 1 (26). pp. 107-110.
[8] Chinakhov D A, Grigorieva E G Osobennosti ispol'zovaniya poroshkovoi provoloki dlya naplavki v CO_2 iznoshennoi poverkhnosti detalei [Special usage of cored wire for welding worn surface parts in CO_2]. Tomsk: Aktual'nye problemy v mashinostroenii [Actual problems in mechanical engineering]. 2015. No.2. pp. 143-147.
[9] R V Chernukhin. Reliability of the steering gear of truck vehicles// Applied Mechanics and Materials. - 2013 - Vol. 379. - p. 36-42.
[10] U S Putilova, Yu I Nekrasov, A A Lasukov Loading of the Manufacturing Systems Elements in the Process of Unsteady Mode Cutting and the Models of their Arrangement Deviations, Applied Mechanics and Materials, Vol. 682 (2014), pp. 192-196

Influence of Filler Metals in Welding Wires on the Phase and Chemical Composition of Weld Metal

N A Kozyrev[1], I V Osetkovskiy[1], O A Kozyreva[1], E A Zernin[2], D S Kartsev[2]

[1]Federal State Budgetary Educational Institution of Higher Professional Education «Siberian State Industrial University», Russia, Novokuznetsk
[2]Yurga Institute of Technology
National Research Tomsk Polytechnic University Affiliate
ul. Leningradskaya 26, Yurga, Russian Federation, 652055

Abstract: The influence of filler metals used in welding wires on the phase and chemical composition of the metal, which is surfaced to mining equipment exposed to abrasive wear, has been investigated. Under a laboratory environment, samples of Mo-V-B and Cr-Mn-Mo-V wires were made. The performed experiments have revealed that fillers of the Cr-Mn-Mo-V system used in powder wire show better wear resistance of the weld metal than that of the Mn-Mo-V-B system; the absence of boron, which promotes grain refinement in the Mn- Mo-V-B system, significantly reduces wear resistance; the Mn- Mo-V- B weld metal has a finer structure than the Cr-Mn-Mo-V weld metal.

1. Introduction

Reconditioning of hoppers and chutes, used in coal transportation, can provide longer operational life and increase a service life of mining machinery. Most loaded chutes are hard surfaced using welding cored wires of 40H3G2MF and 40GMFR types with a higher content of expensive tungsten and the wire discussed in [1-6]. Operating the reconditioned hoppers and chutes shows that the wear of their surfaces is uneven. Chutes during their operation are often exposed to cyclic thermomechanical stresses, corrosion, and abrasive wear. Wear often occurs as stripes that are caused by the presence of weld metal portions having different structures and hardness. The causes of wear that occurs in gutters surfaced using powder wires 40H3G2MF and 40GMFR have been analyzed to indicate that potential of Mo-V-B and Cr-Mn-Mo-V alloys has not been fully realized and is currently under active study. Therefore, it is of interest to study the influence of changes in a chemical composition of expensive components in the system as well the addition of the various components with the purpose of obtaining new properties of weld metal and reducing the cost of cored wire.

2. Research

The work is aimed at studying the possible use of Mn- Mo-V-B and Cr-Mn-Mo-V alloy systems to hard surface hoppers and chutes designed for coal transportation. Under laboratory conditions, standard samples of flux-cored wire with a 3.6 mm wire in diameter were prepared. The wire was coated with a strip St3, and appropriate powdered materials were used as a filler, wherein amorphous carbon was replaced with carbon-fluorine containing dust, formed after purifying waste gases of aluminum production, with a chemical composition, by weight %: Al_2O_3 = 21-46.23; F = 18-27; Na_2O = 8-15; K_2O = 0.4 – 6; CaO = 0.7-2.3; SiO_2 = 0.5-2.48; Fe_2O_3 = 2.1-3.27; Ctotal = 12.5-30.2; MnO = 0.07-0.9; MgO = 0.06-0.9; S = 0.09-0.19; P = 0.1-0.18. The smallest amount of carbon was added to the wire. Previous studies have proved a positive effect of this material to improve the quality of welds [7-8].

Welds were deposited using the AN-26C flux on a plate made of steel St3. X-ray fluorescence spectrometry using a XRF-1800 spectrometer and atomic emission spectrometry using a DFS-71 spectrometer were conducted to determine a chemical composition of the welded layer. The chemical composition of wire samples and similar steels 40H3G2MF and 40GMFR are shown in Tables 1 and 2. An ASAW-1250 welding machine was used for welding with the following characteristics: I = 390-410 A, U = 226-28 B, V = 20cm/min for all samples.

Table 1 - Chemical composition of Cr-Mn-Mo-V wire metal

Types of Wire	C	Mn	Si	Cr	Ni	V	Mo
40H3G2MF	0.35-0.45	1.3-1.8	0.4-0.7	3.3-3.8	< 0.4	0.1-0.2	0.3-0.5
21	0.06	0.48	0.3	1.98	0.15	0.06	0.3
21.1	0.19	0.52	0.74	2.79	0.17	0.14	0.26
71	0.14	0.5	0.57	3.49	0.11	0.05	0.33
72	0.07	0.54	0.57	2.63	0.24	0.08	0.4

Table 2 - Chemical composition of Mn- Mo-V-B wire metal

Types of Wire	C	Mn	Si	Cr	V	Mo	B
40GMFR	0.36-0.44	0.9-1.2	0.17-0.37	0.2-0.5	0.06-0.1	0.08-0.16	0.001-0.003
22	0.06	0.43	0.26	0.36	0.01	0.07	0.003
22.1	0.16	0.32	0.62	0.25	0.19	0.11	0.003
73	0.06	0.4	0.49	0.39	0.04	0.13	-
74	0.04	0.43	0.49	0.3	0.04	0.11	0.002

Note: Boron is not added to sample No.73

Metallographic studies are performed on polished surfaces using an optical microscope OLYMPUS GX-51 in bright field at various magnifications × 100 - 1000 after etching with a solution of alcohol and nitric acid. Longitudinal specimens of the welded layer were analyzed at 100X magnification using a standard technique relating to the presence of non-metallic inclusions. The grain size was evaluated according to the standard test method at 100X magnification. Test specimen No.21 has a ferrite-perlite structure tending to the Widmannstaetten type (Figure 1a). The examined test specimen showed the following non-metallic inclusions: non-deformable silicates (grain size number 2b) and spots of oxides (number 2a) (Figure 2a). The grain size corresponds to numbers 3 and 4 according to the comparison grain size rating chart.

IOP Conf. Series: Materials Science and Engineering **125** (2016) 012027 doi:10.1088/1757-899X/125/1/012027

Table 3 – Wear resistance test results

Number of test specimen	Mass of test specimen before testing, g	Mass of test specimen after testing, g	Loss in mass, (g/ %)	Number of revolutions	Wear rate, g/rev
21	197.059	194.386	2.673/ 1.36	5843	0.00046
21.1	132.768	132.573	0.195/ 0.26	6214	0.00003
71	168.509	168.222	0.287/ 0.17	7685	0.000037
72	134.85	133.548	1.302/ 0.97	7965	0.00016
22	185.317	183.220	2.092/ 1.13	6304	0.00033
22.1	130.281	129.398	0.883/ 1.15	5760	0.00015
73	145.977	143.8537	2.123/ 1.45	7112	0.0003
74	133.4405	131.4593	1.981/ 1.48	7370	0.00027

Note: 21, 21.1, 71, 72 – the test specimens where 40H3G2MF wire was used; 22, 22.1, 73, 74 – the test specimens where 40GMFR wire was used. Test specimens Nos. 21.1 and 71 show the best results in terms of wear resistance.

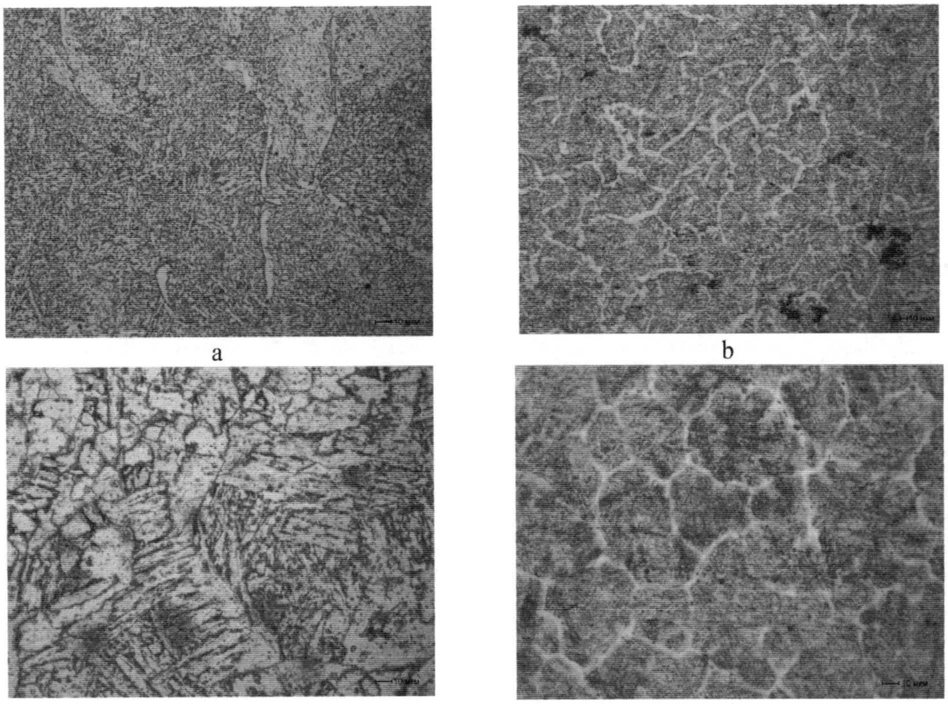

a b

c d

IOP Conf. Series: Materials Science and Engineering **125** (2016) 012027 doi:10.1088/1757-899X/125/1/012027

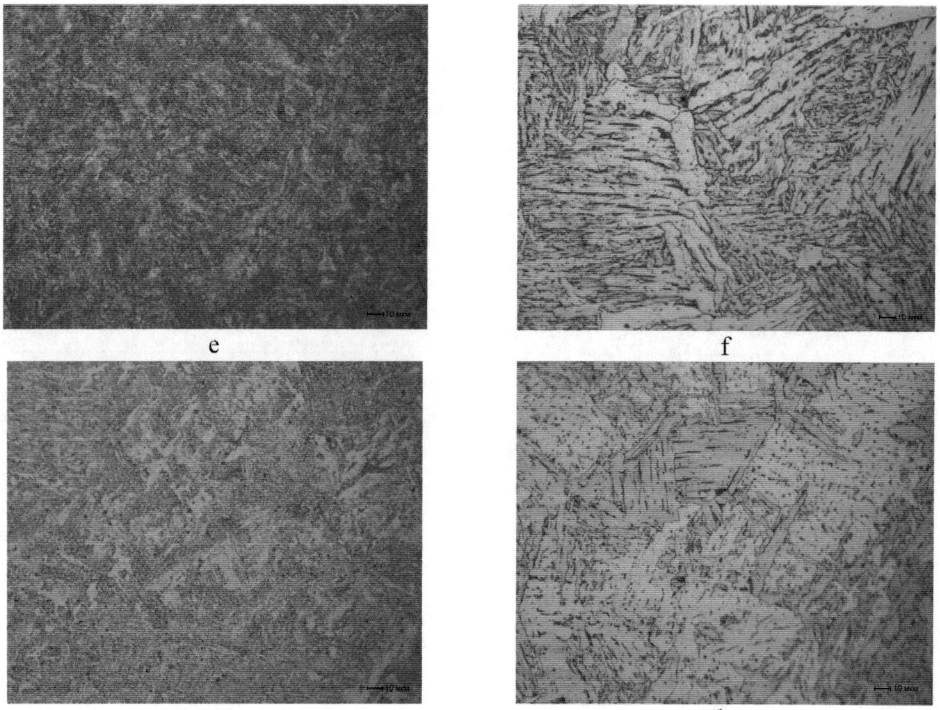

a – test specimen No.21; b – test specimen No.21.1; c – test specimen No.22; d – test specimen No.22.1;
e – test specimen No.71; f – test specimen No.72; g – test specimen No.73; h – test specimen No.74

Figure 1. Weld metal microstructure, × 500

The structure of test specimen No. 21.1 comprises ferrite and pearlite; and ferrite is identified not in the form of a separate structural constituent but a grid (Figure 1b). In this case, there is a high level of content of nonmetallic inclusions. Non-deformable silicates (number 5b and 4b) and spots of oxides (number 2a and 3a) are present as well (Figure 2b). The grain size corresponds to numbers 5 and 4.

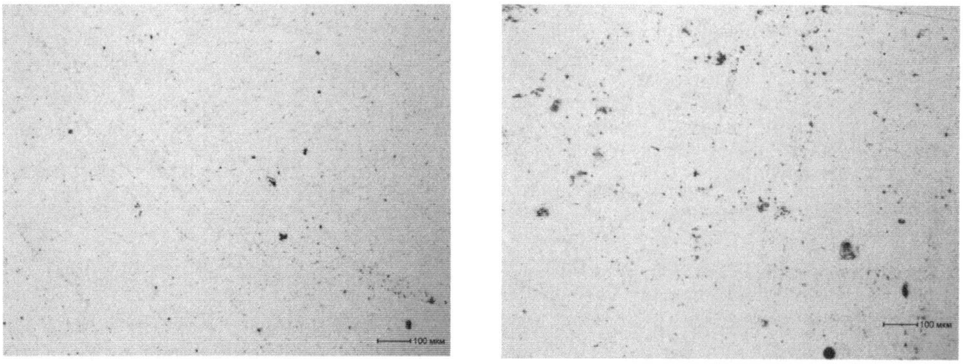

a b

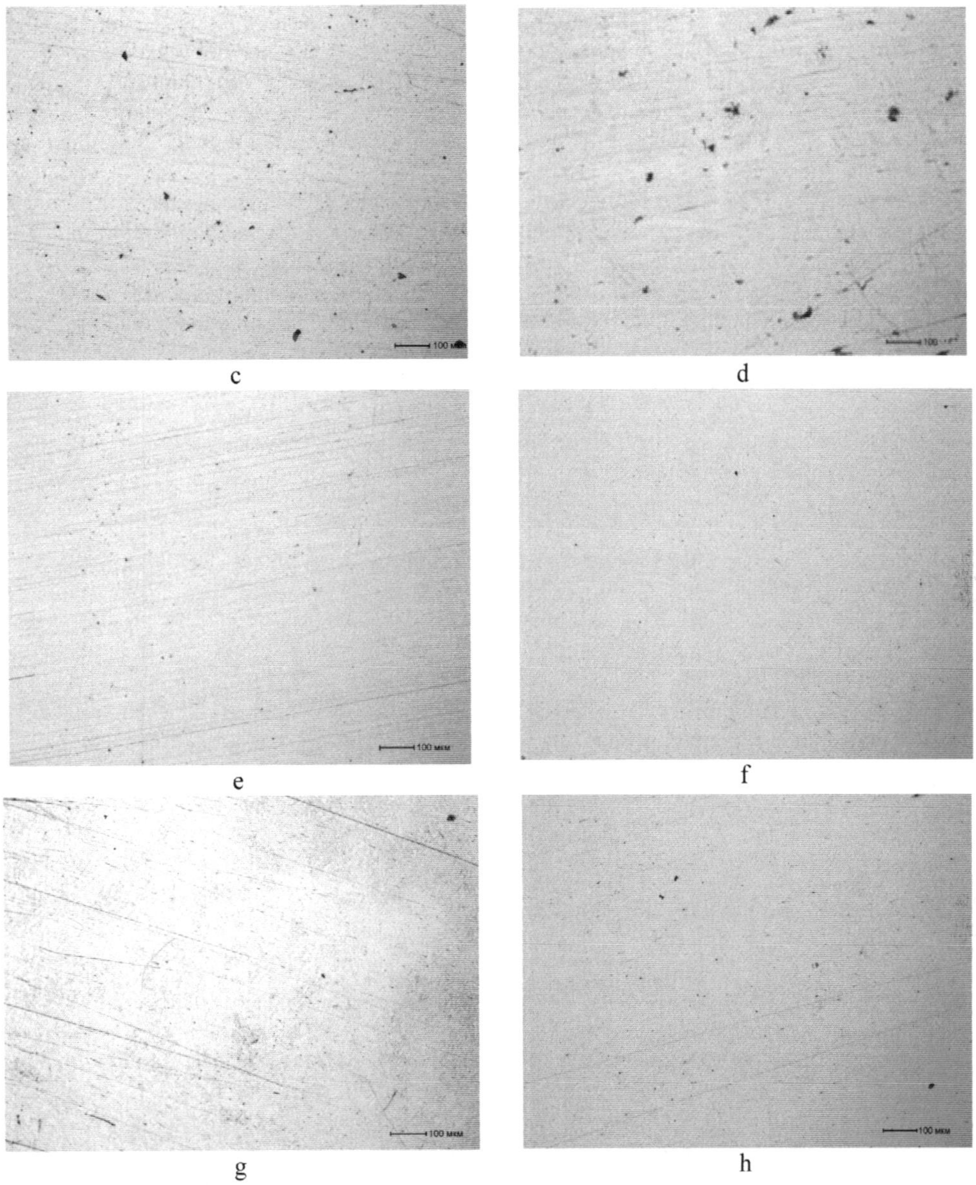

a – test specimen No.21, b – test specimen No.21.1, c – test specimen No.22, d – test specimen No.22.1, e – test specimen No.71, f – test specimen No.72, g – test specimen No.73, h – test specimen No.74

Figure 2. Nonmetallic inclusions in the test specimens

Test specimen No.22 has a ferrite-pearlite structure of the Widmannstaetten type (Figure 1c). This test specimen shows non-deformable silicates (number 2b), spots of oxides (number 1a and 2a) (Figure 2c). The grain size corresponds to numbers 4 and 5 according to the comparison chart used to rate grain sizes.

Test specimen No 22.1 is of a ferrite and pearlite structure, but the ferrite constituent is observed in the grid-like form (Figure 1d). The content of nonmetallic inclusions is higher when compared to the previous specimen: non-deformable silicates (number 3b and 5b), spots of oxides (point 1a) (Figure 2d). The grain size corresponds to number 5 and 6. Thus, 40GMFR steel has a finer grain.

The structure in test specimen No. 71 contains medium acicular martensite (number 5) (Figure 1d) and a little residual austenite. This test specimen shows non-deformable silicates (number2a and 2b) and spots of oxides (number 1a and 2a) (Figure 2e). The grain sizes are 6 and 7 according to the comparison grain size chart.

Test specimen No. 72 has a ferrite-pearlite structure of the Widmannstaetten type orientation (Figure 1f), containing the following non-metallic inclusions: non-deformable silicates (number 2b) and spots of oxides (number 1a) (Figure 2f). The grain size corresponds to numbers 4 and 5.

Test specimen No.73 is of a ferrite-pearlite structure (the Widmannstaetten type) (Figure 1 g) with the non-metallic elements including non-deformable silicates (number 1b and 4b) and spots of oxides (number 1a) (Figure 2g). The grain size corresponds to numbers 5 and 4.

Test specimen No. 74 has a ferrite-pearlite structure tending to the Widmannstaetten type (Figure 1h). This specimen shows a significant amount of non-metallic inclusions: non-deformable silicates (number 2b) and oxides (number 1a) (Figure 2h). The grain size corresponds to numbers 6 and 5 according to the comparison grain size chart.

Conclusions

1. The powder wire of the Cr-Mn-Mo-V system shows higher wear resistance than the powder wire of the Mn- Mo-V-B system.
2. The weld metal of the Mn- Mo-V-B system and 40GMFR steel has a finer grain structure than the weld metal of the Cr-Mn-Mo-V system and 40H3G2MF steel.

1 – This work has been done at Siberian State Industrial University in the framework of government contract No. 11.1531/2014/k with the Ministry of Education and Science RF. The equipment of the Common Use Centre "Material Science" at SibSIU was used in tests, researches and measurement.

References

[1] Lukashov A S, Zernin E A, Kuznetsov M. A. Application of inorganic nanopowders in welding, surfacing and spraying (review) // Applied Mechanics and Materials. - 2015 - Vol. 770. - pp. 299-303

[2] Kuznetsov M A, Kolmogorov D E, Zernin E A Upravlenie strukturoj i svojstvami metallov metodom modificirovanija (obzor) [Controlling a structure and behavior of metals by the process of modification (Summary)] // Journal "Tekhnologiya Mashinostroeniya". - 2012 – No.2 - pp. 5-8

[3] Malushin N N, Valuev D V Measuring stress condition in hard faced work rolls for cold rolling // Applied Mechanics and Materials. - 2013 - Vol. 379. - pp. 71-76

[4] New carbon-fluorine additives for welding fluxes/ N.A. Kozyrev, N.E. Krukov, R. Ev. Krukov, A.V. Roor, L.P. Bashchenko, U.I. Lipatova/ Steel in Translation.-2015. -T. 45. –No. 4. - pp. 251-253.

[5] Kuznetsov M A, Kolmogorov D E, Zernin E A, Lukashov A S Sposoby modificirovanija metalla nanostrukturirovannymi poroshkami dlja uvelichenija dolgovechnosti i jekspluatacionnoj nadezhnosti detalej gornoshahtnogo oborudovanija [Methods of modifying nanostructured metal powders to increase durability and operational reliability of mining equipment]// Journal "Gornoe mashinostroenie". – 2012 – No. OVZ - 552 p., pp. 306 – 309.

[6] Reinhard Rosert, M V Karasev POROShKOVYE PROVOLOKI - TENDENCII, RAZVITIE I IH PRIMENENIE V PROMYShLENNOSTI [Powder wires – tendencies, development, their application in industries] // Int. Sc. Tech. Conf. - St. Petersburg: SPbSPU Publ. 2012. pp. 220-230.

[7] N A Kozyrev, G V Galevsky, D V Valuev, V M Shurupov, O E Kozyreva 2015 IOP Conference Series: Materials Science and Engineering Surfacing with tungsten-containing ores 91 doi:10.1088/1757-899X/91/1/01/012009.

[8] N A Kozyrev, R E Kryukov, G V Galevsky, D A Titov, V M Shurupov 2015 IOP Conference Series: Materials Science and Engineering Development of new materials and technologies for welding and surfacing at research and production center "Welding processes and technologies" 91 doi:10.1088/1757-899X/91/1/01/012013.

IOP Conf. Series: Materials Science and Engineering **125** (2016) 012028 doi:10.1088/1757-899X/125/1/012028

On Quality of a Weld Bead Using Power Wire 35v9h3sf

N A Kozyrev[1], D. Eng. Sc, Prof., G V Galevskiy[1], D. Eng. Sc, Prof., D A Titov[1], PG student, D E Kolmogorov[2] Cand.Eng.Sc., D E Gusarov[2]

[1]Federal State Educational Institute of Higher Professional Education "Siberian State Industrial University", Novokuznetsk

[2]Yurga Institute of Technology (TPU affiliate) National Research Tomsk Polytechnic University, Yurga

Abstract: The effect of introducing carbon-fluorine containing additives and nickel to flux-cored wire 35V9H3SF on its structure, micro hardness of martensite, hardness and wear rate of a welded layer has been investigated. It has been proven that carbon-fluorine additions to a powder mixture increase the hardness of the deposited layer and its wearing strength. Nickel adding increases the hardness of both martensite and the deposited layer, and reduces abrasion wear. Introducing carbon-fluorine containing additives to the powder mixture can reduce the pore formation and lower the level of impurity with nonmetallic inclusions; and nickel adding can lead to austenite having a finer grain size.

1. Introduction

Nowadays, the flux-cored wire PP-Np-35V9H3SF is widely applied in the process of hard surfacing rollers of mill tables and rolls used in hot rolling mills. However, the surfacing process, when using this welding wire, is accompanied by a number of unacceptable defects: pores, cavities, macroscopic nonmetallic inclusions; and in some cases hardness decreases causing a shorter operational life of rolls [1].

It is widely known that sodium fluorosilicate is added to powder wire to reduce pore formation as a result of lower concentrations of hydrogen [1], gas- and slag-forming substances are added to protect against atmospheric gases, and low ionization potential elements are added to ensure stable arcing [2- 4].

2. Results and discussion

Under a laboratory environment, carbon-fluorine containing dust (metallurgical waste) was used instead of amorphous graphite to make samples of flux-cored wire with the following chemical composition, by weight, %: Al2O3 = 21-46; F = 18-27; Na2O = 8-15; K2O = 0.4 – 6%; CaO = 0.7-2.3; SiO2 = 0.5-2.5; Fe2O3 = 2.1-3.3; Ctotal = 12.5-30.2; MnO = 0.07-0.9; MgO = 0.06-0.9; S = 0.09-0.19; P = 0.10-0.18. It has been experimentally established earlier that carbon in a carbon-fluorine containing addition has a higher activity, deoxidizes metals, and reduces oxides in metal and slag to form carbon oxides [5, 6]. For making a powder mixture used for the wire production, the following components and powders were used: amorphous carbon, carbon ferromanganese FMn 78 (A), ferrosilicon FS 75, high carbon ferrochrome FH 900A, ferrovanadium FV50U0,6, nickel PNK-2K9, and iron PZHV1. The surfacing process was conducted using the AN-26S flux.

An X-ray fluorescence spectrometer XRF-1800 and an atomic emission spectrometer DFS-71 were used to determine a chemical composition of the welded layer. The obtained results are given in Table 1.

Longitudinal specimens of the weld deposition were analyzed according to the standard technique relating to the presence of nonmetallic inclusions. A metallographic microscope LaboMet-1I was used to study a polished surface at 100X magnification. Metallographic studies are performed using an optical microscope OLYMPUS GX-51 in bright field at various magnifications after etching with a solution of alcohol and nitric acid. The grain size was evaluated according to the standard test method at 100X magnification. Martensite structure was analyzed using the standard comparison charts to rate sizes of martensite needles. A metallographic test program Siams Photolab 700 was used in determining the length of martensite needles and pore sizes.

Table 1-Chemical analysis of test specimens in 35V9H3SF material

Test No	By wt, %													
	C	Si	Mn	Cr	Ni	Cu	Ti	W	V	Mo	Al	Co	S	P
1	0.14	0.45	1.69	2.60	0.08	0.02	0.005	8.84	0.36	0.01	0.019	0.11	0.028	0.022
2	0.14	0.51	1.73	2.09	0.08	0.11	0.005	7.66	0.14	0.01	0.012	0.11	0.029	0.021
3	0.13	0.45	1.74	2.18	0.38	0.09	0.005	8.11	0.09	0.01	0.008	0.11	0.034	0.019
4	0.13	0.49	1.68	2.27	0.46	0.09	0.004	8.21	0.13	0.01	0.005	0.11	0.034	0.023
5	0.13	0.60	1.54	2.66	0.58	0.09	0.005	7.58	0.15	0.02	0.044	0.11	0.040	0.020
1 – The test specimen with amorphous carbon added. 2 The test specimen with carbon-fluorine containing additives. 3,4, 5 – The test specimens with carbon-fluorine containing additives and nickel.														

According to the study performed, spots of oxides with grain size number 1a (Figure 1 a, c, d, e) and number 2a (Figure 1 a, e), represented in the form of single particle inclusions, are predominantly observed in the test specimens. Almost all the specimens under study have non-deformable silicates in the form of rounded and irregularly shaped inclusions with grain size number 1a (Figure 1, b), 2a (Figure 1, f) and 2b (Figure 1, e, f).

There are spot oxides with grain size numbers 1a and 2a (Figure 1a) and non-deformable silicates with grain size number 1a (Figure 1b) observed in test specimen No.1, with amorphous graphite being used as an additive. Spot oxides with grain size number 1a (Figure 1, c) are observed in test specimen No.2 with carbon-fluorine containing additives. Examining test specimen No.3 shows the presence of oxides with grain size number 1a (Figure 1, d). There are spot oxides with grain size numbers 1a and 2a (Figure 1,a) and a small amount of non-deformable silicates with grain size number 2b (Figure 1,e), observed in test specimen No.4. Specimen No.5 has non-deformable silicates with numbers 2a and 2b (Figure 1, f).

Thus, it has been established that adding carbon-fluorine containing material instead of amorphous graphite to the powder mixture reduces the degree of contamination of steel with nonmetallic inclusions. In this connection, only spot oxides are present in the structure, and non-deformable silicates are not observed. Spot fine carbides are traced in all examined test specimens.

IOP Conf. Series: Materials Science and Engineering **125** (2016) 012028 doi:10.1088/1757-899X/125/1/012028

Metallographic analysis revealed that test specimen No. 1, with using amorphous graphite as an additive, has a structure of a coarse acicular martensite (number 10) and residual austenite that is present in a small amount (Figure 2,a). The grain size corresponds to numbers 6 and 7 in the comparison chart used for grain size rating. In this case, porosity is observed over the entire polish surface. The pore size varies 640 - 1077 micrometers.

After adding carbon – phosphorus containing material instead of amorphous carbon to the powder mixture (test specimen No. 2), ferrite in the form of a separate structural constituent and of a grid in some areas is determined, in addition to a coarse acicular martensite (number 10) and residual austenite in a small amount (Figure 2,b). The grain size corresponds to numbers 5 and 4 according to the comparison grain size chart. In this case, a decrease in porosity is established.

After adding 0.38% nickel to the powder mixture (test specimen No. 3), there is coarse acicular martensite (number 10), residual austenite, and the ferrite component observed in the structure as well (Figure 2,c). The grain size corresponds to numbers 5, 6 and 7 according to the comparison grain size chart. Using carbon-fluorine containing additives and 0.38% nickel does not cause the porosity formation.

The addition of carbon-fluorine containing material and 0.46% nickel (test specimen No.4) to the powder mixture results in the structure, consisting of coarse acicular martensite (number 10), ferrite in the form of separate structural component (Figure 2,d), and a small amount of residual austenite (Figure 2, d, e). The grain size corresponds to numbers 5, 6 and 7 according to the comparison chart; porosity is not observed.

When adding 0.58% nickel (test specimen No.10) to the powder mixture, the ferrite component is not determined in the structure. The structure consists of coarse acicular martensite (number 10) and a small amount of residual austenite with grain size numbers 6 and 7 (Figure 2, f). Martensite has a dispersed structure in comparison with other test specimens; the size of martensite needles is 7-42 microns. In this case, a decrease in porosity is observed.

Thus, it is established, when carbon-fluorine containing additives are used without or with 0.38 % and 0.46% nickel to replace amorphous graphite, the ferrite structure is mainly observed in the form of a separate component as well as coarse acicular martensite and residual austenite. A ferrite–free structure can be obtained when carbon-fluorine containing materials and a larger amount of nickel (0.58%) are used for the wire powder. It should be noted that, when nickel is not used while adding carbon-fluorine material to the wire powder, the austenite grain size increases. The use of carbon-fluorine containing additives separately or together with nickel in an amount of 0.38, 0.46 and 0.58 % reduces porosity in the surfacing material made of steel 35V9H3SF.

The hardness was defined using the Rockwell and Vickers methods. The test method capable to determine the microhardness of structure components was utilized in examining the properties of martensite. A HVS-1000 automatic digital micro hardness tester was used for determining microhardness. 10 measurements were taken to evaluate microhardness of martensite.

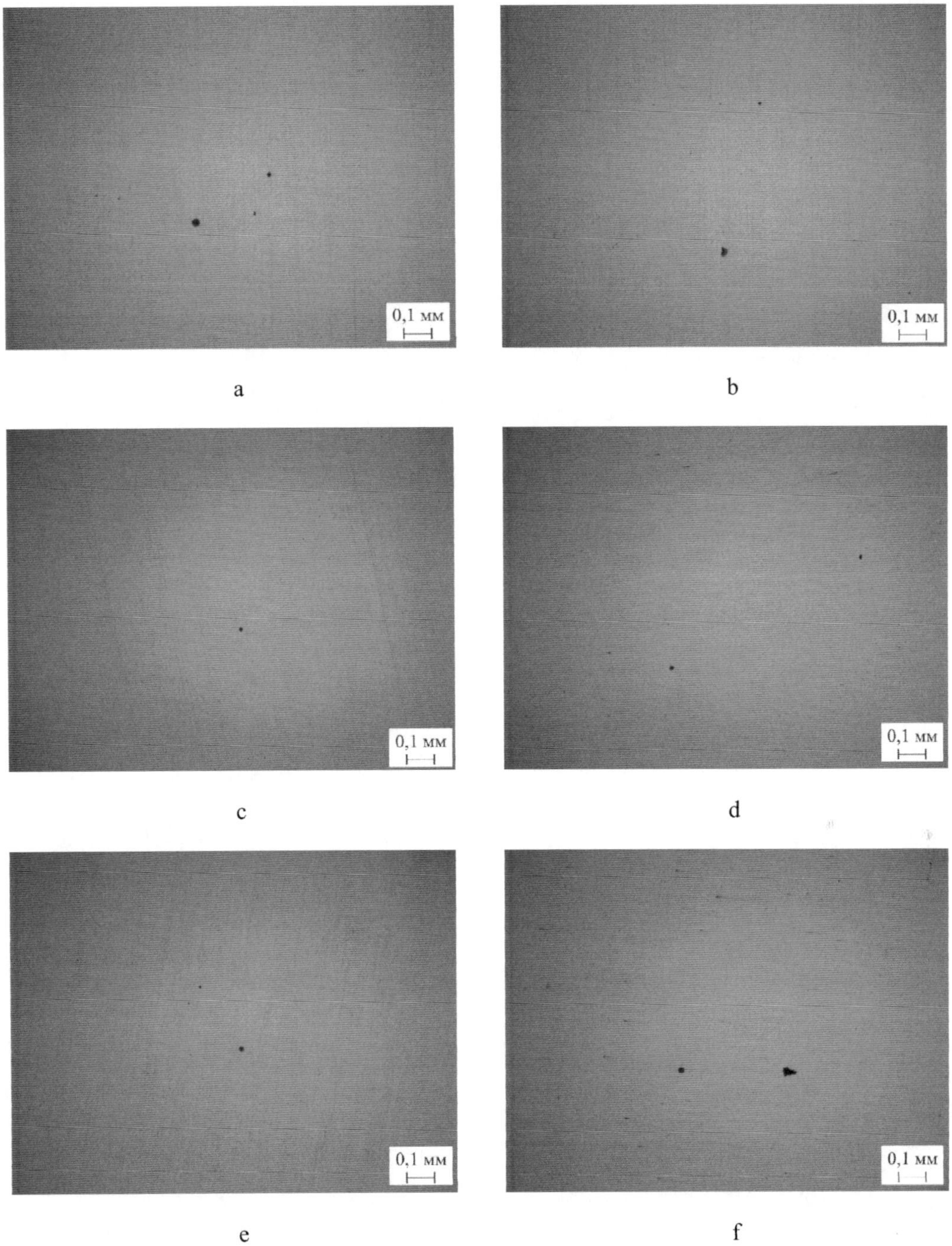

Figure 1- Nonmetallic inclusions: a, b – test specimen No.1; c – test specimen No.2; d – test specimen No.3; e – test specimen No. 4, f – test specimen No. 5

IOP Conf. Series: Materials Science and Engineering **125** (2016) 012028 doi:10.1088/1757-899X/125/1/012028

An etched metallographic polish was prepared for analyzing the structure using a microscope. Within the selected area to be observed, a diamond indenter was pushed to make indention under a load of 1 H; and then the size of the indent on the polished surface was measured using a microscope. Calculating and displaying the hardness number according to the Vickers hardness test method as well as saving images of indents were automatic, enabled by the connection of a CCD-camera with a video capture card to the computer and relevant software to analyze images.

A 2070 SMT – 1 tester was used to measure the wear rate of the deposited layer. In this case, the following parameters were selected: a measurement range for frequency of rotation of shaft with the lower sample (range A), which is 75 -750 min -1 , and measuring range for friction torque (range I), which is 1 - 10 N/m.

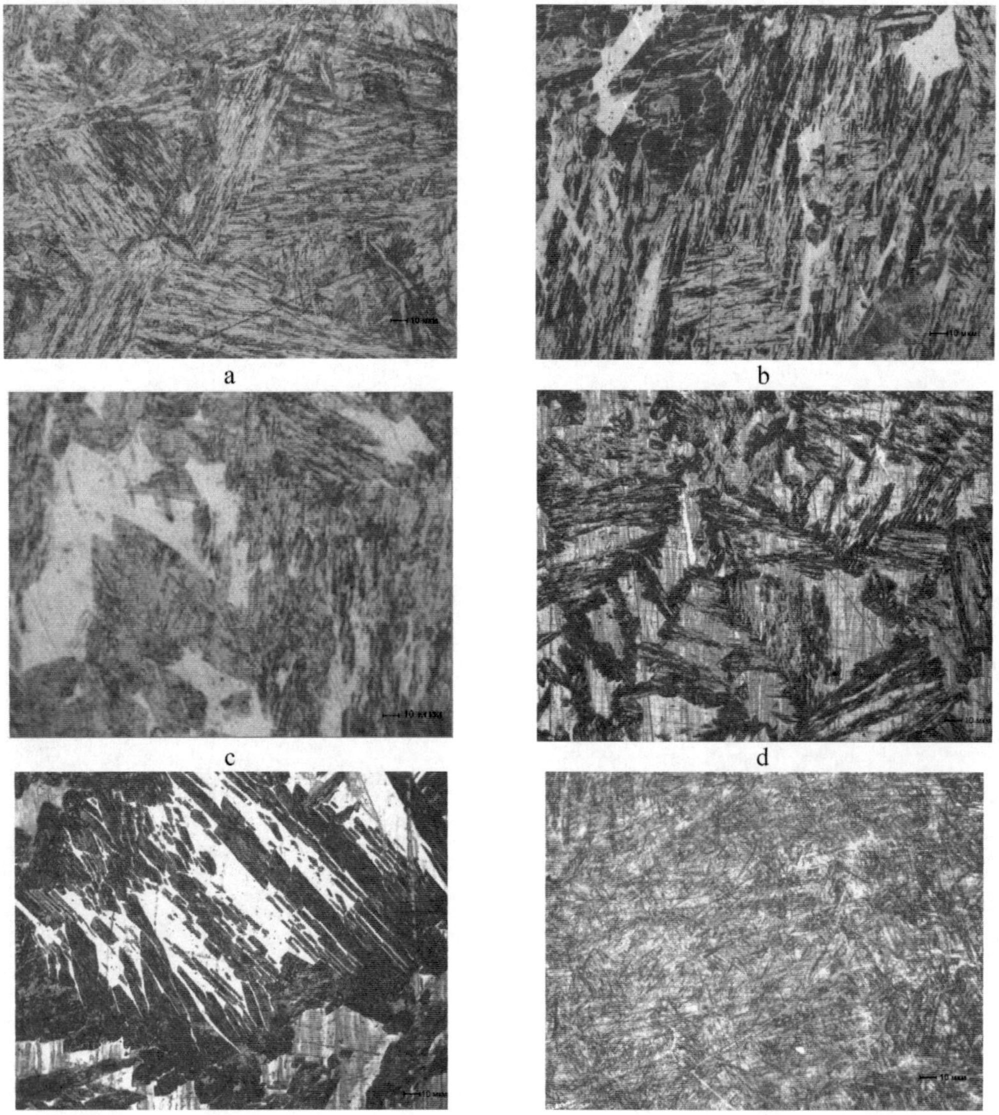

Figure 2 Microstructure, × 500: a, b – test specimen No.1; c – test specimen No.2; d – test specimen No.3; e – test specimen No. 4, f – test specimen No. 5

Metallographic examination of test specimen No.1, where amorphous graphite was used as an additive, revealed that its structure contained coarse acicular martensite (number 10) and a small amount of residual austenite. The microhardness of martensite is 471 HV. Images of the indents for determining microhardness are shown in Figure 3.

The average value of hardness of the deposited layer corresponds to 45,5HRC, and the rate of its wear makes 0.00088 g/min.

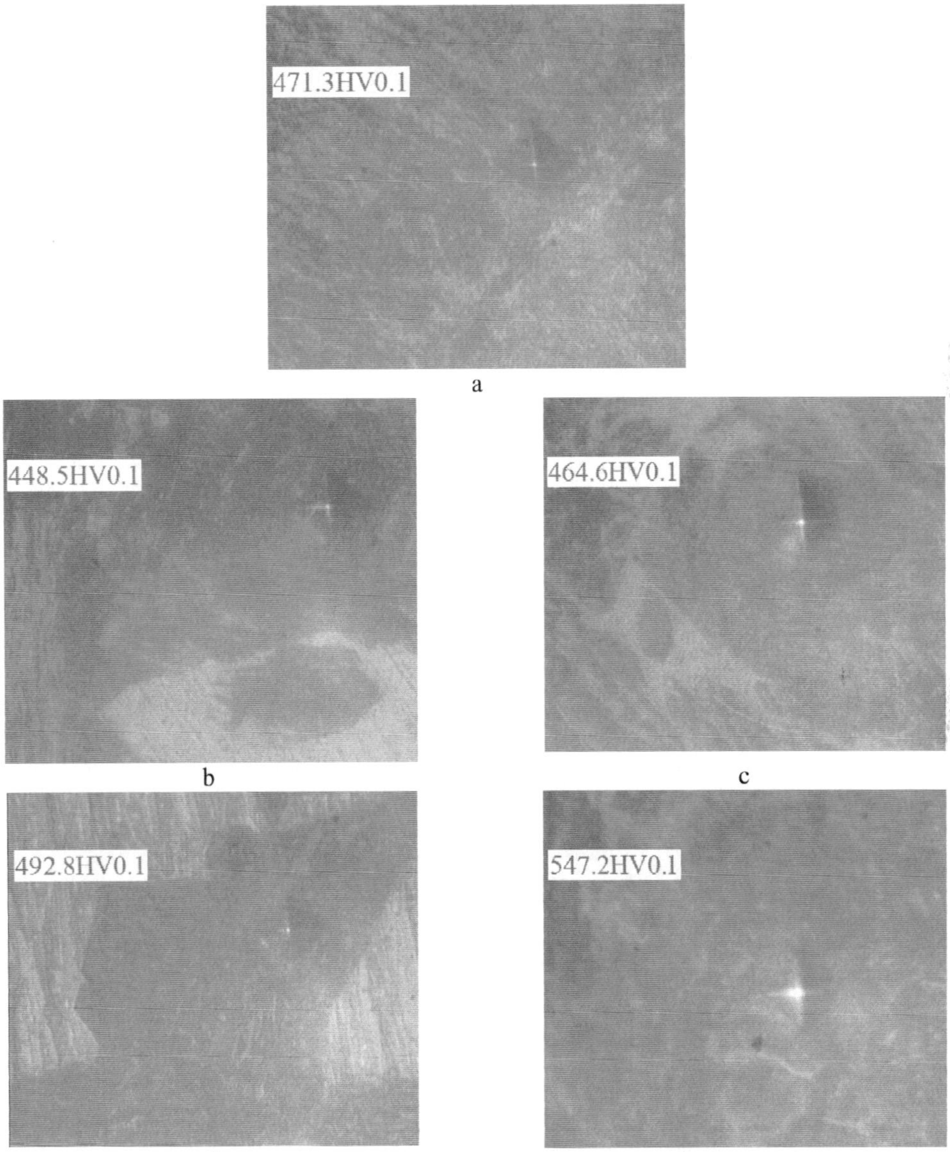

Figure 3 Microhardness of martensite in the specimens under study: a - the specimen with the addition of amorphous graphite; b - the specimen with the addition of carbon –fluorine containing dust; c - the specimen with the addition of carbon –fluorine containing dust and 0.38% nickel; d - the specimen with the addition of carbon –fluorine containing dust and 0.46% nickel; e - the specimen with the addition of carbon –fluorine containing dust and 0.58% nickel

Adding carbon – phosphorus material to the powder mixture (test specimen No.2) to replace amorphous carbon promotes the appearance of ferrite in the form of a structural component and a grid in some areas, besides acicular martensite (number 10) and a little residual austenite. Using carbon-phosphorus additions reduces the microhardness of martensite by 5%, compared with test specimen No.1 (Table 2). However, with this, the hardness of the deposited layer increases, and the rate of wear reduces.

Table 2 – Microhardness of martensite in the test specimen structures, hardness and wear rates of the deposited weld

Test specimens No.	μ_H, HV	HRC average / min/max	Wear rate, g/min
1	471	45,5 / 45,2 / 46,0	0,00088
2	449	46 / 45,2 / 47,0	0,00052
3	464	45,5 / 45,2 / 47,0	0,00052
4	493	47,2 / 45,0 / 49,0	0,00065
5	547	48,0 / 46,5 / 50,0	0,00068

When adding 0.38 and 0.46% nickel (specimens No.3 and 4) to wire, besides needle–like martensite grains (grain size number No.10) and a small amount of austenite grains, the ferrite component is present in the structure. Adding nickel in amount of 0.58% (specimen No.5) provides refinement of martensite needles and makes a ferrite-free structure, consisting of coarse needles of martensite (grain size number No.10) and residual austenite. In specimen No.5 martensite has a more dispersion structure (the size of martensite needles is 7 - 42 microns), if to compare with other specimens, where the size of needles in the martensite structure is not possible to measure. Besides, when comparing with specimen No. 2, adding 0.58% nickel increases the microhardness of martensite by 22% and the hardness of the welded layer by 4%. Additional introduction of nickel in an amount of 0.46 and 0.38% increases microhardness of martensite to a lesser extent (by 19 and 3%, respectively). Adding 0.46% nickel increases the hardness of the welded layer by 2% as compared with specimen No.2. Using carbon - phosphorus containing admixtures to replace amorphous carbon, separately and together with nickel, reduces the rate of wear relating to the welded layer.

The described researches form the basis for making the flux-cored wire, the chemical composition of which is protected by the Russian patent. [7].

Conclusions

1. Adding carbon-fluorine containing material and nickel instead of amorphous graphite to the 35V9H3SF flux-cored wire increases the hardness of the welded layer and reduces the rate of its wear.
2. Adding nickel to the powder mixture improves microstructure due to the removal of ferrite and refinement of martensite that is accompanied by a significant increase in microhardness and a slight increase in hardness of the welded layer.
3. New flux cored wire for surfacing steel is based on the results of this research.

This work has been done at Siberian State Industrial University in the framework of government contract No. 11.1531/2014/K with the Ministry of Education and Science RF. The equipment of the Common Use Centre "Material Science" at SibSIU was used in tests, researches and measurement.

References

[1] I I Frumin, Yu A Yuzvenko, E I Leynachuk.. Tehnologija mehanizirovannoj naplavki [Machine Disposition Technology] . – Moscow : Vysshaya Shkola,1964.- 306p.

[2] Konovalov A V, Kurkin A S, Makarov E L, Nerovniy V M, Yakushin B F Teorija svarochnyh processov [Theory of Welding processes] - Moscow.: Bauman MSTU Publ., 2007. - 752p.

[3] Kuznetsov M A, Zernin E A , Danilov V I , Kartsev D S Application of nanostructured powders to control characteristic of electrode metal transfer and the process of weld structurization // Applied Mechanics and Materials. - 2013 - Vol. 379. - pp.199-203.

[4] N P Aleshin, G G Chernyshov, A I Akulov and colleagues. Svarka. Rezka. Kontrol': Spravochnik.[Welding.Cutting. Control: Handbook in 2 Vol] - Moscow: Mashinostroenie, 2004. Vol.1, 624 p.

[5] Kozyrev N A, Kryukov N E, Kryukov R E, Igushev V F, Koval'skiy I N Tehnologicheskie aspekty ispol'zovanija uglerodftorsoderzhashhej dobavki pri svarke pod fljusom [Technological Aspect of Carbon-Fluorine Adding in flux welding]. Svarochnoe proizvodstvo. – 2015. – No. 4. – pp. 43-47.

[6] N A Kozyrev, R E Kryukov, O A Kozyreva 2015 IOP Conference Series: Materials Science and Engineering 91 New tendencies in development of carbonaceous additives for welding fluxes doi:10.1088/1757-899X/91/1/01/012008.

[7] Kozyrev N A, Igushev V F, Titov D A, Kozyreva O E, Starovatskaya S N Patent RU 2518211, MPK8 V23 K35/368 Poroshkovaja provoloka [Powder Wire].

IOP Conf. Series: Materials Science and Engineering **125** (2016) 012029 doi:10.1088/1757-899X/125/1/012029

Theoretical Foundation for Mechanical Products Service Life Prediction

V V Konovodov[1], A V Valentov[2], T V Lafetova[1], M N Basalaev [2]

[1] Novosibirsk State Agrarian University
147 Nikitin St, 630039 Novosibirsk, Russia; phone: (383) 267-35-07
[2] Yurga Institute of Technology, affiliate of Tomsk Polytechnic University
26 Leningradskaya St, 652050 Yurga, Russia; phone: (38451)-6-05-37

e-mail: valentov@mail.ru

Abstract. The article presents theoretical foundations for prediction of service life of mechanical products, based on the fatigue theory and fatigue limit. Ultimate amplitude and ultimate stress diagrams are presented. Wohler curve, characterizing material durability, is constructed on the results of the tests.

Analytical studies show that about a half of the agricultural machinery failures occur for production reasons. This rate has been maintaining at the same level for the past 15 years. Operation life of complex mechanisms is usually determined by durability of the weakest components and parts.

For example, life of high pressure hoses, composed of high-pressure sleeves and nozzles is determined by durability of sleeve screening braid if careful embedding of nozzles. The service life of internal combustion engines depends on durability of valve springs, specifying the output parameters of engines, namely fuel consumption and capacity.

1. Timeliness of the research

Service life of suspensions is determined by durability of the spring main plates, life of carbide cutting tools - by durability of carbide plates if a reliable connection to holders, life of working bodies of feed mills and tillage machines - by durability of their working bodies, etc.

Thus, to predict service life of machines, a method is to be developed for prediction of durability of components or parts, which are the weakest components in the whole machine.

Durability of any machine element is characterized by the number of load cycles before failure. The problem of determining the durability of components or parts, and hence a service life of machines and mechanisms is being studied constantly; but it has not been resolved definitively.

For components of agricultural machinery, exposed to extreme stresses (in areas of stress concentration, locales of high temperatures and residual stresses) strength calculations (strength of materials) are traditionally used, which can be inadequate and sometimes inappropriate. As a result, for example, cables, having a tenfold safety margin, collapse under loads which are significantly lower than the design loads.

IOP Conf. Series: Materials Science and Engineering **125** (2016) 012029 doi:10.1088/1757-899X/125/1/012029

Modern computational methods allow determining durability (safety factor of durability) of components or parts, under certain magnitude and nature of stress [1], which does not guarantee their uptime.

Therefore, development of methods of forecasting of safety and durability under multiple-cycle fatigue is an urgent task. New techniques of engineering calculations bring about fundamentally new opportunities for lifetime assessment of agricultural machinery at the design stage and during their operation.

Methods of research, authenticity and validity of the results:

Stresses, variable in time, occur in structural elements under load, variable in magnitude or direction. For example, the car axle is bent under the weight load of the car. In the upper part of each axis cross section normal tensile stresses occur. During the motion the wheel and the axis, rigidly connected to the wheel, rotate, and each point of the axis alternately moves from the top (extended) of the half-section, to the bottom (compressed), and vice versa.

Variable stresses arise in shafts of machines, in braided sleeves of high-pressure hoses, in valve springs of internal combustion engines, in spring plates, in suspensions, in working bodies, etc.

Time-scale stress variations can be represented by a graph; time t or the number of cycles of a certain frequency N is plotted on the abscissa axis, a value of normal σ stress or tangential τ stress - on the ordinate axis. It is usually assumed, that time-scale stress variations are characterized by a sine curve (see Figure 1).

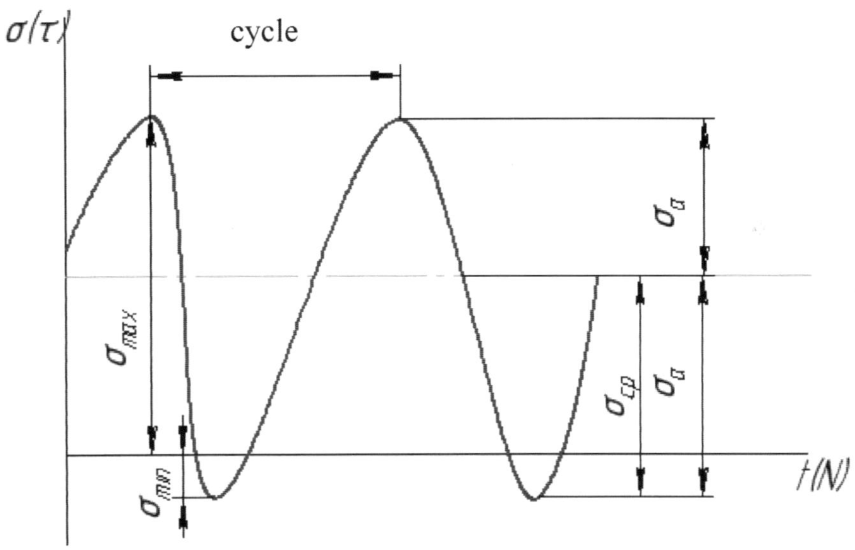

Figure 1. Time-scale stress variations

In fact, as shown by numerous experiments [2], the type of the curve is irrelevant; material strength at variable stresses depends mainly on stress values: σmax and σmin. Population of all successive values of stress variations for one period of change is called the stress cycle. The largest (algebraically) stress of the cycle is called maximum and is denoted σmax (or τmax for tangential stresses). We principally consider normal stresses. For tangential stresses similar conclusions, formulas and calculation procedures are applicable.

Algebraic half-sum of maximum and minimum stress of the cycle is called the average stress.

$$\sigma_{cp} = \frac{\sigma_{max} + \sigma_{min}}{2} \quad (1)$$

Algebraic half-difference of maximum and minimum stresses is called the cycle amplitude.

$$\sigma_{a} = \frac{\sigma_{max} - \sigma_{min}}{2} \quad (2)$$

Average stress of the cycle can be either positive or negative. The amplitude of the cycle is always positive. If stresses σmax and σmin are equal in magnitude and opposite in sign, then the cycle is called reversed. In the reversed cycle σav = 0, and σa=σmax=–σmin

If stress σmax and σmin are not equal in absolute values, the cycle is called irregular. Irregular cycle can be alternating (where σmax is positive and σmin is negative) or constant (when both σmax and σmin have the same sign). In particular cases, when σmin or σmax is equal to 0, the cycle is called fluctuated. Ratio of σmin stress to σmax stress is called ratio R of cycle irregularity.

$$R = \frac{\sigma_{min}}{\sigma_{max}} \quad (3)$$

Stress cycles with the same values of irregularity ratio are called alike. For a reversed cycle R = -1, for fluctuated cycle R = 0, if stresses have positive values.

Results of the research

To determine durability, a special testing of material mechanical properties (fatigue) is carried out. For these tests, a series of identical samples (at least 10) are produced. Pure bending tests with a reversed cycle of stress variations are the most common. They are carried out in the following order.

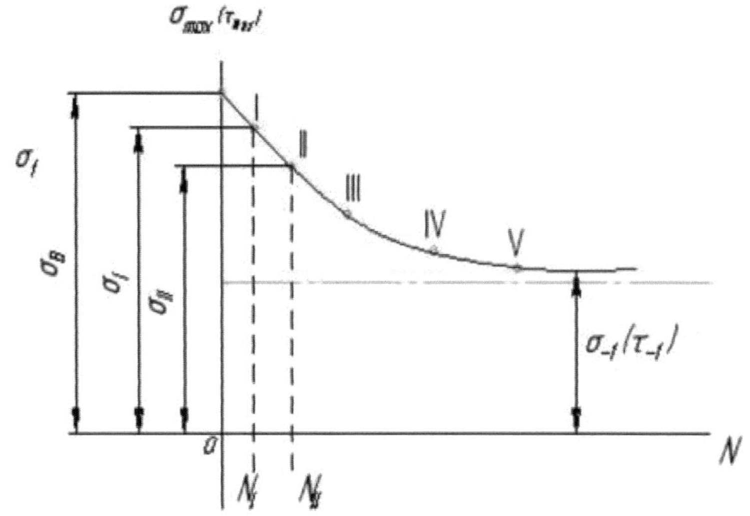

Figure 2. Wohler curve for R = -1

In the first sample a special device produces stress cycles characterized by the values σmax=σI and σmin. Stress value σI is set near to ultimate tensile strength σв of the material so as to obtain a fatigue failure of the sample after a relatively small number of cycles NI. The test result is plotted (in the adopted scale) as a point I, the abscissa is equal to the number of cycles NI, that caused the sample failure, and the ordinate shows the stress value σI (Figure 2). Then, the next sample is bended to failure, if stresses σmax =σII<σI and σmin=–σII. The test result is plotted as a point II.

In the test series, points III, IV, V and etc. are obtained. Joining the points, we plot a so-called fatigue curve or the Wohler curve (A.Wöhler first studied the phenomenon of fatigue in 1852 - 1869.), corresponding to reversed cycles (R = -1) (see Figure 2). By this method, fatigue curves, corresponding to cycles with other values of R, can be obtained.

Limiting stress diagram (LSD) shows the relationship between the limit cycle parameters σ'av-σ'max. LSD is used for functional connections that are used in durability analysis under varying stresses. For a single point of the diagram, a series of identical samples (at least 10) should be tested, and the Wohler curve should be constructed; the latter shows the value of fatigue limit for the cycle with a given R.

For example, tests were conducted at irregular stress cycle. As a result, a value of fatigue limit σ-I is found. The coordinates of this limit cycle are: σ'a =σ'max=σ-1, σ'av=0 (see Eq. (1) - (3)). That is, the point is on the ordinate (point A) (Figure 3).

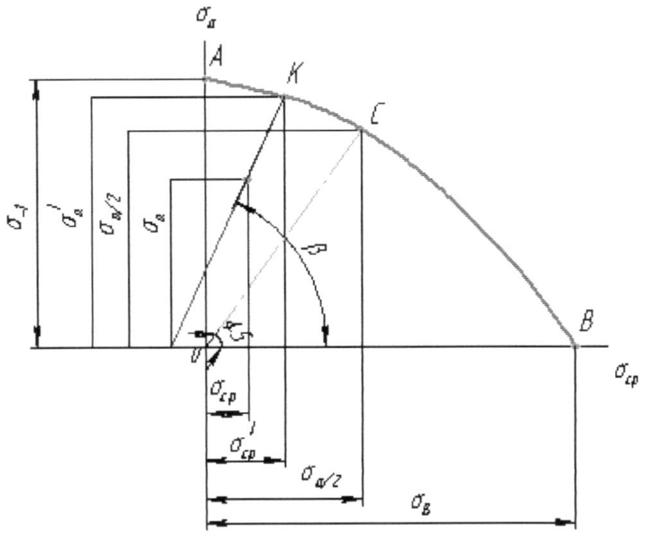

Figure 3. Diagram of limit amplitudes along σ_B

For any irregular cycle, according to the durability limit $\sigma'_{\max} = \sigma_R$, determined by tests, it is easy to find σ'_a and σ'_{cp}.

According to the formula (3), $\sigma'_{\max} = \sigma'_{cp} + \sigma'_a$, but $\sigma'_a = \rho \cdot \sigma'_{cp}$, where ρ – is characteristic of a stress cycle, determined by ratio σ'_a / σ'_{cp}. Consequently,

$$\sigma'_{\max} = \sigma'_{cp} + \rho \cdot \sigma'_{cp} = \sigma'_{cp}(1+\rho)$$

or $\sigma'_{cp} = \dfrac{\sigma'_{\max}}{1+\rho}$ and $\sigma'_{a} = \dfrac{\rho \cdot \sigma'_{\max}}{(1+\rho)}$ (4), (5)

In particular, for a fluctuated cycle (R=0, $\rho = 1$) at durability limit equal to σ_0

$$\sigma'_{cp} = \sigma'_{a} = \frac{\sigma_0}{2}$$

This cycle corresponds to point C in the diagram shown in Fig. 3.

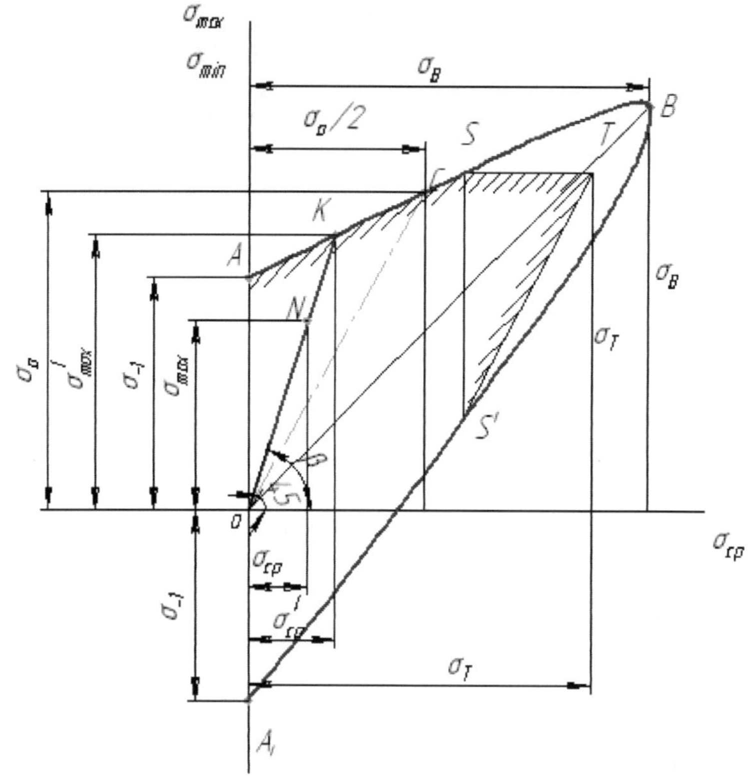

Figure 4. Diagram of limit stresses along σ_B

This diagram shows each cycle with two points. For example, limit reverse cycle is shown with points A and A1: point B corresponds to a limit constant stress $(\sigma'_{\max} = \sigma'_{cp} = \sigma_B)$, limit fluctuated cycle $\sigma'_{\max} = \sigma_0 ; \sigma'_{\max} = \sigma'_{cp} = \dfrac{\sigma_0}{2}$ is shown with points C and F.

To determine fatigue limit for the cycle with ratio R we draw a half-line at angle β to the x-axis from zero [7]. Tangent of the angle is determined by the formula:

IOP Conf. Series: Materials Science and Engineering **125** (2016) 012029 doi:10.1088/1757-899X/125/1/012029

$$tg\beta = \frac{\sigma_{\max}}{\sigma_{cp}} = \frac{2\sigma_{\max}}{\sigma_{\max} + \sigma_{\min}} = \frac{2}{1+R} \qquad (6)$$

K-ordinate of crossing between the half-line and limit stress line gives a value σR= σ'max. To determine areas of cycles safe from fatigue and flow, on the half-line 0B (points of the half-line correspond to time-constant stresses: σmax=σcp, σa=0) we plot a point that corresponds to the cycle with σmax=σcp= σT and draw from this point two straight lines as shown in Fig. 2.5.

Area of safety life cycle is limited by the segment AA1, axes are limited by the curves AS, A1S1 and the broken curve STS1. According to numerous studies [3,4,5,6,7], LSD from Fig. 4 can be simplified as it is represented in Fig. 5,6.

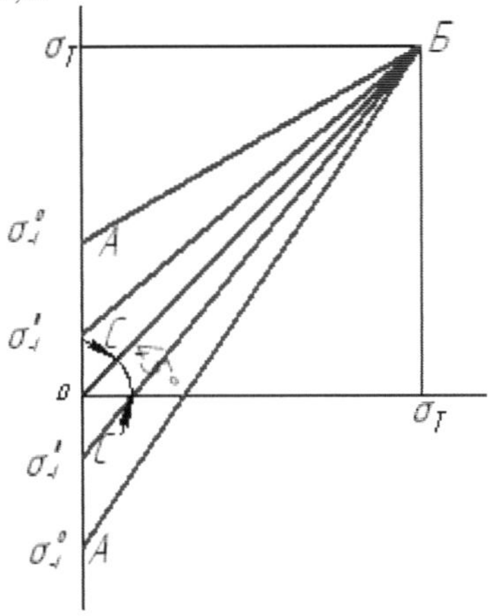

Figure 5. Simplified diagram of the limit stresses along σ_T

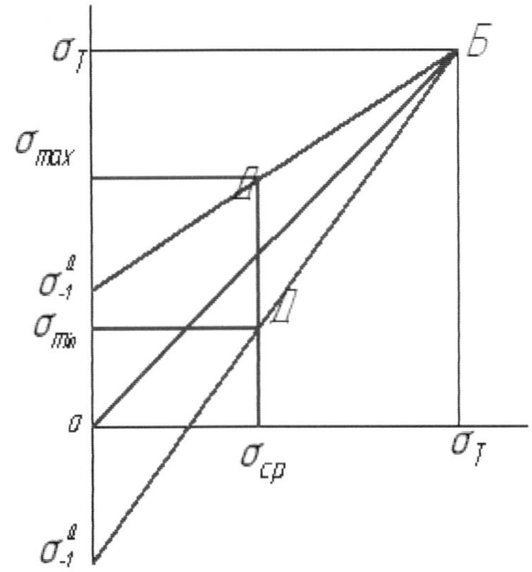

Figure 6. Real diagram of limit stresses

Conclusions

The proposed method is universal. For example, based on testing of a certain helical spring, made of certain steel, LSD can be plotted for any helical springs made from this steel. In tests it is important to find a point of the Wohler curve inflection that characterizes springs durability equal to Nб; that is, to find point Д (see Fig. 6) experimentally, but not by standard specifications.

References

[1] Darkov A.V. Strength of Materials /A.V. Darkov, G.S. Spiro. - M .: Higher School, 1975. - 654 p.

[2] Lasukov, A.A., Mokhovikov, A.A. Influence of tool modified layer on stress - Strain state of cutting wedge, Proceedings - 2012 7th International Forum on Strategic Technology, IFOST 2012

[3] Chernukhin R.V. Reliability of steering gears of truck vehicles// Applied Mechanics and Materials. - 2013 - Vol. 379. - p. 36-42.

[4] Putilova U.S., Nekrasov Yu.I., Lasukov A.A. Loading of Manufacturing System Elements at Unsteady Cutting and the Models of their Arrangement Deviations, Applied Mechanics and Materials, Vol. 682 (2014), pp. 192-196

[5] Retyunsky O.Y. Method of making performs for composite reversible cutting plates // Applied Mechanics and Materials. - 2013 - Vol. 379. - p. 24-27.

[6] Valentov A.V., Retyunsky O.Yu. Use of tungsten carbide cutting tools in processing of mining machinery parts, recovered by surfacing / Mining informational and analytical bulletin (scientific and technical journal). 2012. № S3. Pp 245-249.

[7] Valentov A.V. Retyunsky O.Yu., Konovodov V.V. Increased durability and performance of cutting tools at processing of weld surfaces // Mechanics XXI 2012. № 11. C. 395-399

IOP Conf. Series: Materials Science and Engineering **125** (2016) 012030 doi:10.1088/1757-899X/125/1/012030

Systematic Classifier OF Manufacturing Processes For Medium Size Shafts

D V Lychagin[1], A A Lasukov[2], A V Walter[2] and D A Arkhipova[2]

[1]Tomsk State University, 36, Lenin ave, Tomsk, 634050, Russia
[2]Yurga Institute of Technology, TPU affiliate 26, Leningradskaya street, Yurga, 652050, Russia

E-mail: lasukow@rambler.ru

Abstract. The article considers some issues of increasing efficiency of manufacturing preparation as a part of manufacturing processes design at a machine building enterprise. A tree of routing manufacturing processes for machining shafts of medium size is described as an example of clustering parts according to their structural and technological characteristics. Processing route for a certain part included into a certain group is developed through choosing machining operations for elementary surfaces of a part from the process route developed for a template representative of the group.

1. Introduction

Machining of products under the conditions of small-lot and single-piece production has its unique features. Manufacturing process specifications for an article are not developed in details. Sometimes general processing route is assigned but small operations depend on machine tool operator skills. General purpose equipment and tools are applied. If the procedure specifications are necessary the development process takes a great amount of time which leads to increased manufacturing preparation period. Modern mechanical engineering is characterized by intensive application of new technologies and constantly changing business trends [1]. In this context it is important to provide flexible production [2].

Modern quality requirements and frequent change of manufactured products stipulate the need to improve manufacturing preparation process [3, 4]. Systematization of technologies enables to develop unified manufacturing process specifications. Unification of technologies provides reduction of minor works associated with technological, planning and accounting documentation necessary for preparing and managing a manufacturing process. In view of the above it is reasonable to classify the parts produced at an enterprise according to structural and technological factors and to develop systematic classifier of routing manufacturing processes for a certain kind of parts (here they are shafts). Further on it will allow simplifying development of manufacturing processes and reducing manufacturing preparation time.

The idea to cluster up manufacturing processes belongs to professor S.P. Mitrofanov [5, 6]. Similar concepts can be found in other works [7, 8] including those using graphs [9]. Modern automated design tools allow modular design with which an article is made of structural modules that is unified units of the article. The range of structural modules is much less than the range of products to be assembled of them. Describing a wide range of structural modules is the main trend in the development of the systematic classifier. Such approach enables to cut down the number of the variants being developed. Thus developing a processing route does not take much time and calculation tools are used more effectively [10].

When designing manufacturing process specifications under the conditions of small-lot production there are always a great number of different variants which effect project solutions. As far as the design process is not formalized enough such problems as choosing a blank, developing processing route, choosing a machine-tool and cutting tools can be solved only by means of choosing a variant from some available common representatives. That means that typical solutions make basis for formalization when solving informal tasks in computer aided design of manufacturing processes.

In this account a machine-building enterprise has made a decision to work out a systematic classifier of manufacturing processes for manufacturing standard component parts of power hydraulic systems based on their structural and technological attributes [5, 6, 7] or economic characteristics [11]. The parts are combined into larger groups, which enables to increase series production and reduce process design time due to making template solutions. Unified highly productive machining methods and flexible equipment chosen depending on tools performance figures [12, 13, 14] are assigned for certain groups of parts with similar attributes.

When working out the classifier for an enterprise they aimed to make a description of a set of template solutions to cover the whole range of structural and technological parts and conditions under which each of those solutions can be applied. This work results in reducing manufacturing preparation period, improving data storage safety due to account policy as well as providing easy replication, availability of cooperative work of designers on the same project and easy information exchange between different enterprise departments.

A wide range of products and technological manufacturing processes has been analyzed within the project work. Performance characteristics of the equipment applied at the enterprise as well as jigs, fixtures and tools have been examined. Great attention was paid to fabricability of the group parts which is an important factor of the process design.

The parts were classified depending on their structural and technological characteristics. As a result the parts have been divided into certain types, i.e. classes of parts subjected to common machining operations or manufactured with the same equipment and machine setup, having similar elementary surfaces and installation geometry. Operation-routing sequence for a certain part is formed by selecting appropriate operations for machining the part elementary surfaces from the sequence developed for a template or complex part.

The work on developing the systematic classifier is divide into separate stages according to the parts being considered.

A systematic classifier of medium size shafts is described here to provide an example. Fig. 1 shows parts which were combined into one technological group based on the following main characteristics: similar surfaces or combination of surfaces to be machined, precision of machining, outline dimensions or the possibility to machine differently shaped parts with the same equipment and machine setup. This classification will allow using the same tools and equipment when designing the manufacturing process and accelerate computer-aided design.

After that a complex part which comprises all main structural elements is developed (the part shown in Fig 1 can be considered as a complex part). This part includes all the elements which the parts of the group can contain, for example holes, bevels, notches allowing precise machining.

Then based on major manufacturing process different operation-routeing sequences of machining the parts were developed depending on thermal treatment, some specific operations [15], enterprise special conditions, etc.

Results and Discussion

Modern computer-aided design systems use two methods of manufacturing process design:

IOP Conf. Series: Materials Science and Engineering **125** (2016) 012030 doi:10.1088/1757-899X/125/1/012030

- the first method consists in using a general technology reference guide which contains descriptions of template solutions of manufacturing processes.

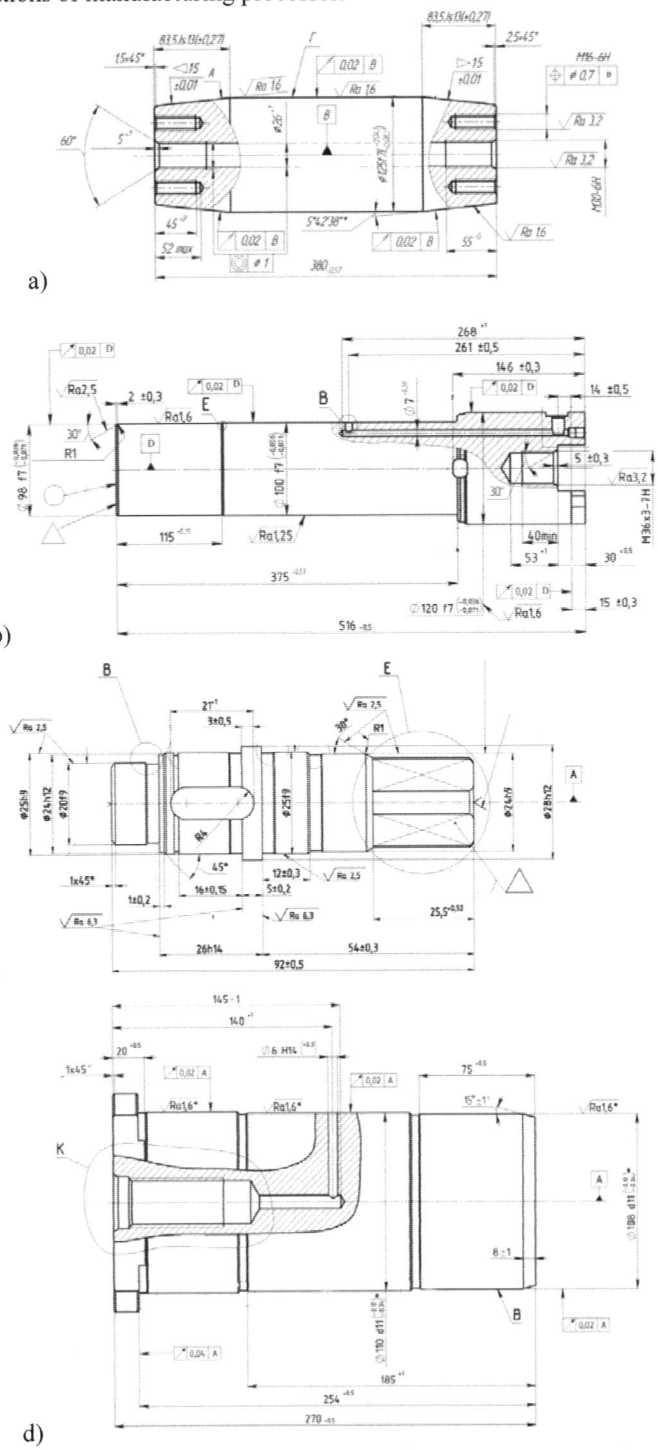

Fig. 1. Parts combined in a group

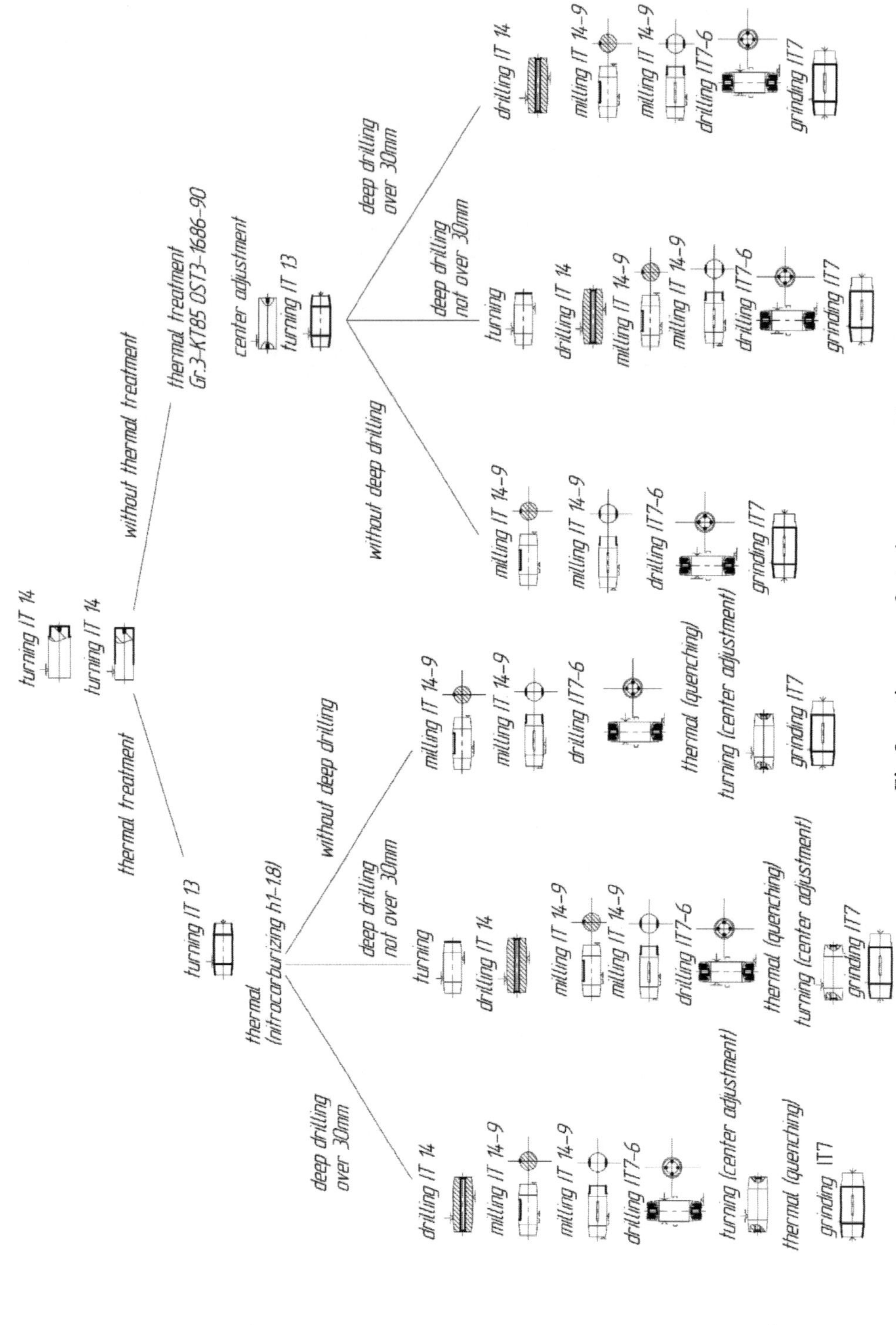

Fig.2 routing manufacturing processes tree

The solutions can form the whole manufacturing process, some of its parts or a separate operation or step.

- the second method uses "Technology Tree" application. With this method the design consists in copying operations from analogue manufacturing processes and their modification to make up a required manufacturing process.

In our case the systematic classifier structurally corresponds to the method using a general technology reference guide. The graphical part consists of a description of template solutions and application conditions.

The operation routeing sequence shown in Fig.2 as a tree-like structure is designed for machining a complex part.

Further on to design a manufacturing process for a new part we should determine its type that is to determine what group of parts characterized by common machining operations or manufactured with the same equipment and machine setup it belongs to. When developing the operation routeing sequence for parts included in some group, the tree branch to be followed is chosen according to the type of thermal treatment, parameters of deep drilling of holes in the parts, equipment to be applied for machining, standard sizes of structural parts.

All parts are divided into two groups depending on whether they are subjected to thermal treatment. Then each group is divided into three subgroups depending on whether deep drilling is applied. Fig.3 shows examples of deep drilling of holes with different diameters. When a hole diameter is less than thirty millimeters drilling is performed by UTB-16 deep drilling machine-tool. Drilling is preceded by end face preparation which consists in cross cutting with minimal deviation of flatness and axis perpendicularity. When a hole diameter is more than thirty millimeters drilling is performed by Rheinmetall machine-tool with relocation of the part. In this case end face preparation is not required but a hole which will serve as a guideway for further machining.

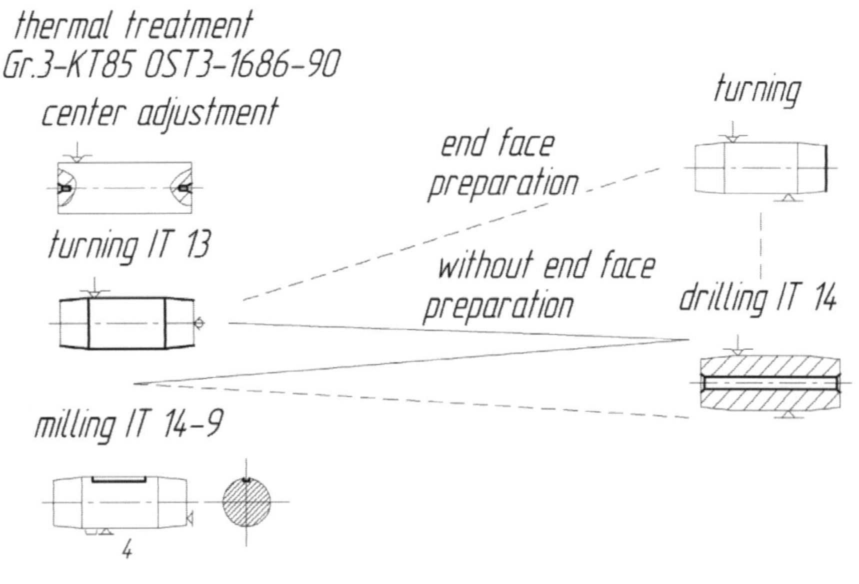

Fig.3 Variants of hole making

3. Conclusion

The example described above shows that based on the developed systematic classifier of manufacturing processes it is possible to deliberately increase production output in small lot and single piece production following the given procedure. It provides the prospects for designing and applying special batch equipment for parts manufacturing to increase accuracy and productivity of the machining process. It reduces the use of multipurpose equipment and the time period required for designing the manufacturing process.

The work can serve the basis for increasing efficiency of manufacturing preparation, extending shop floor material logistics, enlarging regulatory reference data base, improve employees' production skills.

References

[1] Efremenkov A.B. 2011 Forming the subterranean space by means of a new tool (geohod) 6th International Forum on Strategic Technology (IFOST - 2011). – IEEE. – Vol. 1. – pp. 348-350.

[2] Wahaba M.I.M., Stoyanb S.J. 2008 A dynamic approach to measure machine and routeing flexibilities of manufacturing systems *International Journal of Production Economics* Vol. 113 Issue 2 June p.p. 895–913

[3] Skhirtladze A.G., Bogodukhov S.I., Suleimanov R.M., Bondarenko E.V., Manufacturing processes in machine-building. – M:Mashinostroenie, 2009. – 640p.

[4] Waguih ElMaraghya, Hoda ElMaraghya, Tetsuo Tomiyamac, Laszlo Monostorid 2012 Complexity in engineering design and manufacturing *CIRP Annals - Manufacturing Technology* Vol. 61. pp. 793–814.

[5] Mitrofanov S.P. Batch technology of machine-building production: Volume 1. L.: Machinostroenie, 1983. – 403p.

[6] Mitrofanov S.P. Batch technology of machine-building production: Volume 2. L.: Machinostroenie, 1983. – 375p.

[7] Gallagher C.C., Knight W.A. Group Technology, Production Methods in Manufacturing, Ellis Horwood, Chichester, England. 1986.

[8] Muhammad Kutub Uddin, Kripa Shanker 2002 Grouping of parts and machines in presence of alternative process routees by genetic algorithm *International Journal of Production Economics* Vol. 76 pp. 219-228.

[9] Ciurana J., Garcia-Romeu M.L., Castro R., Alberti M. 2003 A system based on machined volumes to reduce the number of routee sheets in process planning *Computers in Industry* Vol. 51. pp. 41 – 50.

[10] Young Choon Lee, Hyuck Han, Albert Y. Zomaya, Mazin Yousif 2015 Resourceefficient workflow scheduling in clouds *Knowledge-Based Systems* Vol. 80 p.p. 153-162.

[11] Aksenov V.V., Walter A.V., Gordeyev A.A., Kosovets A.V. 2015 Classification of geokhod units and systems based on product cost analysis and estimation for a prototype model production *IOP Conference Series: Materials Science and Engineering* Vol. 91 P. 012088

[12] Ovcharenko V E 2012 Evolution of the structure of plasma metal-ceramic coating under pulsed electron-beam treatment *Inorganic Materials: Applied Research* Vol.3 pp. 210-215

[13] Lasukov A.A. 2015 Selection of machining conditions in terms of the temperature dependence of chip formation, *Russian Engineering Research*, Vol. 35, Issue 9, pp. 679–681

[14] Nekrasov R. Yu., Starikov A. I., Lasukov A.A. 2015 Entering the operative correcion machining processes CNC, *IOP Conf. Series: Materials Science and Engineering* Vol. 91 012042

[15] Kovalevskaya Z G, Klimenov V A, Zaitsev K V 2014 Interfacial adhesion between thermal spray coating and substrate achieved by ultrasonic finishing. *Applied Mechanics and Materials*. 682, pp. 459-463.

Integrated Processing: Quality Assurance Procedure of the Surface Layer of Machine Parts during the Manufacturing Step "Diamond Smoothing"

V Yu Skeeba[1], V V Ivancivsky[1], D V Lobanov[2], A K Zhigulev[1] and P Yu Skeeba[1]

[1] Department of Industrial Machinery Design, Novosibirsk State Technical University, 20 Prospekt K. Marksa, Novosibirsk, 630073, Russia
[2] Department of Mechanical Engineering, Bratsk State University, 40 Makarenko st., Bratsk, 665709, Russia

E-mail: skeeba_vadim@mail.ru

Abstract. The present study has found that during the integrated processing after the diamond smoothing, in the surface-hardened sample a cold-worked layer 0.01 ... 0.02 mm in thickness, the microhardness value of which reaches 868 HV, is formed. The intensity of compressive stresses on the part surface increases to σ_τ = -678 MPa. The analysis of the experimental data has shown the relationship between the parameter Ra and the processing modes that can be used during diamond smoothing, based on the high performance and the desired surface roughness. It has been found that the minimum value of roughness Ra = 0.18±0.08 μm is reliably achieved by smoothing processing when the smoothing force P_y ranges from 100 N to 150 N.

1. Introduction

Production of competitive and high-quality products is the main direction for the development of modern engineering. In this regard, in the machine tool industry close attention is focused on developing a new type of processing equipment – hybrid manufacturing systems processing multifunctional capabilities [1 - 7]. Integration of different processes in one machine: laser surface hardening combined with mechanical treatment [8, 9], the abrasive grinding - surface hardening [10 - 12], cutting - hardening [13], turning - quenching - grinding, is caused by the desire of developers to expand technological capabilities of machines and to ensure their autonomy in a flexible mechanical production [14, 15]. This method allows not only to achieve high levels of resource and energy saving, but also to provide the appropriate level of processing performance and quality of the items [16]: the accuracy of the shape, size and arrangement of the surface, roughness and the specified physical and mechanical properties of the surface layers and the material.

The object of this research is the process of manufacturing machine parts, consisting of the following operations: pre-machining (rough turning), surface hardening (high-energy high-frequency heating) and finish machining (finish turning and diamond smoothing).

Since it is the finishing operations which form in the surface layer of parts optimal combination of quality parameters (surface roughness, size and distribution of microhardness and residual stress, microstructure of the surface layer, and others.), determining the performance of products [17 - 21], the aim is to study the quality of the surface layer, achieved during the diamond smoothing in the integrated processing environment.

2. Materials and methods

A cylindrical rod (diameter d = 50 mm, length l = 70 mm), made of steel 45, is used as a test sample. The feed material was checked by the optical emission spectrometer *ARL 3460*. The results are shown in Table 1.

For the integration of mechanical and surface heat treatments a screw-cutting lathe model *UT16PM* has been upgraded, i.e. it has been equipped with additional concentrated source of energy, for which an external quenching circuit implementing high-energy heating by high frequency currents (HEH HFC) and characterized by high thermal coefficient of efficiency, has been used. As an energy source generator *HFG 6 - 60 / 0.44* with the current operating frequency ω = 440 kHz has been used.

When turning, HEH HFC hardening and diamond smoothing, the sample was fixed in a three-jaw gripper with the center in abutting relation to the screw-cutting tailstock.

Table 1. Chemical compositions of steel 45

Elements	C	Si	Mn	S	P	Cr	Ni	Cu
Weight, [%]	0.44	0.23	0.61	0.013	0.019	0.11	0.15	0.17

Rough turning was carried out with a replaceable insert straight turning tool (the material of the insert T15K6) under the following conditions: cutting speed $V_d \approx 92$ m min^{-1} (work rotational speed $n_d = 588$ min^{-1}); feed per revolution $S_{rev} = 0.35$ mm; cutting depth t = 1 mm.

When surface hardening, the loop-type inductor, equipped with brand ferrite N87, was used. The heating process was carried out on a progressive feed method (the thickness of the hardened layer did not exceed the depth of penetration of the current into the hot metal - 0.6 ... 0.8 mm) [22, 23]. The studies were conducted using heavy water spray cooling of the surface within the following range of processing modes: the specific source power $q_s = (1.5 - 4.0)\cdot 10^8$ W m^{-2}, the travelling speed of the parts under the inductor $V_p = (0.05 ... 0.1)$ m s^{-1} (corresponding to $n_d = 19.9 ... 39.8$ min^{-1}). The width of the active conductor of the inductor B_i was 2 mm, the processing was carried out with a gap Δ = 0.1 ... 0.2 mm.

Finish turning was carried out with an indexable-insert straight turning tool (the material of the insert - oxycarbide ceramic insert based on the composition Al_2O_3-TiC). Cutting modes: cutting speed $V_d \approx 133$ m min^{-1} (work rotational speed $n_d = 882$ min^{-1}); feed per revolution $S_{rev} = 0.025$ mm; cutting depth t = 0.01 ÷ 0.015 mm. When rough and finish turning, a cooling lubricant was sulphonated mineral oil "*Sulfofrezol*".

Diamond smoothing was based on the two-pass scheme using designed and manufactured insert holder with elastic fixing of the diamond indentor (Technical Regulations 2-037-631-88) with the radius *R* = 1 mm. The radial component of the smoothing force P_y considering the stiffness of the processing equipment, the surface layer hardness of the processed part after the HEH HFC surface hardening (HV = 700 ... 800) and the radius of the diamond areas, respectively, was equal to 50, 100, 150, 200 N. In this the surface meters per minute of the processed part was V_{ds} = 25.3; 33.2; 41.9 m min^{-1} (work rotational speed n_d = 168; 220; 278 min^{-1}); and the feed per revolution - S_{ds} = 0.018 ... 0.08 mm. A cooling lubricant for the diamond smoothing was industrial oil *I-20A*.

Structural studies of the samples were performed on an optical microscope *Carl Zeiss Axio Observer Z1m* and on a scanning electron microscope *Carl Zeiss EVO 50 XVP*, equipped with energy dispersive analyzer *INCA X-ACT (Oxford Instruments)*. The microstructure of the samples was revealed by etching with a 5% ethanolic solution of nitric acid and a saturated solution of picric acid in ethanol with surfactants [24].

Microhardness of the hardened surface layer of parts was evaluated on the instrument *Wolpert Group 402MVD*. Residual stress research was carried out with application of the X-ray method on

IOP Conf. Series: Materials Science and Engineering **125** (2016) 012031 doi:10.1088/1757-899X/125/1/012031

the high-resolution diffractometer *ARL X`TRA* and the mechanical destructive method: the layered electrolytic etching of the sample [25, 26]. For the detection of defects in the surface layer in each manufacturing operation we used: visual and optical method using a microscope *Carl ZeissAxio Observer A1m*, capillary method, eddy current method using eddy current flaw detector *VD - 70*. In the study for the simultaneous measurement of form deviations, waviness and roughness of the surface we used profilograf - profilometer *Form Talysurf Series 2* of *Taylor Hobson Company*. The evaluation of the surface topography was carried out on a laser profilograph-profilometer *Zygo New View 7300*.

Statistical analysis of the experimental studies results was performed in software products *Statistica, Table Curve 2D* and *Table Curve 3D*.

3. Results and discussion

Studies of surface quality after the first manufacturing operation of the integrated processing showed that in the process of rough turning we get a defect-free surface the roughness value of which according to the parameter *Ra* is 3.3±0.7 μm.

At the second manufacturing operation we carry out HEH HFC surface hardening, maintaining the required depth of hardening h = 0.6 mm and rational nature of the residual stresses distribution; the specific power (q_s [W m^{-2}]), and the speed of parts movement (V_p [m s^{-1}]) was carried out by solving the equations system h (q_s, V_p) and Ψ (q_s, V_p) [27, 28]:

$$\begin{cases} h(q_s, V_p) = a + bV_p + cq_s + dV_p^2 + eq_s^2 + fV_p q_s + gV_p^3 + hq_s^3 + iV_p q_s^2 + jV_p^2 q_s, \\ \Psi(q_s, V_p) = k + lV_p + mq_s + nV_p^2 + oq_s^2 + pV_p q_s + rV_p^3 + sq_s^3 + tV_p q_s^2 + uV_p^2 q_s, \\ 0{,}25 \le \Psi(q_s, V_p) \le 0{,}35. \end{cases} \quad [1]$$

where a = 0.426008, b = 2.827121, c = $3.025072 \cdot 10^{-9}$, d = -301.591960, e = $-4.694423 \cdot 10^{-18}$, f = $3.600666 \cdot 10^{-8}$, g = 1953.668810, h = $3.216427 \cdot 10^{-27}$, i = $1.375401 \cdot 10^{-17}$, j = $-3.779403 \cdot 10^{-7}$, k = 0.087564, l = -7.429933, m = $1.062284 \cdot 10^{-8}$, n = 235.19293, o = $-3.424286 \cdot 10^{-17}$, p = $-8.850919 \cdot 10^{-8}$, r = -1309.3045, s = $2.9423 \cdot 10^{-26}$, t = $1.403793 \cdot 10^{-16}$, u = $1.010925 \cdot 10^{-7}$ - coefficients of functional dependencies for 45 steel.

Figure 1 is a graphical solution of this problem. When hardening Steel 45 to a depth of 0.6 mm the range of the recommended conditions is limited within the points A and B on the curve (black solid line): while q_s = (3.0 ... 3.4) 10^8 W m^{-2}, V_d = (0.072 ... 0.081) m s^{-1}. The determined processing modes (shaded area) guarantee a required depth of hardening and rational value of the transition zone.

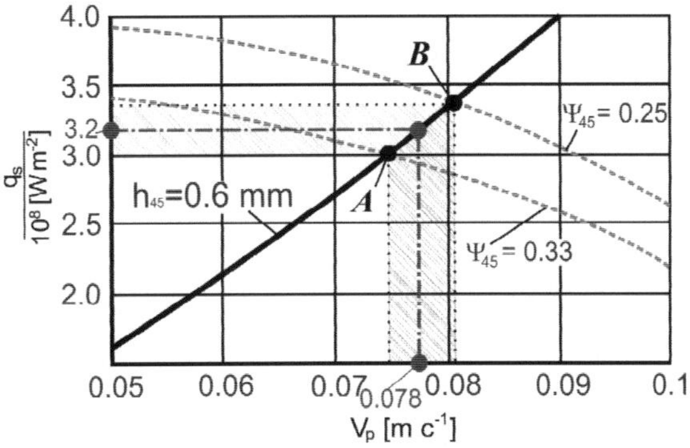

Figure 1. The dependence of the source specific power on its travelling speed during the HEH HFC hardening of Steel 45 to a depth h_{45} = 0.6 mm

The job setting was carried out in the following modes: $q_s = 3.2\ 10^8$ W m^{-2}, and $V_p = 0.078$ m s^{-1} (work rotational speed $n_d \approx 31$ min^{-1}). To ensure the required spindle rotational speed the main drive has been upgraded, that is equipped with the frequency converter *HF Inverter model F1500-G0015S2B.*

Microhardness pattern of samples in the cross section is shown in Figure 2. The mean level of the hardened after HEH HFC layer microhardness was 739 HV, wherein the base material microhardness was 204 HV.

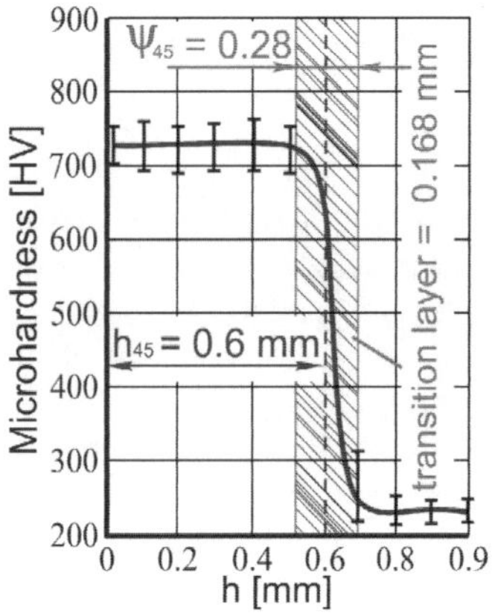

Figure 2. Microhardness distribution in the surface layer of Steel 45 after HEH HFC hardening: the mode $q_s = 3.2 \cdot 10^8$ W m^{-2}, $V_p = 0.078$ m s^{-1}, $B_i = 2$ mm

The maximum level of the surface compressive residual stress (RS) was σ_τ = -560±20 MPa. The peak tensile stress reached σ_τ = 75±45 MPa and is at a depth of ~ 0.8 mm.

Finish turning was carried out in the part size d = 48 mm, the surface roughness value of the parameter Ra was 1.2±0.2 μm. Surface microhardness and residual stress were at the level achieved in the transition zone HEH HFC.

The process of diamond smoothing forms a plastically deformed surface, topography of which is characterized by a rounded shape of the microrelief without scoring and snatching.

Figure 3 shows the dependence of the effect of speed V_{ds} and feed S_{ds} on roughness parameter *Ra* for different values of the smoothing force P_y. As it can be seen from the figures, the increase in V_{ds} and S_{ds} in the studied ranges of mode parameters results in the increase of the surface roughness. To a greater extent it is the change of the feed S_{ds} that influences the surface pattern.

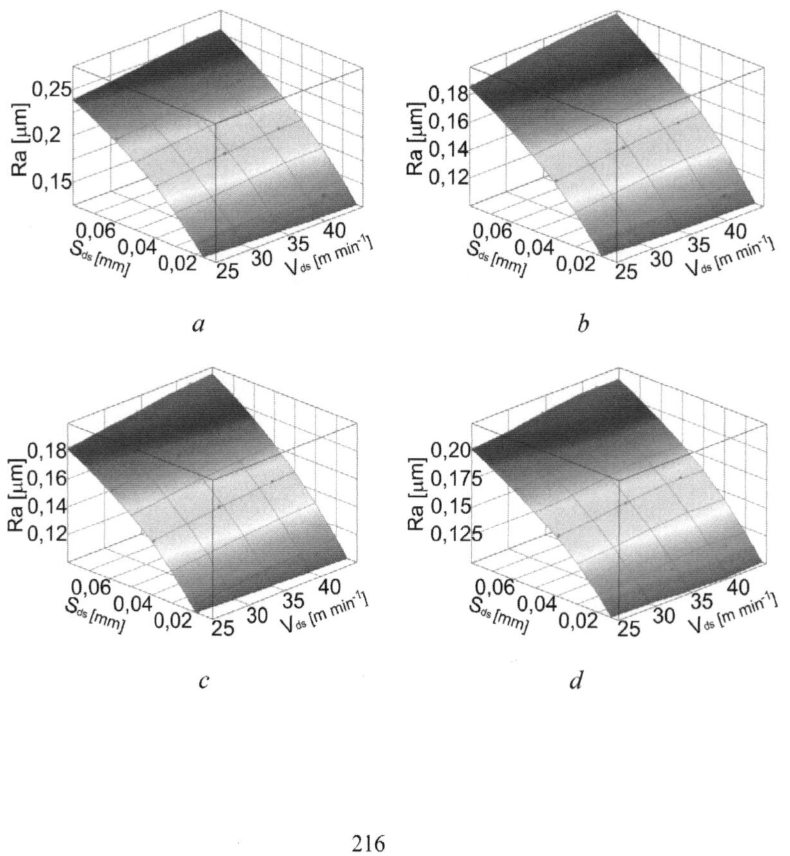

In the study of the smoothing effect it was revealed that in the section from 50N to 100N we observed a high reduction of the roughness parameter Ra (Figure 4). In the interval [100N; 150N] the value of the Ra stabilized at the level of 0.18±0.08 μm: initial surface pattern was almost completely leveled, and the surface roughness depends on the parameters V_{ds} and S_{ds}. For values Py = 100N and Py =150N the functional relationship *Ra* (V_{ds}; S_{ds}) was determined:

$$Ra(V_{ds},S_{ds})=a+bV_{ds}+cS_{ds}+dV_{ds}^{2}+eS_{ds}^{2}+fV_{ds}S_{ds}+gV_{ds}^{3}+hS_{ds}^{3}+iV_{ds}S_{ds}^{2}+jV_{ds}^{2}S_{ds}, \quad [2]$$

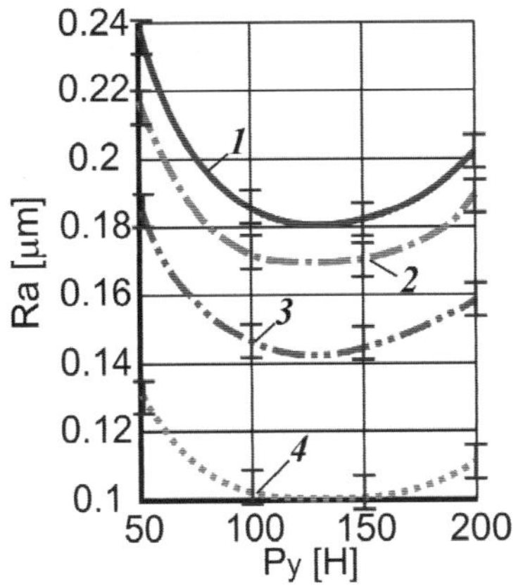

Figure 4. The dependence of the roughness parameter *Ra* on smoothing force P_y : 1 – V_{ds} = 25.3 m min^{-1}, S_{ds} = 0.08 mm; 2 – V_{ds} = 33.2 m min^{-1}, S_{ds} = 0.06 mm; 3 – V_{ds} = 33.2 m min^{-1}, S_{ds} = 0.04 mm; 4 – V_{ds} = 25.3 m min^{-1}, S_{ds} = 0.018 mm.

where

for P_y = 100 N:

a = 0.098931, b = -0.00422, c = 2.512088, d = 0.000135, e = -23.857743, f = 0.020124, g = -1.326322·10^{-6}, h = 105.49039, i = -0.028432, j = -0.000193;

for P_y = 150 N:

a = 0.070639, b = -0.002061, c = 2.729073, d = 7.897542·10^{-5}, e = -23.406582, f = 0.005446, g = -8.396753·10^{-7}, h = 92.631383, i = -0.001955, j = -1.421271·10^{-5}.

A further increase in the values of P_y enhances the parameter *Ra.* These results are coherent with the materials presented in [29 - 32].

The minimum value of the roughness parameter *Ra*, which had been recorded while processing in the modes V_{ds} = 25.3 m min^{-1}; S_{ds} = 0.018 mm; P_y = 150 N, was ~ 0,1 μm (Figure 5). Structural studies carried out on a scanning electron microscope showed that in the process of diamond smoothing in a surface hardened sample a hardened (cold-worked) layer with a thickness of 0.01 ... 0.02 mm was formed (Figure 6). Thus, microhardness and compressive stress in the surface layer increase to the values of 868 HV and σ_τ = -678±20 MPa, respectively (Figure 7).

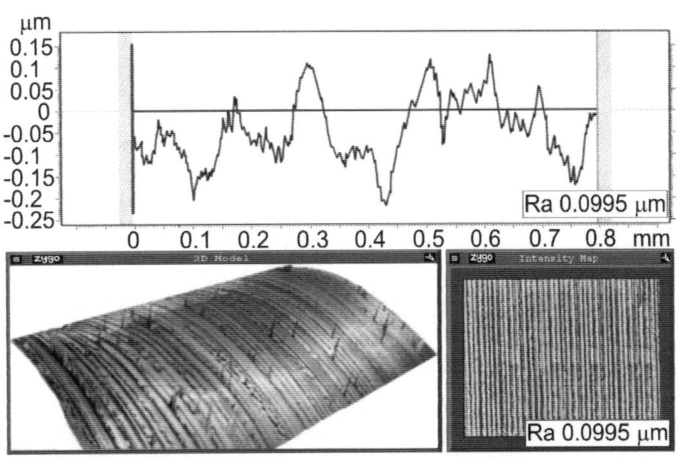

Figure 5. Surface profilogram and topography of the part after diamond smoothing: V_{ds} = 25.3 m min^{-1}; S_{ds} = 0.018 mm; P_y = 150 N; R = 1 mm; $Ra_{initial}$ = 1.2±0.2 μm; $HV_{initial}$ = 739.

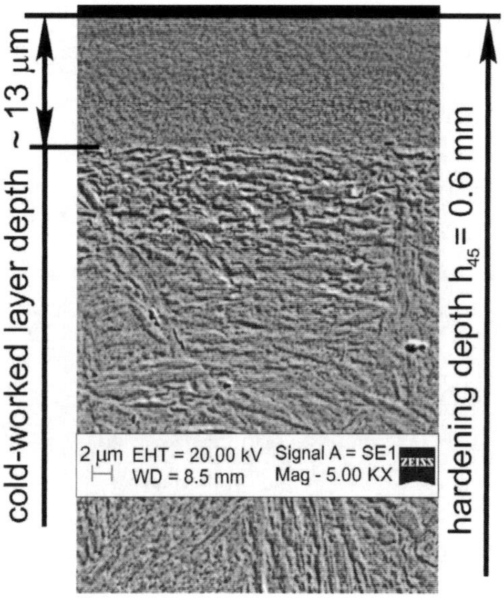

Figure 6. The microstructure of the surface layer after diamond smoothing

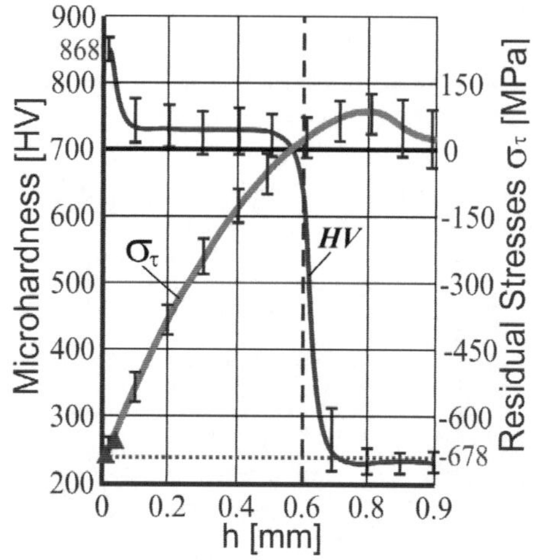

Figure 7. Microhardness and residual stress distributions in the surface layer of Steel 45 after diamond smoothing.
▲ - Residual stresses, obtained by X-ray determination method.

4. Conclusions

Studies have shown that diamond smoothing performed as part of an integrated processing can significantly improve the quality of the surface layer of machine parts. It was established experimentally that when completing the final transition with respect to the HEH HFC transition it becomes possible to increase the surface microhardness and the level of residual compressive stresses in the surface layer of the product by ~ 15 ... 20%. It was revealed that after diamond smoothing in the surface hardened sample a cold-worked layer with a thickness of 0.01 ... 0.02 mm, the microhardness of which is ~868 HV, is formed, while the level of compressive stress in the surface layer increases to σ_τ = -678±20 MPa. The rational range of the smoothing force $P_y \in [100N; 150N]$ has been determined, wherein the minimum value of the roughness parameter Ra = 0.18±0.08 µm is guaranteed. The resulting functional relationship $Ra(V_{ds}, S_{ds})$ allows determining the modes of diamond smoothing, based on the high performance and the desired surface roughness.

Acknowledgements
The study was partially supported by RFBR, research project No. 13-08-01102 a.

References

[1] Garro O, Martin P and Veron M 1993 CIRP Annals – Manufacturing Technology 42 433–6
[2] Nagae A 2004 Proc. 11th Int. Conf. on Machine Tool Engineers (Tokyo) pp 312-23
[3] Moriwaki T 2008 CIRP Annals – Manufacturing Technology 57 736–49
[4] Makarov V M 2011 RITM: Repair. Innovation. Technologies. Modernization 64 20-3
[5] Zhu Z, Dhokia V G, Nassehi A and Newma S T 2013 International Journal of Computer Integrated Manufacturing 26 596-615
[6] Lauwers B, Klocke F, Klink A, Tekkaya A E, et al 2014 CIRP Annals – Manufacturing Technology 63 561-83
[7] Skeeba V Yu, Pushnin V N, Erohin I A and Kornev D Yu 2015 Materials and Manufacturing

Processes 30 1408-11
[8] Brecher C, Hoffmann F, Karlberger A and Rosen C J 2008 41st CIRP Conf. on Manufacturing Systems (Tokyo) (London: Springer) pp 425-8
[9] Ding H T, Shin Y C 2010 International Journal of Machine Tools and Manufacture 50 106-14
[10] [Ivancivsky V V, Skeeba V Yu 2010 Obrabotka metallov (tekhnologiya, oborudovanie, instrumenty) 49 15-21
[11] Skeeba V Y, Ivancivsky V V, Zub N P and Turevich S V 2010 Innovative Activities 10-1 66-9
[12] Skeeba V Y, Ivancivsky V V, Zub N P and Turevich S V 2010 The World of Scientific Discoveries 2-3 91-3
[13] Salonitis K, Chondros T and Chryssolouris G 2008 The International Journal of Advanced Manufacturing Technology 38 48-58
[14] Bushuev V V 2008 Bulletin of MSTU "Stankin" 1 8-13
[15] Pushnin V N, Erokhin I A, Kornev D Yu and Skeeba V Yu 2014 Actual problems in machine building 1 245-55
[16] Ptitsyn S V, Skeeba V Yu, Chesov Yu S and Merezhko E V 2013 Obrabotka metallov (tekhnologiya, oborudovanie, instrumenty) 59 33-8
[17] Martyushev N V, Petrenko Y N 2014 Applied Mechanics and Materials 682 231-5
[18] Martyushev N V 2014 Advanced Materials Research 880 163-7.
[19] Yanyushkin A, Lobanov D, Arkhipov P and Ivancivsky V 2015 Applied Mechanics and Materials 698 351-4
[20] Arkhipov P, Yanyushkin A, Lobanov D and Petrushin S 2013 Applied Mechanics and Materials 379 124-30
[21] Lobanov D V, Yanyushkin A S 2011 Russian engineering research 31 236-9
[22] Skeeba V, Pushnin V, Erokhin I, Kornev D 2014 Obrabotka metallov (tekhnologiya, oborudovanie, instrumenty) 64 90-102
[23] Skeeba V, Ivancivsky V and Pushnin V 2015 Applied Mechanics and Materials 698 288-93
[24] Vander Voort G F 2004 ASM Handbook. Vol. 9: Metallography and microstructures (Ohio: ASM International)
[25] Totten G E, Howes M and Inoue T 2002 Handbook of residual stress and deformation of steel (Ohio: ASM International)
[26] Sharpe W N 2008 Springer handbook of experimental solid mechanics (Leipzig: Springer Science)
[27] Ivancivsky V V, Skeeba V Yu and Stepanova N P 2005 Obrabotka metallov (tekhnologiya, oborudovanie, instrumenty) 28 22-4
[28] Ivancivsky V V, Skeeba V Yu and Stepanova N P 2005 Obrabotka metallov (tekhnologiya, oborudovanie, instrumenty) 33 17-9
[29] Koval'nogov V N 2009 Russian Engineering Research 29 747-9
[30] Shvetsov A N, Skuratov D L 2014 Vestnik of the Samara State Aerospace University 47 62-7
[31] Abul'khanov S R, Goryainov D S, Skuratov D L and Shvetsov A N 2015 Russian engineering research 35 147-9
[32] Kuznetsova V P, Tarasovc S Yu and Dmitriev A I 2015 Journal of Materials Processing Technology 217 327-35

IOP Conf. Series: Materials Science and Engineering **125** (2016) 012032 doi:10.1088/1757-899X/125/1/012032

Nanosized Borides and Carbides for Electroplating. Metal-Matrix Coatings: Specifications, Performance Evaluation

G V Galevskiy [1,a], V V Rudneva [1], S G Galevskiy [2,b], D P Il'yashchenko [3,c], D S Kartsev [3]

[1] Siberian State Industrial University, Russia, Kemerovo Region, Novokuznetsk,
[2] National Mineral Resources University, St. Petersburg region, St. Petersburg,
[3] Yurga Institute of Technology, TPU Affiliate, Kemerovo region, Yurga.

e-mail: [a]kafcmet@sibsiu.ru, [b]feanor1985@mail.ru, [c] mita8@rambler.ru

Abstract. This paper summarizes experience of application of nano-sized carbides and borides of titanium and chromium, silicon carbide as components of electro-depositable coating compositions based on nickel, zinc, and chromium. Basic physical and mechanical properties of the coatings are determined. Technological and economic evaluation is completed; practicability of high-cost nano-diamonds substitution for nano-sized borides and carbides is justified.

1. Introduction.

Rise of operational requirements for materials is constantly encouraging developments in the field of surface hardening. Therefore, currently an urgent problem is development of environmentally friendly new techniques of material hardening by means of changing structural-phase state and physical-mechanical specifications of surface layers.

These techniques have some advantages in comparison with conventional electroplating, thermo-mechanical and chemical-thermal processes as well as with methods of concentrated energy flux exposure. Developers of innovative hardening techniques turn to high-temperature ultra-hard materials (diamond, borides, carbides, nitrides and others). Analysis of scientific and applied studies on various methods of surface hardening and cooperation with authors and developers, confirms technological and economic practicability of use of nanoscale materials, i.e. powders with a particle size less than 100 nm.

The objective of this work is analysis and synthesis of experience and results of use of nano-dispersed borides and carbides of chromium, titanium and silicon as components of galvanic composite metal-matrix coatings.

2. Manufacture of nanopowders borides and carbides

Electroplating is extensively used in modern engineering technology. Electroplating conventional techniques and equipment have almost reached the peak of development. Modern industry requires new techniques of surface hardening of machine parts and mechanisms, giving them increased mechanical, corrosion-resistant, anti-friction and aesthetic properties. This problem can not be solved by conventional electroplating. Therefore active developments of innovative hardening methods are conducted.

Galvanic composite coating (GCC) is one of the innovative techniques. These coatings are deposited as a thin electrolyte layer, containing a dispersed phase, to items with conductive surface. New composite coatings are based on matrices of chromium, nickel, zinc, copper, gold, silver and others. Composite plating technology has been developed by scientific Schools of professors R.S. Saifullin (KSTU), G.V. Khaldeev (PSU), T.E. Tsupak (UCTR) V.Yu.Dolmatov (SPbSPU).

1 - 10 micron micropowders of diamonds, carbides, borides, nitrides, and silicides of transition metals are conventionally used for composite electroplating. Powders of such granularity significantly increase hardness and wear resistance of composite coatings. However, relatively large particles are unevenly distributed in the matrix volume; that causes increased composite porosity and, as a consequence, reduces thermal resistance, heat resistance and corrosion resistance. These weaknesses have questioned practicability of nano-dispersed diamonds, borides and carbides use for electroplating.

[1-5] represent nanodiamond synthesis technology and application. Manufacturing involves detonation synthesis, product chemical cleaning and conditioning. A nanodiamond is a complex object with three-layer structure consisting of 4-6 nm diamond core, a transition shell of 0.4-1.0 nm X-ray amorphous carbon structures, a surface layer consisting of carbon, nitrogen, oxygen and hydrogen. Acceptable content of the main impurity, oxidized carbon, is 0.5-1.0%. Impregnared in materials, nanodiamonds are structure-forming agents, providing precipitation hardening of the composition.

Relatively low cost of nanodiamonds, compared with static synthetic diamonds, aids optimal "quality-to-price" ratio. A positive effect is achieved by impregnating as much as 1.0% of nanodiamonds into the material. There are many areas of application of nanodiamonds; these are three main ones: finish polishing (approximately 70%), electroplating (25%) and oil compositions (about 5%). Nanodiamonds are used in electrochemical composite coatings based on chromium, nickel, tin, zinc, copper, gold, silver, aluminum, iron, various alloys.

Nanomaterials, impregnated into electroplated coatings, greatly increase micro-hardness, wear resistance, corrosion resistance, improve exterior view and reduce porosity. Along with the advantages of nanodiamonds in GCC technology there is a number of significant weaknesses. Professionals-electroplaters note an extremely high level of dispersion, specified by the manufacturing method, because of which the nanodiamonds are supplied as an aqueous suspension; a significant tendency to coagulation; rapid loss of sedimentation stability; multioperational complex technology of electrolyte preparation and operation; chrome-diamond coating limited lifetime, caused by oxidation of nanodiamond particles, impregnated to the surface layers of chromium matrix, at temperatures above 473-573 K.

Manufacturing technology for nanopowders of borides and carbides of titanium, chromium and silicon is described in [6]. The procedure includes plasma synthesis and, if appropriate, refining.

Table 1 shows parameters of boride- and carbide-formation in conditions of the three-jet vertical continuous-flow reactor with 150 kW capacity as well as general properties of borides and carbides. It can be seen that basic phase content and dispersion grade of the materials comply with specifications for components of metal-matrix composite coatings:

- 30-60 nm dispersion grade, wherein particles in the matrix are distributed micro-uniformly, stability of depositing suspensions, and the most complete activity of the particles (chemical interaction with the matrix, diffusion processes, material hardening and corrosion resistance improving);
- homogeneity of phase composition, which determines the identity of particle behavior in the electrolyte and coating, local reproducibility of performance properties of the coating;
- homogeneity of chemical composition, with a minimum amount of impurities, primarily free carbon (there are some technical difficulties when using nano-carbides with a high content of free carbon);
- controlled state of particle surfaces (gas content, presence of oxide films, etc.), which significantly impact on the development of liquid coalescence in the electrolyte-suspension;

- high corrosion resistance provides constant suspension compositions and electroplating parameters;
- high thermal stability provides scaling resistance.

Metal-matrix composite coatings with nano-components

The following are basic specifications of coatings with nano-components: chrome, nickel - silicon carbide, nickel - carbonitride, boride, chromium, nickel - carbide, titanium boride, zinc - chromium boride (Table. 2). Electroplating conditions, structure and properties of coatings are described in detail in [6-11]. Analysis of the results reveals identity of procedures of composite chromium plating, nickel plating and galvanizing as well as some basic factors of improving of coating service properties:

– boride and carbide nanopowder particles are spherical or oval without sharp edges; have high chemical adsorption activity, form electrolyte-suspensions, stable to sedimentation and coagulation; the particles, because of their small size and weight, are effectively transferred to the surface;

– during electroplating, suspended boride and carbide particles interact with a growing surface sediment through hydrodynamic, molecular and electrostatic forces; that leads to formation of the composite coating;

– when electroplating of chromium, nickel and zinc, nanoparticles are centers of metal crystallization; in consequence of large number of particles, crystallization has a multicelled nature; and the resulting coatings have small dimensions of structural fragments, specific matt appearance and are almost non-porous;

- aggregate of almost non-inertial particle mass transfer and mass crystallization ensures uniform depositing of metal coatings on equipotential surfaces;

– a small size of metal crystallite particles provides an exact copy of the surface micro-topography; that increases the general surface and composite coating adhesion to the substrate;

– coating quality, namely corrosion resistance and micro-hardness, is achieved at a low boride and carbide content in the coating (0.5 - 1.0)%, thereby reducing production costs.

Therefore, unlike micropowders, boride and carbide nanopowders are not only filling matter and second phase, but strong modifiers or structure-forming agents in the metal electro-crystallization procedure. They form a superfine non-porous coating structure with improved corrosion resistance, wear resistance and micro-hardness.

Table 1. Parameters of boride- and carbide formation and general specifications of nanopowders

Synthesis parameters and Specifications	CrB_2	$Cr_3(C_{0,8}N_{0,2})_2$	TiB_2	TiC	SiC
Composition of gas - heat-transfer agent, % vol.: - Nitrogen/hydrogen/methane	74/25/1	99/–/1	74/25/1	99/–/1	99/–/1
Technology option of synthesis	$Cr+B+H_2$	$Cr+CH_4$	$Ti+B+H_2$	$Ti+CH_4$	$Si+CH_4$
Productivity on raw materials, kg / h	3.6	3.1	3.6	3.2	3.0
Boron amount in charge, % of stoichiometric	100-120	–	100-120	–	–
Carbide constituent amount, % of stoichiometric	–	120-140	–	120-140	120-140
Plasma flow initial temperature, K	н.м. 5400	н.м. 5400	н.м. 5400	н.м. 5400	н.м. 5400
Hardening temperature, K	2600-2800	2000-2200	2600-2800	2600-2800	2800-3000

IOP Conf. Series: Materials Science and Engineering **125** (2016) 012032 doi:10.1088/1757-899X/125/1/012032

Synthesis parameters and Specifications	CrB_2	$Cr_3(C_{0,8}N_{0,2})_2$	TiB_2	TiC	SiC
Phase content	CrB_2	$Cr_3(C_{0.8}N_{0.2})_2$	TiB_2	TiC	β-SiC
Basic phase content, %	92-93	92-93.5	92-93	93-93.5	91-92
Basic phase output, %	91-92	90.5-91.5	91.5-92.5	92-92.5	87-90
Productivity kg/h	3.0	3.4	3.4	3.7	4.05
Intensity kg/ $\cdot m^3$	1364	2010	1980	2105	2200
Specific surface, m^2/kg	33000-35000	31000-35000	46000-48000	33000-35000	40000-44000
Size * of particles, nm	42.0	34.0	36.0	35.0	55.0
Particle shape	spherica l	spherical	spherical	faceted cube	faceted
Oxidation ** of nanopowder $x10^7$, kg O_2/m^2	9.0-9.7	8.0-10.0	5.8-7.6	8.5-9.5	6.5-8.0

* - calculated on the specific surface value;
** - measured after exposure in air for 24 h.

Table 2. Basic physical and mechanical specifications of composite coatings with nanocomponents

Coatings	Specifications					
	Micro-hardness ±0,3 hPa	Adhesion , MPa	Wear resistance* *	Oxidation rate ** T=1173K	Inner stress, MPa	Corroding current, mcA/cm^2
Chrome	5.4	27.9-28.8	1			
Cr – NP SiC	9.7	29.3-30.4	2.5	*	*	*
Cr – MP SiC	8.9	28.4-29.2	2.2			
Nickel	2.2	29.1-30.3	1	2.55	1.36	0.17
Ni – NP	4.5	29.8-32.1	1.72	1	0.50	0.02
$Cr_3(C_{0,8}N_{0,2})_2$	3.2	28.9-30.2	1.54	1.34	0.78	0.11
Ni – MP Cr_3C_2	4.6	31.2-33.3	1.69	1	0.41	0.02
Ni – NP CrB_2	3.5	29.8-30.6	1.52	1.17	0.72	0.10
Ni – MP CrB_2	4.4	30.7-32.8	1.70	1	0.52	0.03
Ni – NP TiC	3.3	29.2-29.6	1.50	1.31	0.78	0.12
Ni – MP TiC	5.3	29.5-32.2	1.76	1	0.54	0.02
Ni – NP TiB_2	3.3	29.1-31.2	1.47	1.24	0.79	0.11
Ni – MP TiB_2						
Zinc	1.0	*	*	*	*	1.0***
Zn – NP CrB_2	1.2					2.3***

* - not specified
** - relative values
*** - corrosion resistance according to State Standard 9.308-85, relative values

NP – nanopowder
MP – micropowder

IOP Conf. Series: Materials Science and Engineering **125** (2016) 012032 doi:10.1088/1757-899X/125/1/012032

A comparison of characteristics of composite coatings based on chromium, nickel and zinc, with diamond, carbide and boride nano-components has been performed (Table. 1,2,3) to assess the possibility of the carbide and boride nanopowders substitution by nanodiamonds in chrome plating, nickel plating and galvanizing. It can be seen (Table 3) that chromium carbide coatings have wear resistance, micro-hardness, corrosion resistance, service life comparable to chrome-diamond coatings when exploited in temperatures above 473-573 K.

If nanodiamonds are substituted by nanoscale boride and chromium carbide when nickel plating and galvanizing, fine-grained coatings substantially nonporous will be obtained with similar corrosion resistance. Consequently, carbide and boride nanopowders can be used in galvanic composite chromium and nickel electroplating and galvanizing.

Numerous studies confirm technological and economic practicability of substitution of nanodiamonds by nano- borides and carbides in the composite nickel plating, chrome plating and galvanizing.

Under similar performance and service life of reinforced parts, such substitution results in the following manufacturing and economic advantages: the electrolyte-suspension preparation procedure is significantly simplified and accelerated; 2.5-fold deposition rate increase;1 m^3 electrolyte-suspension cost diminishes by a factor of 8-10. Substitution of 1 ton of nanodiamonds by boride and carbide nanopowders brings economic efficiency of 135 million rubles.

Table 3. Comparative Specifications of coatings based on chrome with silicon carbide SiC and diamond nanopowders

Electroplating conditions and achieved results		Technological options					
		1	2	3	4	5	6
Electrolyte composition, kg/m^3	Chromic anhydride	250	250	250	250	250	260
	Sulphuric acid	3	3	3	3	3	0.3 4
	trivalent Chromium	8	8	8	8	8	-
	Potassium fluorosilicate	-	-	-	-	-	19
	Barium sulfate	-	-	-	-	-	7
	NP SiC	4	6	7	8	10	-
	NP diamond	-	-	-	-	-	20
Deposition mode	Temperature, K	328	328	328	328	328	331
	Current density, A/dm^2	55	55	55	55	55	55
Results	Wear resistance*	0.96	1.0	1.05	1.07	1.08	1.0
	Micro-hardness, hPa	8.9	9.0	9.1	9.15	9.15	9.0
	Corrosion resistance *	0.96	1.05	1.07	1.10	1.10	1.0
	Service life at T above 473-573 K*	1.1	1.50	1.70	2.00	2.06	1.0
	1 m^3 electrolyte suspension cost *	0.13	0.16	0.18	0.20	0.23	1.0

Note: * Wear resistance, corrosion resistance, service life and 1 m^3 electrolyte suspension cost are given in relation to option 6, for which values of these indicators are taken as 1.0.

3. Conclusion

The results of the comparative analysis of nanoscale diamonds, borides and carbides use in galvanic metal-matrix composite coatings show that use of nanoscale borides and carbides with formed set of specifications instead of nano-diamonds ensures achievement of stable parametric, concentration, structural and economic advantages; that testifies their high competitiveness in the domestic and international electroplating

IOP Conf. Series: Materials Science and Engineering **125** (2016) 012032 doi:10.1088/1757-899X/125/1/012032

This work is executed in SibSIU within the scope of government assignment of Ministry of Education and Science of the Russian Federation № 11.1531/2014/K.

References

[1] Dolmatov V. Yu. 2005 Current Industrial Production Technology for Detonation Synthesis Nanodiamonts and Basic Fields of Their Application Proc. 4th International Conference on Materials Processing for Properties and Performance (MP3-IV), Tsukuba Science City, Ibraki, Japan. Singapore: Institute of Materials (East Asia) pp.284-286.

[2] Burkat G.K., Dolmatov V.Y., Jo S.K., Suh B.S., Kwon M.T., Orlova E.A. 2005 Appl. for Pat. Republic of Korea 10-2005-0037999. Composite Chrome-diamond coating, coating layer and manufacturing method Appl. 06.05.05.

[3] Loubnin E.N. Pimenov S.M. A. Blatter, Schwager F. and Detkov P.Ya. 1999 New Diamond Frontier Carbon Technology. № 4. Electroplating of gold-nanodiamond composite coatings pp. 273—282.

[4] Pyzyr A.P., Dolmatov V.Yu., et.al. 2004 Diamond and Related Materials. Destruction of cells in interaction with detonation nanodiamonds in experiments in vitro. №13 pp. 2020-2023.

[5] Ilyushin M.A., Tselinsky I.V., Ugryumov I.A., Dolmatov V.Yu., Shugalei I.V.. 2015 Central European Journal of Energetic, Warsaw, Materials 2, №1. Study of sub-micron structured energetic coordination metal complexes for laser initiation systems. pp. 21-33.

[6] Nozdrin I.V., Shiryaeva L. S., Galevskii G. V., Rudneva V. V.2011 Steel in Translation. Reactor for plasmometallurgical production of refractory borides and carbides. Vol. 41. №. 8 pp. 644 – 648.

[7] Efimova K.A., Galevskii G. V., Rudneva V. V., Kozyrev N.A., Orshanskaya E.G. 2015 Nikel-based Composition Coatig with Titanum Nanoboride Applied Mechanics and Materials. Vol.788. pp. 75-81.

[8] Malushin N.N., Osetkovsky V.L., Valuev D.V., Valueva A.V., Serikbol A.; 2014 J. Applied Mechanics and Materials Selection criteria parameter thermal cycle during surfacing heat resistance high hardness parts of metallurgical equipment 682 p. 69-74.

[9] Valuev D.V., Valueva A.V., Serikbol A.; 2014 J. Advanced Materials Research An apparatus for implementing the thermal cycle for plasma surfacing of metallurgical equipment,. 1040 p. 673-677.

[10] Zemenkov Yu D, Shalay V V and Zemenkova M Yu 2015 Immediate analyses and calculation of saturated steam pressure of gas condensates for transportation conditions Procedia Engineering 113 254-8.

[11] Zemenkov Yu D, Shalay V V and Zemenkova M Yu 2015 Expert systems of multivariable predictive control of oil and gas facilities reliability Procedia Engineering 113 312-5.

Effect of Initial Microtopography and Ultrasonic Treatment Mode on Steel Surface Layer Quality

K V Zaitsev [1, a], D V Lychagin [2], D A Arkhipova [1]

[1]Yurga Institute of Technology, TPU affiliate, 26, Leningradskaya Str., 652050, Yurga, Russia
[2]Tomsk State University, 36, Lenin ave, Tomsk, 634050, Russia

e-mail: [a]zaitsev81@rambler.ru

Abstract. The article presents results of studies on the effect of pre-lathed surface micro-relief on surface micro-hardness after ultrasonic plastic treatment, as well as the effect of ultrasonic treatment on structure and properties of surface layers of steels 20 and 40X. The effect of ultrasonic treatment processing modes on roughness and micro-hardness of the surface layers was studied. It is shown that roughness values as well as form of ridges and grooves, obtained by pre-lathing, effect the growth of micro-hardness values after ultrasonic plastic treatment.

1. Introduction

Ultrasonic plastic working (ultrasonic treatment) is a method of high impact to metal surfaces; the essence of the method is that a component surface after machining is treated with the instrument (a ball) vibrating with ultrasonic frequency [1-5].

Ultrasonic plastic treatment causes formation of a modified layer, with a high concentration of dislocations and diffuse interface with the base material, on the surface; formation of dangerous stress concentrators and emergence of fatigue cracks slow down significantly in the surface layer; that blocks a development of fatigue processes [6-8, 17, 18].

For practical application of ultrasonic treatment special ultrasonic equipment is used, that is the ultrasonic head, the ultrasonic generator and tools. Machining of outer and inner surfaces is performed by metal-working machines (turning, milling) of normal accuracy.

The main parameters of ultrasonic hardening are: tool static pressure on the detail surface; rates of vibration movement, longitudinal and lateral feed; a number of strokes; tool dimensions and shape [1, 2, 3, 5].

Auxiliary movement (feeding) and the number of working strokes both determine the number of strokes per unit of in-process surface per unit of time. The ultrasonic instrument size and shape both define the impact force and, thereby, the pressure on the in-process surface [4, 5].

Ultrasonic treatment has several features that distinguish this method from other types of surface deformation. The first feature is impact stress, which is repetitive and is distributed on a relatively small surface area. The second feature is rapidly alternating compression and shear deformations on the workpiece surface.

Specific process factors, namely magnitude of tool static pressure to the workpiece, amplitude and frequency of vibrations, have a great effect on physical-mechanical condition of the surface layer (roughness, micro-hardness, internal residual stresses). Static pressure force ensures contact between the tool (ball) and the surface. Both vibration amplitude and ball static pressure determine surface roughness, degree and depth of hardening, and magnitude of residual stresses in the metal surface layer as well as processing performance, feeding and workpiece rotation rate and a stroke number. Static pressure force is inversely proportional to the workpiece material elasticity; and is directly proportional to the initial roughness, diameters of in-process workpiece and the ball; feeding rate and workpiece rotation velocity.

This paper represents researches on effect of ultrasonic plastic processing modes on roughness and micro-hardness of steel surfaces. The authors investigated impact of initial roughness, indenter feed rate and in-process sample rotation velocity on values of micro-hardness and roughness after ultrasonic treatment.

2. Materials and methods

To determine the effect of lathe-cut initial surface micro-relief on surface hardening magnitude after ultrasonic plastic treatment, test samples were initially processed by lathing. Longitudinal support feed, setting the distance between individual micro-relief surface ridges and cut depth, setting the height of the irregularities, were variable parameters. Geometric parameters of the turning tool were also varied.

Ultrasound plastic treatment of the samples from 20 and 40X steels, 30 mm in diameter, was performed using ultrasound process set UZTK-02 [9, 20, 21]. Modes and processing parameters are as follows: generator output is 300 watts, indenter vibration amplitude is 20 microns, indenter static contact force is 50 N; vibration frequency is 24 kHz, indenter diameter is 5 mm.

Processing (Fig. 1) was carried out as follows: before ultrasonic plastic treatment, pre-lathing of the sample surface was performed by a machine-tool.

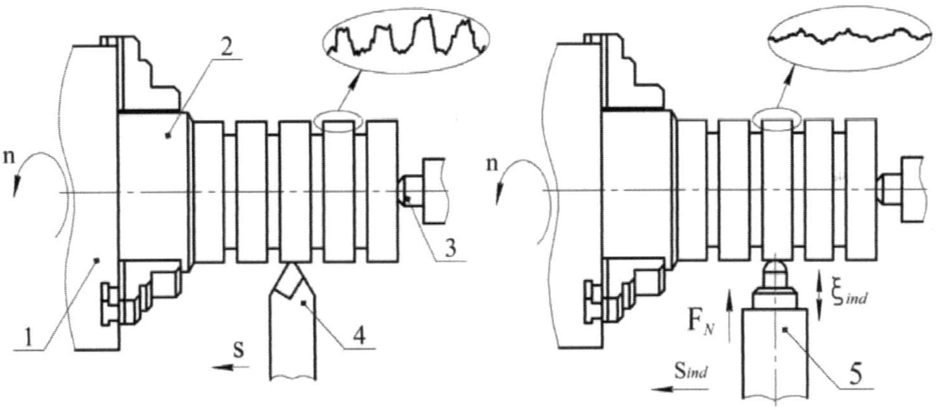

Figure 1. Ultrasonic treatment scheme: 1 - Three-jaw chuck;
2 – Workpiece (Sample); 3 - Rear Center; 4 - Cutter; 5 - Ultrasonic Tool

The sample 2 was placed in the three-jaw chuck 1 of the lathe and pressed by the rear center 3. Lathing of sample necks was performed in the following modes: support longitudinal feed was from 0.04 to 0.28 mm/rev; cutting depth - from 0.5 to 1 mm; lathing tool plan angle φ - from 40° to 85°, processing velocity - from 17 to 50 m/min. Then the cutter 4 was replaced by the ultrasonic impact tool 5. Sample necks were processed with the hard-metal indenter vibrating at an ultrasonic frequency. Sixteen necks of each workpiece were subjected to treatment

To determine the effect of ultrasonic treatment on indentation hardness of the metal surface, surface treatment was carried out without longitudinal feed by two modes: with ultrasonic action and without ultrasonic action.

3. Results and discussion

Tool stroke ultrasonic frequency modifies surfaces to a much greater extent than conventional methods of surface plastic deformation. A complex surface topography is formed by rotating the in-process sample and movement of the tool traverse manipulator. The surface topography is formed by waves of deformable metal, extruded under the indenter stroke displacements at an ultrasonic frequency.

In a local area of the indenter contact with the in-process surface, a plastic deformation zone occurs, which shifts with the indenter. Roughness Rz in the range of 3.5-5.5 mc was formed on sample surfaces after treatment with the ultrasound tool. Thus, decrease of surface roughness is achieved by indenter levelling impact on the thin surface layer, namely, on the roughness ridges formed after pre-lathing. The magnitude of surface hardening after ultrasonic treatment was evaluated by increase in micro-hardness. Table 1 shows the results of micro-hardness measurement on sample surfaces after ultrasonic treatment, comparing the highest (sample 2) and the lowest (sample 1) values of hardness.

Table 1

Micro-hardness value after ultrasonic treatment

	Steel 20			Steel 40X		
	Initial	Sample 1	Sample 2	Initial	Sample 1	Sample 2
Micro-hardness, MPa	1800 ± 200	2500 ± 250	3300 ± 250	2500 ± 150	3100 ± 200	4300 ± 200

Ultrasonic metal treatment significantly improves structure of surface and near-surface layers and increases micro-hardness values [6, 7, 10, 11, 23, 22]. Metallographic structure image (Fig. 2) shows a non-pickling 2-3 micron thick surface layer; non-equiaxed elongated grains, following machine direction, are located depthward, at a depth of 3-12 microns. Even below the treated layer transits gradually to the initial steel structure [3, 5, 12].

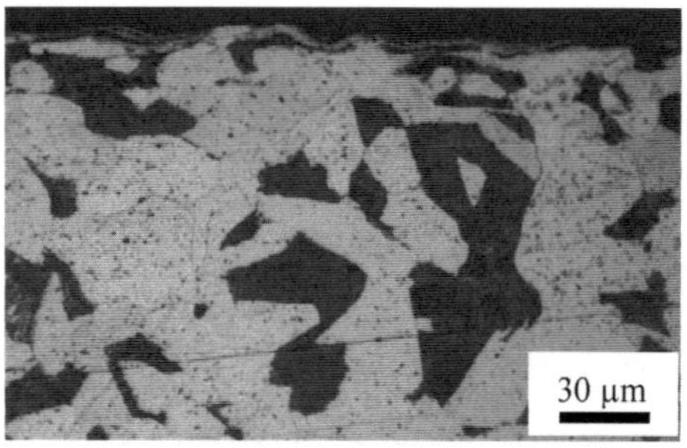

Figure 2. Metallography steel 20 after ultrasonic treatment

The results of micro-hardness value distribution of the most hardened samples (samples 2 of Table 1) are shown in Fig. 3.

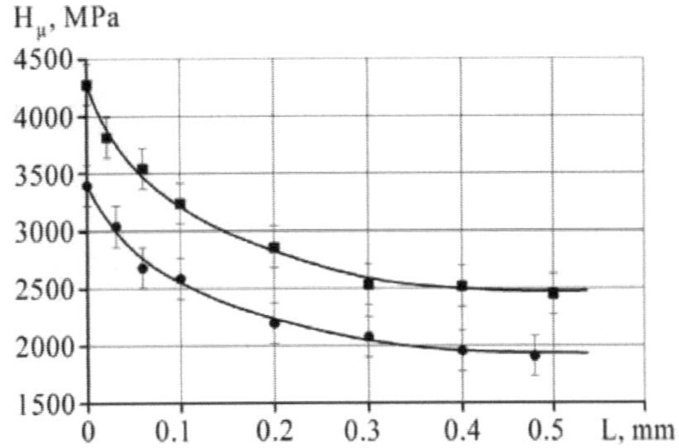

Figure 3. Depthward distribution of micro-hardness after ultrasonic treatment:
● - Steel 20; ■ - steel 40X

After ultrasonic treatment micro-hardness has a maximum value at the sample surface. Below the surface, micro-hardness monotonically diminishes to the initial level of material micro-hardness. Monotonic micro-hardness diminishing is resulted from changes in macro-, micro- and sub-microstructure of the steel, as well as changes of residual compressive stress conditions during the ultrasonic treatment [1, 2, 13, 14, 19]. Both absolute maximum value of surface micro-hardness and a hardened layer depth depend on carbon concentration in steel, and increase with carbon content [4, 6, 7, 8, 15].

In this part of the study it is found, that maximum surface hardening is achieved on surfaces with micro-relief as shown in Figure 4, and minimum hardening - on surface with micro-relief as shown in Figure 5.

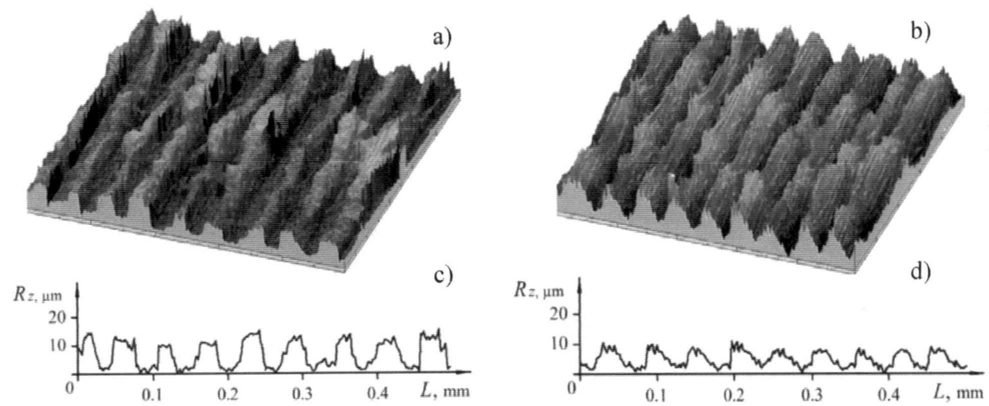

Figure 4. 3D-image of the surface (a, b) and profilogram (c, d) after pre-turning:

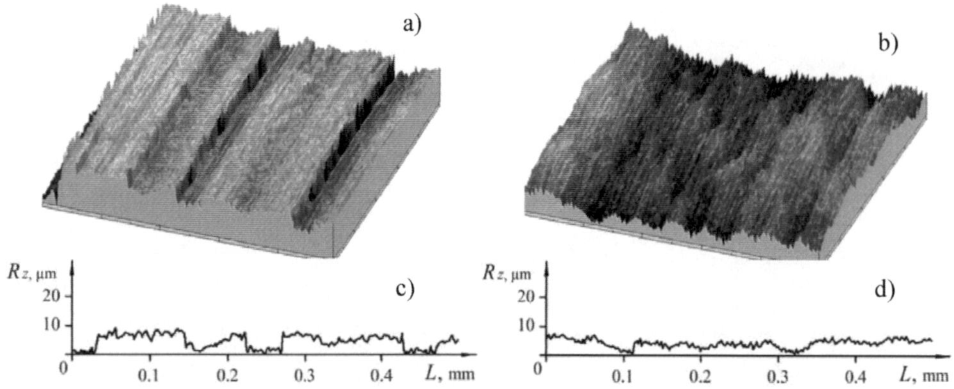

Figure 5. 3D-image of the surface (a, b) and profilogram (c, d) after pre-turning: a, c - Steel 20; b, d - steel 40X

The surfaces shown in Figure 4 are formed by lathe cutting, with the following modes: 0.05 mm/rev feed; 1 mm cutting depth. Surface roughness has a regular mode; micro-relief is an alternation of ridges and grooves [16] of trapezoidal shape with roughness value Rz 10-15 mc.

The surfaces shown in Figure 5 are formed by lathe cutting, with the following modes: 0.02 mm/rev feed; 1 mm cutting depth. Micro-relief is irregular; ridges and grooves vary in shape and size; roughness value after treatment is Rz 7-12 mc.

Subsequent ultrasound treatment of surfaces shown in Figures 4 and 5 was conducted with the following modes: feed - 0.2 mm / rev; ultrasonic action power - 300 W.

Figure 6 shows the surface, shown in Fig. 4, after ultrasonic plastic treatment, which has maximum hardening value. Figure 7 shows the surface, shown in Fig. 5, after ultrasonic plastic treatment, which has minimum hardening value.

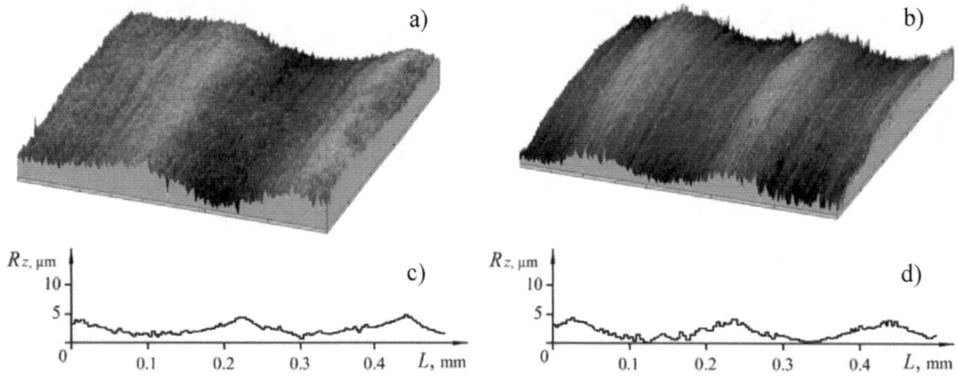

Figure 6. 3D-image of the surface (a, b) and profilogram (c, d) after pre-turning: a, c - Steel 20; b, d - steel 40X

Surface micro-relief (Fig.6) has a regular mode formed by alternation of ridges and grooves with roughness value Rz 5–6 mc. Surface micro-relief (Fig.7) is irregular; roughness value is Rz 5–6 mc.

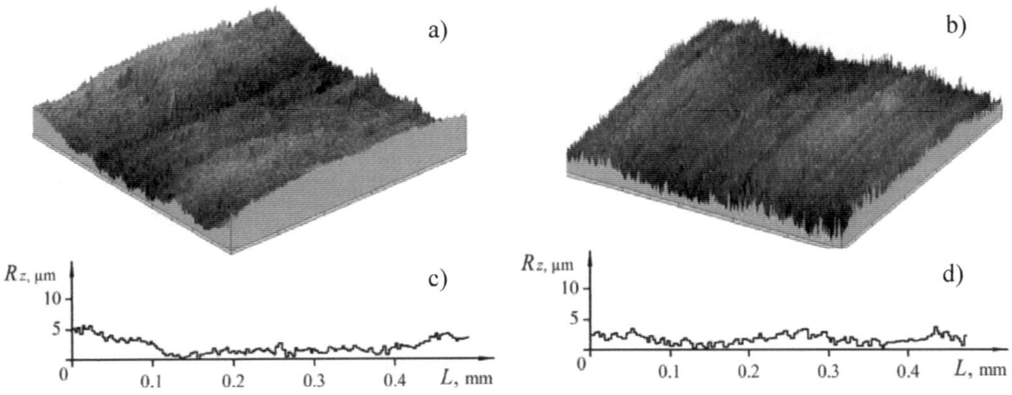

Figure 7. 3D-image of the surface (a, b) and profilogram (c, d) after pre-turning: a, c - Steel 20; b, d - steel 40X

Figure 8 shows a diagram of material surface layer deformation if contact of the hard-alloy indenter with the sample surface without both longitudinal feed and ultrasonic action. When loading the indenter by static force FN it is impressed in the in-process material. With the increase of pressing force, first elastic deformation of the surface occurs, then plastic deformation. After unloading, residual plastic deformation is expressed by indent sizes, namely, H groove width and h groove depth. Indentation depth was determined using the following formulas:

$$h = BC = OC - OB \qquad (1)$$

$$OB = \sqrt{r^2 - \left(\frac{H}{2}\right)^2} \qquad (2)$$

AO = OC = r; AB = H/2

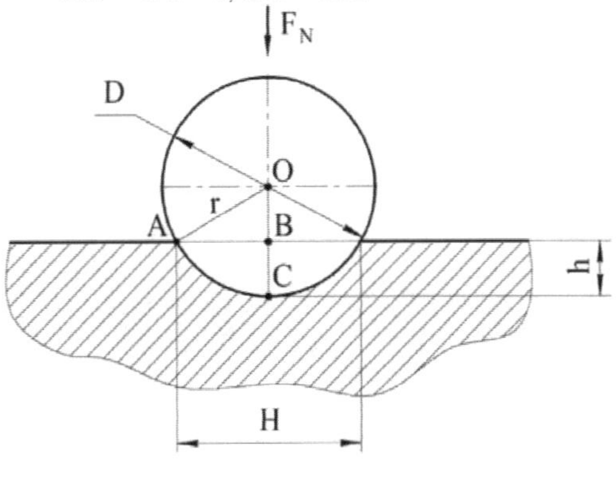

Figure 8. Scheme of surface deformation with indenter: F_N - static force; D - 5 mm indenter diameter; h - groove depth; H - groove width

Table 2 shows the results of contact of the hard-alloy indenter with the sample surface without both longitudinal feed and ultrasonic action. H groove width and h groove depth were determined using formulae (1) и (2). Table 2 shows that indentation width and depth are significantly larger if ultrasound treatment.

Table 2

Groove dimensions after contact of indenter with sample surface

Sample material	With ultrasonic action		Without ultrasonic action	
	Indentation groove width H, mm	Indentation groove depth h, mm	Indentation groove width H, mm	Indentation groove depth h, mm
Steel 20	1.6	0.13	0.8	0.03
Steel 40X	1.0	0.06	0.6	0.02

It was studied how indenter feed rate and in-process sample rotation velocity both affect roughness and micro-hardness values at ultrasonic treatment. Graph (Fig. 9) shows that heights of surface micro-roughness enhance with indenter feed increase. Simplistically we can assume that surface micro-relief is formed by arcs of equal radii; centers of the arcs are spaced from each other by an indenter feed advance. Accordingly, the larger indenter feed rate is, the smaller is groove interpenetration formed by deformation; hence, a value of machined surface roughness is higher.

The graph (Fig. 10) shows that surface roughness is diminished with increasing of rotation velocity and at 23 m/min has a minimum value; if a further increase in rotation velocity, roughness enhances. Roughness enhance is caused by reduction in number of indenter strokes per unit area because of increase of sample rotation velocity. At lower rotation velocity roughening enhances due to increasing number of indenter strokes per unit area. There is a phenomenon of over-hardening and metal particles flaking thereby roughness enhances.

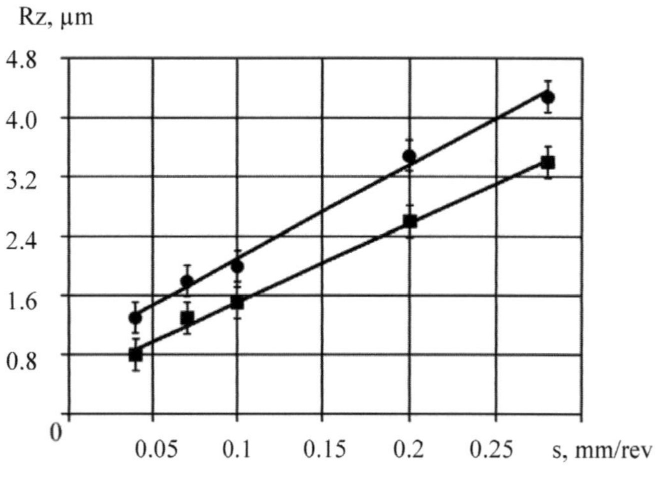

Figure 9. Influence of the supply amount of the indenter to a height of asperities after ultrasonic treatment: • - Steel 20; ▪ - Steel 40X. Processing velocity 23

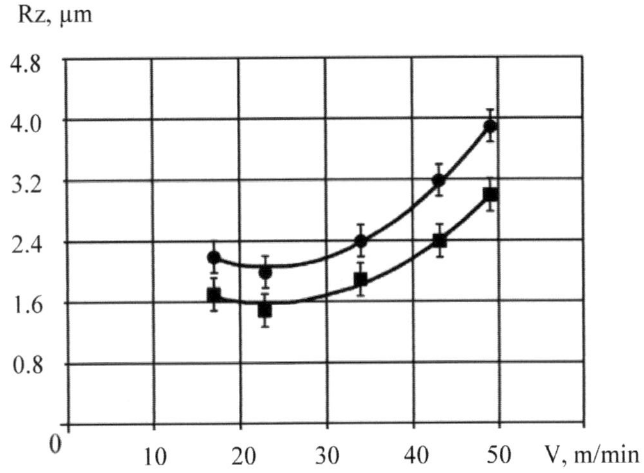

Figure 10. Effect of workpiece rotation velocity on micro-roughness heights after ultrasonic treatment: ● - Steel 20; ■ - Steel 40X. Indenter feed 0.1 mm/rev

The graphs (Fig. 11) shows that indenter feed increase diminishes surface microhardness at ultrasonic treatment. The larger indenter feed rate is, the smaller are both number of strokes per unit area and groove interpenetration formed by deformation; hence, values of plastic deformation and micro-hardness go down.

Graph (Fig. 12) shows that with increase of ultrasonic treatment rate, surface micro-hardness enhances and takes a maximum value at a speed of 23 m/min; then it gradually diminishes. At lower rotation velocities a number of indenter strokes per unit area increases; hence, a phenomenon of over-hardening occurs; thereby micro-hardness degrades.

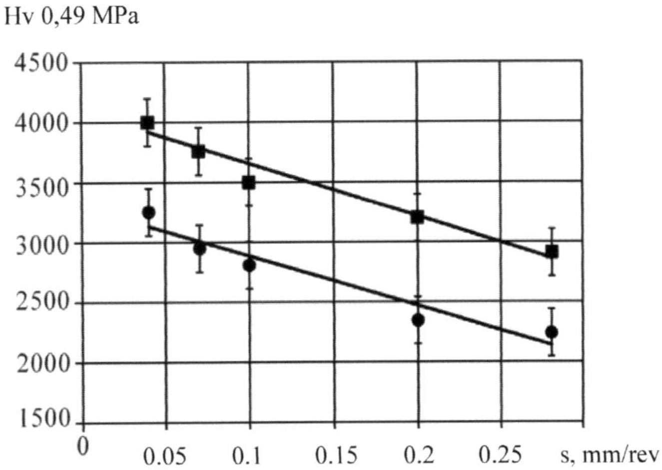

Figure 11. Effect of indenter feed on micro-hardness after ultrasonic treatment: ● - Steel 20; ■ - steel 40X. Processing velocity 23 m/min

Further increase in processing velocity leads to micro-hardness enhance to a maximum value for given treatment conditions and materials. From the turning-point a gradual micro-hardness decrease begins. This is caused by reduction in number of indenter strokes per unit area and, therefore, a degradation of surface plastic deformation.

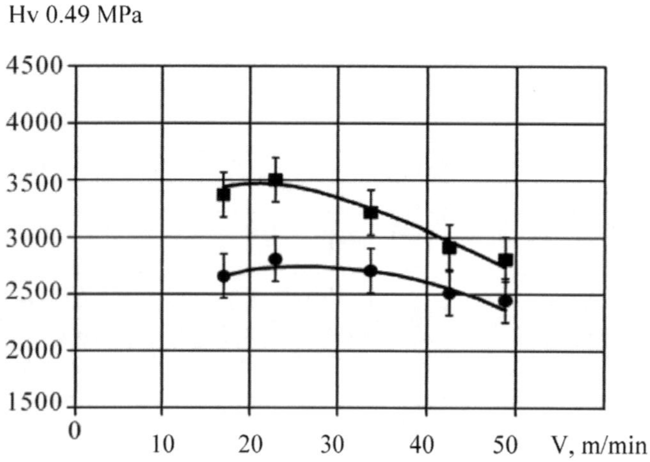

Figure 12. Effect of processing velocity on surface micro-hardness after ultrasonic treatment: ● - Steel 20; ■ - steel 40X. Indenter feed 0.1 mm/rev

Surface roughness-feed relation is linear within the research range. Roughness-peripheral speed relation is clearly nonlinear and approximated satisfactorily by a second order polynomial.
The following empirical relations are obtained by mathematical processing, for calculation of micro-relief Rz (3), (4), (7), (8) and micro-hardness HV (5), (6), (9), (10) of material.

For steel 20: $R_z = 12.64 \cdot s + 0.834$, (mc) (3)

$$R_z = 0.002 \cdot V^2 - 0.129 \cdot V + 3.558, \text{(mc)} \quad (4)$$

$$HV = -4039 \cdot s + 3277, \text{(MPa)} \quad (5)$$

$$HV = -0.627 \cdot V^2 + 32.698 \cdot V + 2315.4, \text{(MPa)} \quad (6)$$

For steel 40X:

$$R_z = 10.59 \cdot s + 0.458, \text{(mc)} \quad (7)$$

$$R_z = 0.002 \cdot V^2 - 0.096 \cdot V + 2.694, \text{(mc)} \quad (8)$$

$$HV = -4279 \cdot s + 4060, \text{(MPa)} \quad (9)$$

$$HV = -0.633 \cdot V^2 + 20.997 \cdot V + 3241.6, \text{(MPa)} \quad (10)$$

Where: s – indenter feed, mm/rev; V – surface velocity, m/min.

The graphs (Figs. 11 and 12) contain processing velocity extreme points providing minimum roughness values and maximum micro-hardness values.

Conclusion

1. It is found that ultrasonic plastic processing ensures 2-2.5 times maximum micro-hardness increase, if roughness (Rz 15 mc for steel 20 and Rz 10 mc for steel 40X) with trapezoidal ridges and grooves, obtained after turning.

2. Parameters of indenter feed and rotation velocity at ultrasonic plastic treatment are determined, which enable forming of minimum roughness Rz 3.5-5.5 mc and maximum micro-hardness 3200-4200 MPa for given processing conditions.

3. Modes of turning and geometric parameters of cutting tool are determined, which allow forming of surface micro-topography with the most effective form of ridges and grooves for subsequent ultrasonic treatment.

4. Modes of ultrasonic plastic deformation are determined, which allow forming a regular micro-topography with a maximum surface layer hardening and ensuring gradual micro-hardness transition to initial material.

References

[1] Odintsov L.G. Hardening and finishing by surface plastic deformation; Handbook. - M.: Engineering, 1987. - 328 p.
[2] Abramov O.V., Horbenko I.G., Shvegla Sh. Ultrasonic treatment of materials / Ed. O.V. Abramov. - M.: Engineering, 1984. - 280 p., Ill.
[3] Klimenov V.A., Nekhoroshkov O.N., Uvarkin P.V., Kovalevskaya J.G., Ivanov Yu.F. Structure, phase composition and properties of steel 60 after ultrasonic finishing // Physics. - 2006. - v.9. - Spec. Issue - pp. 173-176.
[4] Markov A.I. Use of ultrasound in machining and surface hardening of intractable materials. // Use of ultrasound in industry. Ed. A. Makarov. M: Mechanical engineering. 1975. - pp. 157-180.
[5] Konstantin V. Zaitsev, Vasiliy A.Klimenov, Marina A. Loshchilova, Cergey C. Polovnikov Ultrasonic plastic deformation of steels // YIT-ITEE IOP Publishing IOP Conf. Series: Materials Science and Engineering 91 (2015) 012056 doi:10.1088/1757-899X/91/1/012056
[6] Kulemin A.V., Kononov V.V., Stebelkov I.A. On choice of optimal mode for ultrasonic strengthening of metals. // Physics and Chemistry of Materials Processing. 1982. - №2. - pp.93-97.
[7] Kulemin A.V., Kononov V.V., Stebelkov I.A. Improving of fatigue strength by ultrasonic surface treatment. // Problems of strength. - 1981. №1 - S. 70-74.
[8] Nekhoroshkov O.N. Effect of ultrasonic treatment on structure, properties, and degradation of compositions forming while coating and welding. // Thesis for a scientific degree of candidate of technical sciences. - Tomsk, 2006. - 157 p.
[9] Ultrasonic processing set UZTK-02. Manual.
[10] Klimenov V.A., Panin V.E., Bezborodov V.P. and others. Substructure and phase transformations during ultrasonic machining of martensitic steel // PCMP - 1993. - № 6. - pp. 77-83.
[11] Klimenov V.A., Ivanov Yu.F., Perevalova O.B. and others. Structure, phase composition and mechanisms of austenitic steels hardening after ultrasonic indenting // PCMP. 2001. - pp. 90-97.
[12] Klimenov V.A., Kovalevskaya J.G., Zaitsev K.V., Borozna V.Y., Tolmachev A.I. Procedure of hardening of structural alloys. Patent № 2354715. Reg. 10.05.2009. Publ. 10.05.2009
[13] Kolomeets N.P., Mikhailov V.S. Use of ultrasonic technology for hardening of welds and superfinishing of friction units // Shipbuilding. 2001. - № 4. pp 32-33.
[14] Kulemin A.V. Ultrasound and diffusion in metals. M.: Metallurgy, 1978. - 200 p.
[15] Mukhanov I.I. Pulse hardening-finishing of machine parts with ultrasonic tools. M: Mechanical Engineering, 1978 - 44.
[16] Dyachenko P.E. Quantifying of surface microtopography. M.: Engineering, 1963. - 140 p.
[17] Aksenov V.V., Walter A.V., Gordeyev A.A., Kosovets A.V. Classification of geokhod units and systems based on product cost analysis and estimation for a prototype model production // IOP Conference Series: Materials Science and Engineering. – 2015. – Vol. 91. – P. 012088.
[18] Petrova V.A., Bakanov A.A., Walter A.V. Crack Resistance of Weld Seals, Hardening Methods and Composite Tools Quality // Applied Mechanics and Materials. – 2014. – Vol. 682. – P. 431-437.
[19] Zhanna G. Kovalevskaya, Vasiliy A. Klimenov, Konstantin V. Zaitsev Interfacial Adhesion Between Thermal Spray Coating and Substrate Achieved by Ultrasonic Finishing. AppliedMechanicsandMaterialsVol. 682 (2014) pp 459-463.
[20] U.S. Putilova, Yu.I. Nekrasov, A.A. Lasukov Loading of the Manufacturing Systems Elements in

the Process of Unsteady Mode Cutting and the Models of their Arrangement Deviations, Applied Mechanics and Materials, Vol. 682 (2014), pp. 192-196
[21] R. Yu. Nekrasov, A. I. Starikov, A.A. Lasukov Entering the operative correcion machining processes CNC, IOP Conf. Series: Materials Science and Engineering Vol. 91 (2015) 012042
[22] Klimenov, V., Borozna, V., Zaitsev, K. Influence of ultrasonic superficial processing on properties of titanic alloys. METAL 2009 - 18th International Conference on Metallurgy and Materials, Conference Proceedings 2009. pp. 273-278.
[23] Klimenov, V.A., Kovalevskaya, Z.G., Ul'yanitskii, V.Y., Zaitsev, K.V., Borozna, V.Y. Effect of ultrasound treatment of the substrate on the formation of the coating in detonation spraying. Welding International. 2010. 24 (10), pp. 803-807.

Production of Welding Fluxes Using Waste Slag Formed in Silicomanganese Smelting

N A Kozyrev[1, a], R E Kryukov[1], O E Kozyreva[1], U I Lipatova[1], A V Filonov[2, b]

[1]Siberian State Industrial University
654007, Novokuznetsk, ul.Kirova 42
[2]Yurga Institute of Technology, National Research Tomsk Polytechnic University Affiliate
652050, Yurga, ul. Leningradskaya 26

e-mail: [a]kozyrev_na@mtsp.sibsiu.ru, [b]a.filonow@gmail.com

Abstract: The possibility in principle of using slag, which is formed in the silicon-manganese smelting process, in producing welding fluxes is shown. The composition of and technology used for a new fused flux has been designed. A comparative evaluation of the new flux and the widely used AN-348 type flux was done. It has been proved that the new flux has high strength properties.

1. Introduction

Currently, much attention is paid to the development of new fluxes and additions, with analyzing the effect of flux chemical compositions on mechanical properties of a weld, and the content of oxygen and nonmetallic elements in welds. [1-6].

Today, the AN-348 type fused flux produced in Ukraine is widely used in welding and surfacing low alloy steels in Russia [7, 8].

It should be noted that such fluxes are oxidative, and their production is based on silicomanganese redox reactions, and in this connection the products of these reactions are the oxides of silicon and manganese. As a result, using such fluxes in welding promotes impurities, such as nonmetallic inclusions contained in weld metal and, as a consequence, physical and mechanical properties of a weld are negatively affected. Carbon-fluorine additives of a FD-UFS type were suggested for the removal of impurities in weld metal to improve its mechanical properties [5, 6]; this allows de-oxidation of the weld metal with carbon, causing significant lowering in the content of non-metallic elements represented by oxides.

Another significant disadvantage of using fused fluxes is their high cost, because natural materials are expansive and costs associated with mixtures, prepared for smelting flux in specially designed melting facilities, are high.

The use of metallurgical wastes, including the slag formed in the silicomanganese smelting process, for the production of welding fluxes is one of the ways to reduce the cost of welding fluxes. Analysis of published data shows that, when smelting silicon-manganese alloys, the slag, which is dumped, has a chemical composition generally in line with the requirements to the chemical composition of welding fluxes. Thus, according to work in reference [9], the slag contains: 14–16% MnO; 45–60% SiO_2; 7–8% Al_2O_3; 12–15% CaO; 3–4% MgO with ratio of CaO/SiO_2 = 0.52–0.58.

According to the data given in [10]: 47–49% SiO_2; 18–20% MnO; 12.2–14% CaO; 7–8% Al_2O_3; 2.9–3.1% MgO. Following [11]: 6.2–8.5% MnO; 45–47% SiO_2; 18–23% CaO; 9.2–11.6% Al_2O_3; 7.6–12.1% MgO; 0.3–0.7% FeO; ≤ 3% C. In the work referenced in [12]: 47–49% SiO_2; 18–20% MnO; 12.5–14% CaO; 7–8% Al_2O_3; 2.9–3.1% MgO. In the work referenced in [13]: 3.2–4.5% MnO; 43–47% SiO_2; 22–30% CaO; 12–16% Al_2O_3; 6–10% MgO; 0.3–0.7% FeO; ≈ 3,5% C. It should also be noted that, when silicomanganese alloys are obtained using the carbothermic process, up to 20.3% of manganese contained in raw materials may pass to the slag [13].

Based on these assumptions, we discussed that slag wastes formed in silicomanganese processing could be used in the production of a flux suitable for welding and surfacing of low-alloy steels.

2. Methods of Research

When producing the flux intended to be used in welding, the slag was used that was formed by the silicomanganese smelting in ore-thermal furnaces, employing the carbothermic continuous process. Table 1displays the chemical composition of slag. The product was smelted according to the conventional process scheme. The charge mixture consisted of manganese ore, quartzite, iron chips and coke fines. Ferroalloys (silicomanganese) were tapped with the slag into the ladle. After pouring silicomanganese, the slag was cooled while pouring into the ladle.

Table 1 – Chemical composition of silicomanganese slag

Material	Weight, %											
	Al_2O_3	CaO	SiO_2	FeO	MgO	MnO	F	CaF_2	Na_2O	K_2O	S	P
Si-Mo slag	10.61	18.62	50.55	1.55	8.03	9.63	0.38	-	0.41	0.61	0.13	0.05
AN-348	≤6	≤12	40-44	0.5-2.0	≤7	31-38	-	3-6	-	-	≤0.12	≤0.12

Under a laboratory environment, the flux was made by crushing, screening and sieving into fractions. Optimal fractions and their proportions were taken into consideration. Submerged arc welding was performed on samples of metal sheet in 09G2S steel using Sv-08GA wire and an ASAW-1250 welding machine. Of the welded plates, test specimens were cut to analyze micro- and macrostructure, and mechanical behaviour (ultimate tensile strength 6_B, MPa yield strength 6_T, MPa, elongation δ,%; impact strength KCV at -20 °C, J/cm2).

For the examination, the following fractions were used: less than 0.45 mm; 0.45-2.5 mm; 2.5-5 mm; 5-10 mm. Welding was performed with the following characteristics: Isv = 700 A; Ud = 30 V; Vsv = 35 m/h, and surfacing was done with: Isv = 410 A; Ud = 27; Vsv = 30 m/h. The tests have proved that 2.5-5 mm and 5-10 mm fractions do not ensure the quality of a weld surface (slag inclusions and high porosity of the weld), the fraction of less than 0.45 mm causes the formation of some kind of pitted surface of on the weld surface; the optimal fraction is 0.45-2.5 mm (Figure 1-4).

IOP Conf. Series: Materials Science and Engineering **125** (2016) 012034 doi:10.1088/1757-899X/125/1/012034

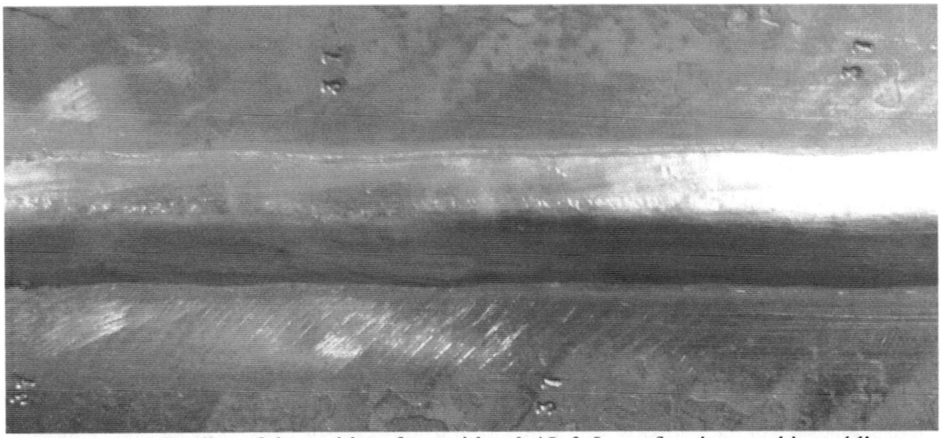

Figure 1 – Quality of the weld surface with a 0.45–2.5 mm fraction used in welding

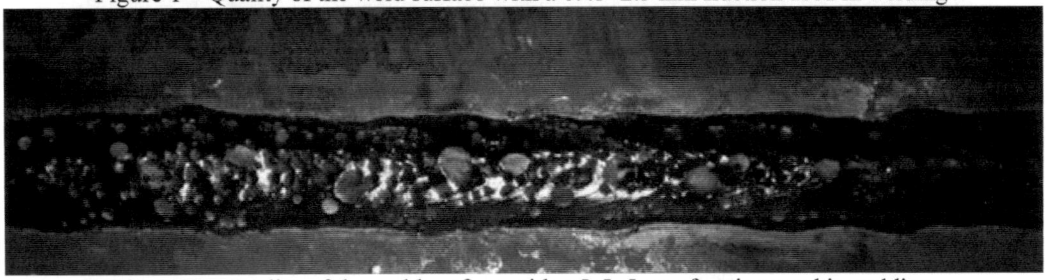

Figure 2 – Quality of the weld surface with a 2.5–5 mm fraction used in welding

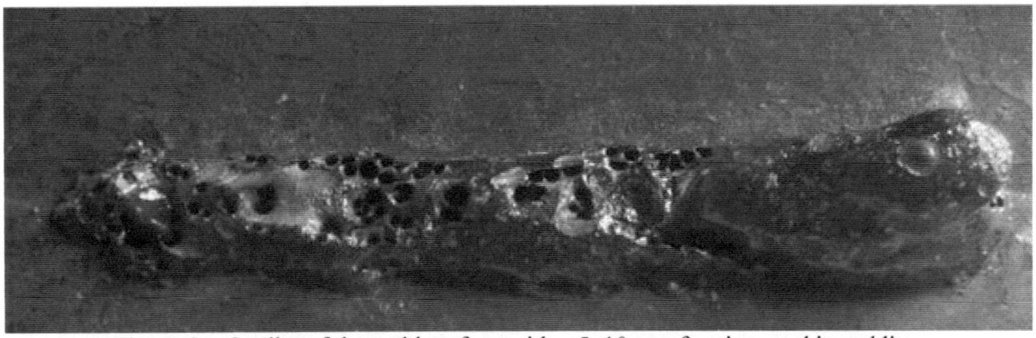

Figure 3 – Quality of the weld surface with a 5–10 mm fraction used in welding

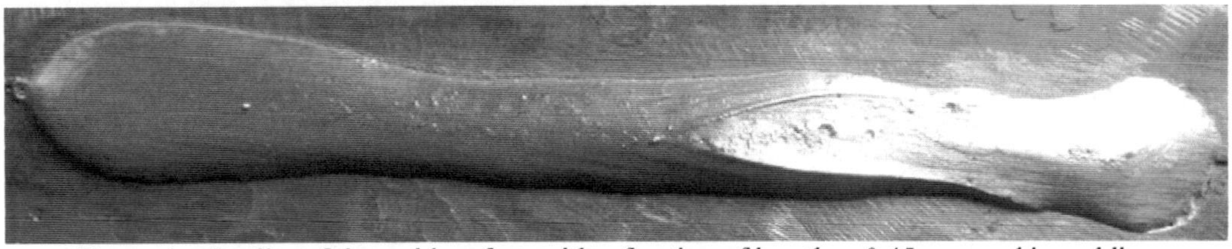

Figure 4 – Quality of the weld surface with a fraction of less than 0,45mm used in welding

The chemical composition of the crust (slag) on top of the weld after flux cored welding, with the flux obtained using the slag from silicomanganese processing, is given in Table 2.

Table 2 – Chemical composition of the slag crust

Wt, %										
Al_2O_3	CaO	SiO_2	Fe_2O_3	MgO	MnO	F	Na_2O	K_2O	S	P
10.01	18.02	47.7	2.83	7.09	9.42	0.34	0.39	0.61	0.12	0.02

3. Results and Discussion

Welds made using the flux that was produced using silicomanganese slag and the widely used AN-348 flux were analyzed relating to their mechanical properties. The comparative evaluation has showed that the strength of the welded joints when using the new flux is significantly higher, but the value of impact strength at negative temperatures is not satisfactory; the same is observed using the AN-348 flux. When adding a FD-UFS fluxing agent in different proportions (1-5%) to the experimental flux, the values of impact strength KCV at -20 °C are considerably higher. The results of mechanical tests are given in Figures 5-8 and Table 3.

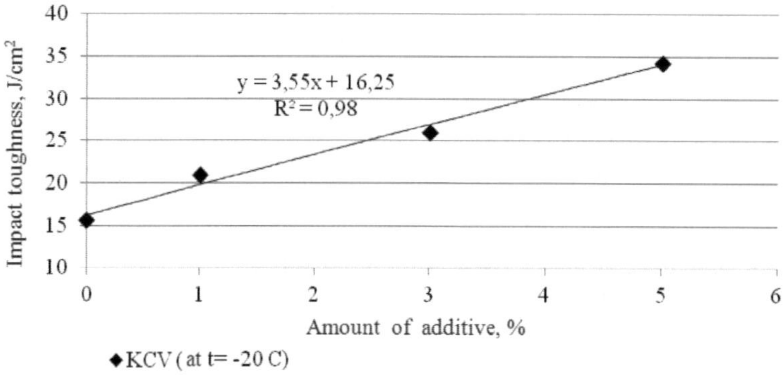

Figure 5 – Change in impact strength KCV of the weld metal at T = -20 °C with the amount of carbon-fluorine additions to flux

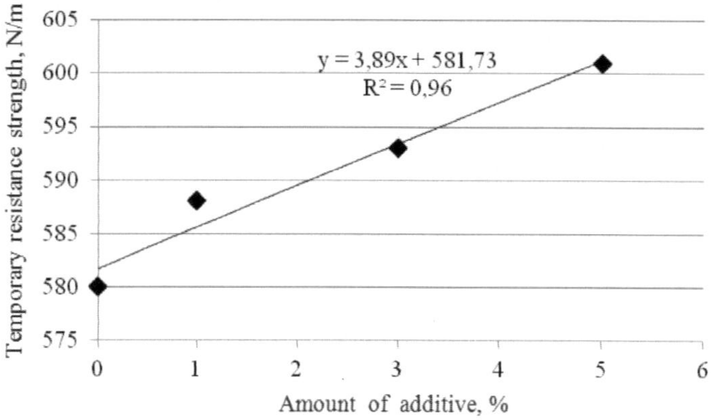

Figure 6 – Change in ultimate tensile strength with the amount of carbon fluorine additions to flux

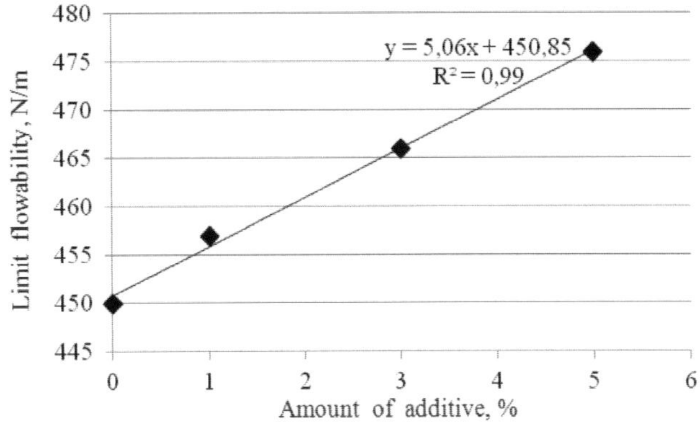

Figure 7 – Change in yield strength with the amount of carbon-fluorine additions to flux

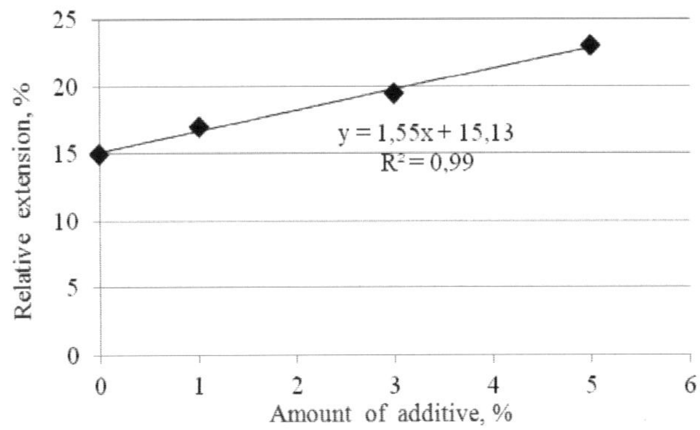

Figure 8 – Change in elongation with the amount of carbon-fluorine additions to flux

Table 3 – Mechanical properties of weld metal

Weld; Flux-cored welding	Ultimate tensile strength (Ϭв), MPa	Yield strength (Ϭт), MPa	Elongation (δ), %	Impact strength KCV at T= –20 °C, J/cm² (weld)
AN-348	535* 530-543**	360 355-368	25 24-26	18 16-21
Silicomanganese slag	580 576-583	450 447-452	15 14-16	16 15-17
Silicomanganese slag + 1 % FD-UFS	588 585-592	457 451-463	17 15,5-18	21 18-23
Silicomanganese slag + 3 % FD-UFS	593 590-596	466 463-469	19,5 19-20	26 20-32
Silicomanganese slag + 5 % FD-UFS	601 597-604	476 472-479	23 22-24	35 26-42

* - average values; ** - minimum and maximum values.

Conclusions

1. As laboratory experiments have shown, the use of slag, formed in the silicon-manganese smelting process, for the production of welding fluxes is possible in principle. Currently, the welding technology with using the experimental flux is under testing in the pilot-scale manufacturing conditions.

2. Based on the experiments performed, the dependences, describing the effect of carbon-fluorine additions to flux on the mechanical properties of welds, were obtained.

This work has been done at Siberian State Industrial University in the framework of government contract No. 11.1531/2014/к with the Ministry of Education and Science RF. The equipment of the Common Use Centre "Material Science" at SibSIU was used in tests, researches and measurement.

References

[1] Study of the relationship between the composition of a fused flux and its structure and properties/ Amado Cruz Crespoa, Rafael Quintana Puchola, Lorenzo Perdomo Gonzáleza, Carlos R. Gómez Péreza, Gilma Castellanosa, Eduardo Díaz Cedréa & Tamara Ortíza / Welding International. – 2009. – Volume 23. – No.2. – pp. 120-131.

[2] Using a new general-purpose ceramic flux SFM-101 in welding of beams/ Yu. S. Volobueva, O. S. Volobueva, A. G. Parkhomenkob, E. I. Dobrozhelac & O. S. Klimenchukd / Welding International.– 2012.- Volume 26. – No.8. – pp. 649-653.

[3] Special features of agglomerated (ceramic) fluxes in welding / V. V. Golovko & N. N. Potapov / Welding International – 2011.– Volume 25. – No.11. – pp. 889-893.

[4] New carbon-fluorine additives for welding fluxes / N.A. Kozyrev, N.E. Krukov, R. Ev. Krukov, A.V. Roor, L.P. Bashchenko, U.I. Lipatova / Steel in Translation. – 2015. – Vol. 45. – No. 4. – pp. 251-253.

[5] Development of additions to welding fluxes in welding of low-alloy steels / N.A. Kozyrev, V.F. Igushev, R Ev Krukov, A V Roor, I N Kovalsky / Welding International. – 2014. – Vol. 28. – No.5. – pp. 403-405.

[6] R E Kryukov, N A Kozyrev, G V Galevsky, Y V Bendre, V F Goryushkin, D V Valuev 2015 IOP Conference Series: Materials Science and Engineering 91 Some aspects of oxidation-reduction under carbon-bearing flux welding.

[7] Svarochnye materialy dlja dugovoj svarki: Spravochnik. 2 t. T. 1. Zashhitnye gazy i svarochnye fljusy: spravochnoe posobie [Welding consumables for arc welding: Handbook in 2 volumes. Vol.1. Shielding gases and welding fluxes: Handbook]// ed. N.N. Potapov. – Moscow: Mashinostroenie, 1989. – 544 p.

[8] Podgaetskiy V V, Rabkin D M Fljusy dlja avtomaticheskoj i poluavtomaticheskoj svarki. [Fluxes for automatic and semiautomatic welding] – Kiev: the USSR Academy of Sciences Publ., 1954. – 56 p.

[9] Povolotskiy D Ya, Roshchin V E, Mal'kov N V Elektrometallurgija stali i ferrosplavov [Electrometallurgy of steels and ferroalloys] – Moscow: Metallurgiya, 1995. – 592 p.

[10] Gasik M I, Lyakishev N P, Emlin B I Teorija i tehnologija proizvodstva ferrosplavov [Theory and technology of production of ferroalloys] – Moscow: Metallurgiya, 1988. – 784 p.

[11] Povolotskiy D Ya, Roshchin V E, Ryss M A, Stroganov A I, Yartcev M A Elektrometallurgija stali i ferrosplavov.[Electrometallurgy of steels and ferroalloys] – Moscow: Metallurgiya, 1984. – 586 p.

[12] Gasik M I, Ljakishev N P Teorija i tehnologija jelektrometallurgii ferrosplavov [Theory and technology of electrometallurgy ferroalloys] – Moscow: Intermet Engineering, 1999. – 764 p.

[13] Ryss M A Proizvodstvo ferrosplavov [Production of ferroalloys] – Moscow: Metallurgiya, 1985. – 344 p.

YIT-MT 2015 IOP Publishing
IOP Conf. Series: Materials Science and Engineering **125** (2016) 012035 doi:10.1088/1757-899X/125/1/012035

Analysis Of The Retailoring Methods And The Workability Of Deposited Surfaces

V V Konovodov[1], A V Valentov[2], E G Grigoryeva[2], K A Abdrasulov[2]

[1]Novosibirsk State Agrarian University
147 Nikitin St, 630039 Novosibirsk, Russia; phone (383) 267-35-07
[2]Yurga Institute of Technology, affiliate of Tomsk Polytechnic University
26 Leningradskaya St, 652050 Yurga, Russia; phone: (38451) 6-05-37

e-mail: valentov@mail.ru

Abstract. This work is devoted to the conduction of the analysis of methods and means of retailoring of parts in the conditions of technical repair facilities as well as of the workability of deposited surfaces. The results of the research show that the most important processing methods of reconditioning of component parts severely worn in the course of exploitation are various ways of deposit welding with subsequent mechanical handling. Furthermore, low efficiency of mechanical handling of retailored parts' surfaces is conditioned on their low workability that results from the specific status of weld pad.

Component parts are reconditioned by means of a technological influence which changes the geometric shape, the dimensions, or the internal state of material.

As a rule, the procedure of component parts reconditioning is a combination of operations of various means of building-up with their subsequent mechanical processing.

For the purpose of rebuilding of the geometric shape, the dimensions, and the surface of component parts they use following process operations: building-up of surface layers of material onto the worn surface; soft mud shaping; replacement of a component part as well as mounting of additional elements with subsequent removal of a material part using a certain method of mechanical processing.

1. Timeliness of the research

In comparison with other reconditioning methods, deposit welding helps make a layer of necessary thickness and chemical composition, of high hardness and wear resistance on the surface of a component part, while the original cost of building-up is quite low. In the course of subsequent mechanical handling, they achieve the necessary geometrical shape, prescribed dimensions of the part as well as required parameters of surface coarseness.

Currently, at the repair facilities of agricultural business, they use a wide range of building-up methods of worn surfaces, the listing of which is presented in table 1 [1].

The deposit welding is used for the reconditioning of component parts of agricultural machinery which are severely worn in the course of exploitation. They include the parts of following types: "single-diameter and stepped shafts", "disc plates, flywheels, and sheaves", "spider center caps, plugs", "splined shafts" [2, 3]. In particular, it is typical for undercarriage parts of cat tractors as well as work tools of agricultural machinery [3], while the building-up can be executed up to 10 mm.

In order to correct analogous defects of standard parts, at TRF of the agricultural business systems, they use various technological methods [3]. For instance, lower track wheels of cat tractors are reconditioned using 12 methods; crank axles - 8; rolling axles - 3; journals - 7; splined surfaces - 15, etc.

Component parts of the type "splined shaft" are also reconditioned through the use of a range of processes [3]: plastic working; short-circuit arc surfacing with plastic working; CO2 spline deposition on side surfaces; CO2 deposit welding or deposit welding under a flux layer through the helical line, while all of these methods are followed by mechanical handling of splined surfaces.

Table 1 — Methods of components parts reconditioning by means of building-up in the conditions of technical repair facilities (TRF)

Methods of building-up	The rate of reconditioned component parts, %
Deposit welding under a flux layer	32
Short-circuit arc surfacing	12
Deposit welding in an atmosphere of shielding gas	20
Flux-cored wire welding	10
Plasma spraying	1,5
Electric contact deposit welding	6
Galvanic methods	5
Electromechanical deposit welding	1
Electroslag facing	1,5
Pouring with supernatant liquid	2
Reconditioning by means of polymers	5
Other building-up methods	5

2. Methods of research, authenticity and validity of the results

The effectiveness of the defined reconditioning process which meets the requirements of GOST 19677-74, GOST 17524-80 is defined by the correlation of the price of a new component part as well as the original cost of reconditioning taking into account resource recovery factor and specific capital costs [3]. Furthermore, the original cost of reconditioning includes the expenses for building-up of worn surface, its mechanical handling, and other operations directed to the recovery of resource of a component part. Thus, the effectiveness of the reconditioning of worn component parts in a TRF system depends on technological characteristics of the applied building-up method of worn surface and its subsequent processing. For instance, the usage of high-alloy wear-resistant depositions, which significantly raise the wear-resistance of a component part, increases the original cost of the building-up operation as well as notably raises the expenses for subsequent mechanical handling of reconditioned surfaces. In such cases, technological variants of reconditioning are chosen in conjunction with mutual technological and economic restrictions of building-up methods and sizing, taking into account their impact on the recovery factors of component part resource.

As follows from table 1, currently, at the repair facilities of agricultural business, they mostly use deposit welding under a flux layer, deposit welding in an atmosphere of shielding gas, short-circuit arc surfacing, and flux-cored wire welding. These methods are widely applied because of their flexibility which allows to deposit a wide listing of recovered component parts, as well as because of the good quality of built-up surfaces with relatively low original cost.

Deposit welding under a flux layer gives following advantages: high productiveness, especially when multielectrode head sections and fluctuating electrodes are used; process stability; high-quality alloying with basic metal; the possibility to produce deposited layers of large thickness (over 8 mm) and to control chemical composition. Herewith, as a result of significant warming of the deposited component part, we observe the change of physical and mechanical characteristics and possible deformation of the component part.

The restrictions of this method include the complexness of depositing of a small diameter (up to 60 mm), the impossibility to gain a thickness of deposited layer less than 2 mm, while deposit welding under a flux layer is highly effective when the layer is thicker than 3 mm [4, 5]. The characteristics of deposited surfaces under a flux layer depending on the flux brand and the used wire are presented in table 2.

Table 2 — The characteristics of deposited metal under a flux layer

Flux	Electrode	Hardness, HRC_3	Relative wear resistance
AN 348 A	Welding (Sv) – 08G2S	20…22	1,02
AN 348 A	Sv – 18HGSA	31…33	1,18
AN 348 A	Nonconsumable (Np) – 20	17…27	1,08
AN 348 A	Np – 80	34…35	1,24
AN 348 A	Np – 30HGSA	34…36	1,17
ZHSN - 1	Sv – 08	47…51	2,75
ZHSN - 2	Sv – 08	36…40	1,65
ZHSN - 5	Sv – 08	35…38	1,62
ANK - 18	Sv – 08	38…47	2,62

Short-circuit arc surfacing is highly productive, up to 2.6 kg/h; in contrast to deposit welding under a flux layer, the heat-affected zone is much smaller, from 0.6 to 4 mm, the detail is warmed insignificantly. The welding bead is intensively cooled down through heat removing into the detail and into the environment, and becomes tempered. The weld pad is heterogeneous by its structure and hardness.

In the repair industry, they widely use deposit welding in an atmosphere of shielding gas, in particular, in an atmosphere of carbon dioxide, an inactive gas, or steam vapor as well as with gas-flame shielding. The productivity of this method of building-up is 25-33 per cent larger than the productivity of deposit welding under a flux layer, the heat-affected zone is smaller which helps produce the depositing of beads with a diameter over 10 mm. For the deposit welding in an atmosphere of shielding gases, they use wires Sv-08G2, Sv-30HGSA etc., the hardness of deposited surfaces equals respectively to HB 220...390; as well as flux-cored wires PP-3H2V8T, PP-30H5G2SM, PP-30H12SMT etc. with the hardness HRC3 40-46. The productivity of deposit welding by means of flux-cored wires equals to 10 kg/h; this method is used for the reconditioning of component parts of agricultural machinery with intense abrasive wear; increased wear resistance is provided by means of a high-carbon weld pad.

Table 3 — Hardness of surface deposited by means of short-circuit arc and flux-cored electrodes

Short-circuit arc surfacing		Deposit welding using flux-cored wires	
Surfacing wire	Hardness of the weld pad, HRC_3	Surfacing wire	Hardness of the weld pad, HRC_3
Sv – 08	12…32	PP – 3H63F-0	40…46
Np – 20	13…35	PP – U15H12M-0	40…42
Np – 40	15…45	PP – U15H7T-0	40…44
Np– 60	25…60	PP – R18T	61…64
Np – 80	25…65	PP – 3H2V8T	40…44
Sv – 30HGSA	15…60	PP – 3H13-0	52…54

Deposit welding by means of flux-cored materials is not so widely used because of the high cost of deposit materials, their cost is 5-10fold higher [6] than the cost of full-metal wires and ribbons; as well as because of low workability of such layer, especially in the course of reconditioning of complex profiled surfaces.

In order to recover huge wears of massive component parts of the type "wheel", at the repair facilities, they use electroslag facing which is characterized by low special original cost and high wear resistance of deposited surfaces. However, this method is not widely used due to the following technological restrictions: minimal thickness of the applied surface is 6-8 mm, and the analysis shows that the rate of such component parts, e.g. lower track wheels of cat tractors, does not exceed 25-30%; large heat-affected zone that hinders selective reconditioning of a worn component part.

Most component parts are delivered for reconditioning when their work surfaces are worn to a maximum of 0.2-0.3 mm, i.e. their working capacity is recovered by heating of an insignificant thickness of worn surface. Thus, electroslag facing is effective within large repair programs and high concentration and specialization of manufacture. In a real economic situation, the methods are most appropriate which are characterized by high flexibility of technological processes as well as the possibility to recondition a higher listing of component parts.

For component parts with small wear, the reconditioning by means of electric contact deposit welding is suitable. The essence of this method consists in the heating of welding filler by high-density current as well as in the deposition of pressure onto in; as a result, metallic bond is created because of partial surface melting and diffusion. The productivity of this method equals to 1000 mm2/min, the coating thickness is up to 1 mm, the waste of welding filler does not exceed 5%, the heat-affected zone does not exceed 0.3 mm.

For component parts with relatively low wear (up to 0.3 mm), the methods of electrolytic build-up are used: chromium coating (the hardness of the surface amounts to 1200 HB, the coating thickness amounts to 0.3 mm2/h); steeling results in the coating thicknesses up to 3 mm; electrolytic polishing of component parts of the type "shaft, hornet".

The essence of mechanical handling consists in the removal of allowance which was formed in the course of reconditioning process. The size as well as physical and mechanical features of the allowance are defined by technological parameters of the preceding building-up method.

The size of the built-up layer is determined with account of component part wear and the allowance for subsequent sizing [4]:

$$A_l = (P_n - P_w) + Z_{al} \quad (1)$$

where P_n and P_w are the sizes of a new and a worn component part respectively, mm;

Z_{al} is the allowance for machining, mm.

The size of the allowance is conditioned by the parameters of irregularities and geometric properties of deposited surface [7]. At this, the authors have noticed that the height of irregularities significantly grows when using multielectrode and multi-layer deposit welding. According to the research, the size of the allowance (fig. 1) depends on geometric parameters of deposited surfaces.

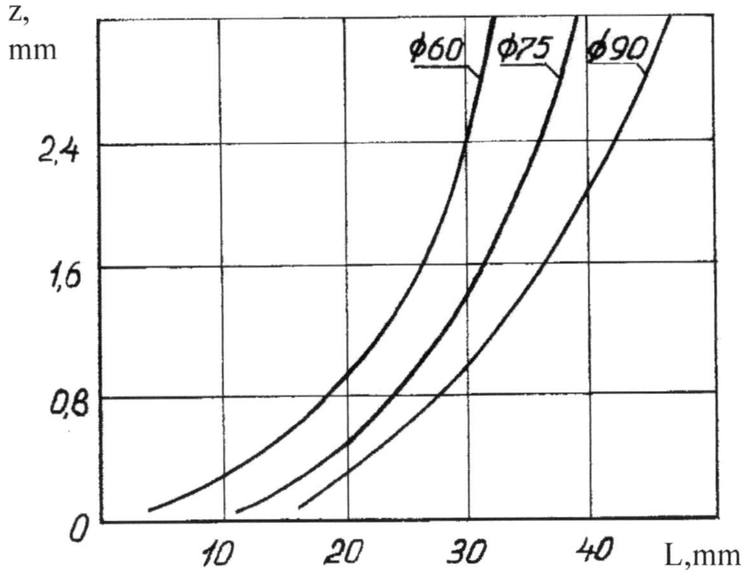

Figure 1 — Dependence of the allowance for machining on the diameter and the length of deposited surface

3. Results of the research

Specific particularities which define the workability of deposited surfaces, according to the researches [7, 8, etc.] are the following: significant macroirregularities, lappings, interstices, abscesses which exceed 2 mm; unequal distribution of alloying elements, heterogeneity of the micro-macrostructure in its thickness and cross-section which leads to significant hardness fluctuations up to 1.5-2fold; the presence of non-metallic impurities and a large allowance. The above listed factors result in periodic blows, vibration, high temperature and strength in the cutting zone which are 1.5-1.8fold larger than in the course of steel turning with close mechanical features, as well as their significant fluctuations: Q = 623-1523°K, Pz=150-3000 H. Hard processing conditions destroy the cutting edge of the tool to 68-94%, and lead to intense wear which 2-4 times exceeds general mechanical standards, while the cutting speed decreases to 15-25%.

Despite the development and the recommendation to implement promising methods of building-up of worn component parts, currently, due to economic and technological reasons, various methods of short-circuit arc surfacing are mostly used. The highest productivity and the lowest labor intensity and energy output are provided by edge cutting machining of built-up surfaces, in the course of which processed surfaces are hardened by means of cold working.

Based on the presented research and practical experience of repair facilities, we can conclude that the most important technological methods of reconditioning of component parts of agricultural machinery intensively worn within the process of exploitation are various methods of deposit welding with subsequent mechanical handling. Furthermore, low effectiveness of mechanical handling of retailored surfaces is conditioned by their low workability which results from the specific state of weld pad.

4. Conclusions

Accordingly, direct transfer of technological parameters and technical means of mechanical processing, especially edge cutting machining, from automotive-tractor and agricultural mechanical engineering into the technological process of reconditioning results in an ineffective use of machining facilities, overexpenditure of cutting tools, high labor intensity of processing by its low productivity. The solution of the problem of mechanical handling and of the increase of general effectiveness of reconditioning, apart from the quality improvement of deposit welding, lies in the development and implementation into repair industry of progressive cutting tools which provide productive cutting modes and condition high sustainability and durability.

References

[1] Metodika opredelenia economicheskoi effektivnosti tekhnologii i selskokhoziastvennoi tekhniki (Methods of determination of economic effectiveness of technologies and agricultural machinery). P.1. - Moscow: VNIESKH, 1998. - 219 pages

[2] Valentov A.V., Retyunskiy O.Yu., Konovodov V.V. Povyshenie stoikosti i proizvoditel'nosti reztsov pri obrabotke poverkhnostei, vosstanovlennykh naplavkoi// [Increased durability and performance of cutting tools used to machine surfaces parts reconditioned using deposit welding] Bratsk: Scientific Conference Mekhaniki XXI veku[Mechanics to XXI century].. 2012. No.11. pp. 395-399.

[3] Lasukov, A.A., Mokhovikov, A.A. Influence of modified layer of tool on stress - Strain state of cutting wedge, Proceedings - 2012 7th International Forum on Strategic Technology, IFOST 2012

[4] M.E. Porter. On competition / M.E. Porter: Translation from English: Uch. Pos. - Moscow: Publishing house "Williams", 2001. - 495 pages

[5] U.S. Putilova, Yu.I. Nekrasov, A.A. Lasukov Loading of the Manufacturing Systems Elements in the Process of Unsteady Mode Cutting and the Models of their Arrangement Deviations, Applied Mechanics and Materials, Vol. 682 (2014), pp. 192-196

[6] Konovodov, V. V.; Valentov, A. V.; Kukhar, I. S.Analysis of the influence of warming on the quality of soldered instruments/VI INTERNATIONAL SCIENTIFIC PRACTICAL CONFERENCE ON INNOVATIVE TECHNOLOGIES AND ECONOMICS IN ENGINEERING Серия книг: IOP Conference Series-Materials Science and Engineering: 91; №012053; 2015

[7] R.V. Chernukhin. Reliability of the steering gear of truck vehicles// Applied Mechanics and Materials. - 2013 - Vol. 379. - p. 36-42.

[8] Retyunsky O.Y. Method of making performs for composite reversible cutting plates // Applied Mechanics and Materials. - 2013 - Vol. 379. - p. 24-27.

IOP Conf. Series: Materials Science and Engineering **125** (2016) 012036 doi:10.1088/1757-899X/125/1/012036

Computer Simulation of Stress-Strain State of Pipeline Section Affected by Abrasion Due to Mechanical Impurities

P V Burkov[1,2], R G Afanas'ev[1], S P Burkova[1]

[1] National Research Tomsk Polytechnic University, 30, Lenin Ave., Tomsk, Russia
[2] Tomsk State University of Architecture and Building, 2, Solyanaya Sq., Tomsk, Russia

e-mail: [1]burkovpv@mail.ru

Abstract. The paper presents the effect of abrasive wear of the pipeline section occurred due to mechanical impurities in the transported gas flow. The approaches to the detection of the maximum specific wear of the pipeline wall and the geometry of abrasion are the main problems of computer simulation described in this paper.

1. Introduction

Although the natural gas undergoes the required purification before its transportation, investigations of abrasive wear in main pipelines are rather relevant since purification systems fault sometimes or the natural gas composition contains a large number of particulates. In this connection, it becomes necessary to estimate the consequences of gas transportation and, possibly carry out a feasibility study of transportation with the original gas flow instead of its purification allowing for the abrasion effect produced by mechanical impurities.

Moreover, in the light of recent discoveries in the field of nanoparticles that have an effect on the intensity of turbulent fluctuations in moving media fluxes [1], the practical use of the interaction between solid particles and transported gas flow becomes possible in gas production. Thus, it seems to be indispensable to investigate all possible aspects of implementing research achievements in this field. One of these aspects is abrasive wear of the pipeline material induced by solid particles. Undoubtedly, abrasive wear is a negative phenomenon in the use of solid particles for the reduction of pressure losses in gas flow. A computer simulation of this process allows detecting all negative facts and making a proper decision concerning the use of this specific natural effect.

The experience in gas transportation with suspended solids shows that in absence of the pipeline bends, its surface wears in the bottom area. This is because solid particles are directed towards the bottom due to gravity and demonstrate the intensive interaction with the bottom surface of the pipeline [2].

[1]Correspondent author

2. Results and discussion

Since relations that describe the abrasive wear in the main pipelines are absent, it is advisable to employ Putilov model designed for the related field, i.e. hydraulic pulverizing and ash/slag-discharge machinery at thermal power stations [3]. Since abrasive wear usually propagates uniformly along the pipeline, its model has two dimensions. The main parameters of abrasion in this work are accepted to be the maximum depth H_{max} and maximum width B_{max} as shown in figure 1.

The maximum width B_{max} of abrasion is obtained for geometrical reasons. In case the bulk concentration is known, the relation between transverse squares of the spot and the pipeline are known as well due to the uniform longitudinal distribution of impurities. At the same time, the maximum width of the spot equals to the chord of the circular segment formed by the spot of particulates (see figure 2).

Thus, the problem is reduced to finding the chord of the circular segment by the known square. The maximum width B_{max} is to be obtained as the third side of triangle formed by the given side and two radii:

$$B_{\max} = R\sqrt{2\cdot(1-\cos\alpha)} \tag{1}$$

To obtain B_{max}, it is necessary to get the angle α that forms this segment. This angle can be obtained from the segment square, because it is known as well as the square S_p of the pipeline:

$$\gamma = \frac{S_1}{S_p}, \tag{2}$$

Therefore, we have

$$S_1 = \gamma \cdot S_p = \frac{R^2}{2}(\alpha - \sin\alpha) \tag{3}$$

From here

$$\alpha - 2\pi \cdot \gamma = \sin\alpha \tag{7}$$

Substituting the angle value for the cosine theorem, we get B_{max} value.

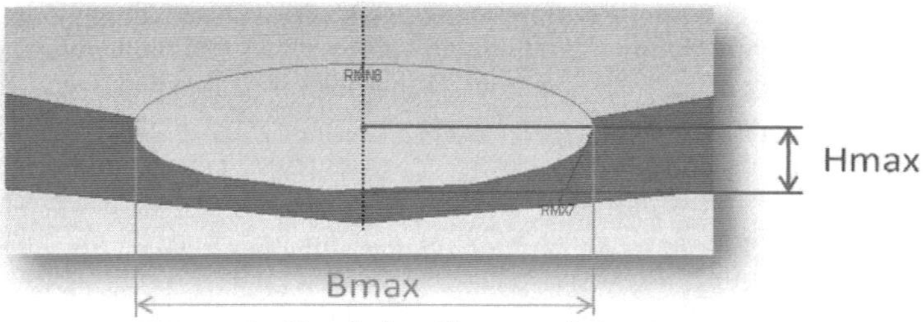

Figure 1. Simulation diagram of abrasive wear.

Having the maximum values of abrasion depth H_{max} and width B_{max} only, it is impossible to obtain the geometry of defects that satisfies the conditions of accurate simulation. These values allow obtaining only square, triangular or ellipsoid defects. Thus, the abrasive wear simulated in the form of annular groove, as

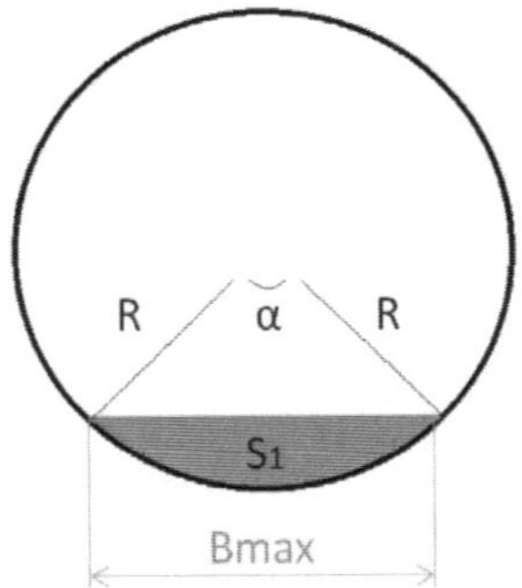

Figure 2. Geometrical substantiation of maximum abrasion width estimate

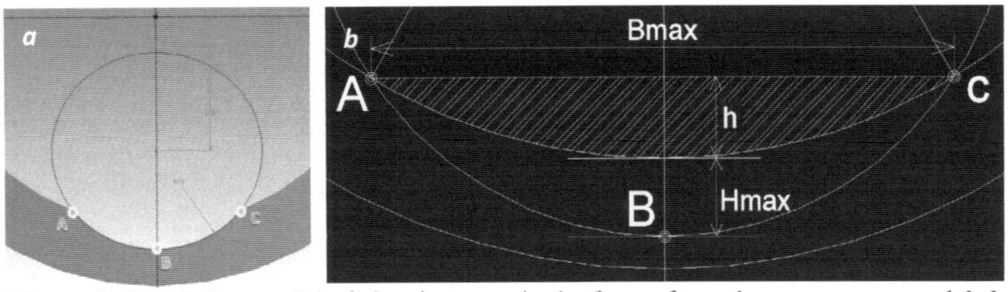

Figure 3. Simulation model of abrasive wear in the form of annular groove: *a* – model; *b* – schematic drawing

shown in figure 3 seems to be the most suitable for the pipe body. The values H_{max} and B_{max} should be transferred to X_c and Y_c coordinates of the circle centre with radius R for the correct simulation of the geometric model. Since the coordinates centre can be easily placed into the centre of the simulation model, then $X = 0$ and Y is the distance from X axis to the centre of the annual groove.

Having got points *A, B, C*, the distance between the centres of the pipeline and the circle that forms the spot of particulates can be obtained to properly model the geometry. Let us use the coordinate translation formulas for three points of the circle into the centre coordinates and radius value. Thus, having obtained the geometry of the pipeline (length, wall thickness, inner diameter) and the annular groove (coordinates of the circle midpoint and radius), it is possible to simulate the abrasion as a geometric object cut out of another one, and further analyze the stress-strain state of the defective pipeline.

A typical design was used to represent initial data of the stress-strain state of the pipeline based on OOO 'Gazprom transgaz Tomsk' technical requirements for the reconstruction of Parabel' - Kuzbass main pipeline (see table 1).

Table 1. General parameters of the pipeline

Internal diameter, (mm)	Wall thickness, (mm)	Length, (m)	Pipeline depth, (m)	Natural gas density, (kg/mi)	Operating pressure, (MPa)
1000	15 mm	11 m	1.5 m	0.770	5.4
		Soil properties			
Specific weight, (N/mi)	Angle of internal friction				
19000	20°				

Material properties				
Type	Density, (kg/mi)	Yield stress, (MPa)	Vickers number	
17G2SF steel	7850	372	171	
Particulate properties				
Origin	Density, (kg/mi)	Average size, (мm)	Bulk concentration	Particle velocity, (m/s)
Oxides, corrosion	4500	600	$1 \cdot 10^{-4}$	1

Having got the geometrical parameters of the pipeline section, it is necessary to determine loads induced by both the soil from above and below and transported product, insulating coating, metal and also pressure that has the effect on the inner surface of the pipeline wall [4].

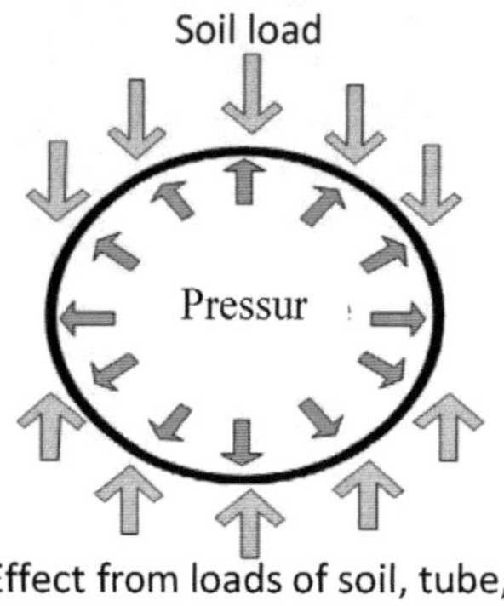

Figure 4. Pipeline wall exposed to loads

These calculations are characterized by the use of relations that generate results in Newtons. This is because the computer simulation of the stress-strain state of the pipeline is carried out in ANSYS software package in which loads are suitable to express in unit of force. The theoretical force acts on the half-area of the pipeline. As shown in figure 4, the force acting from soil above comes to 443130 N, while force produced from soil below, transported product, insulating coating, and pipeline material achieves 445611 N.

The geometry of the pipeline and abrasive wear is then simulated with ANSYS Workbench [5-8]. The suggested design techniques facilitate the simulation of this type of defects in CAE programs due to visualization of the interaction between two bodies having the simple geometry (figures 5).

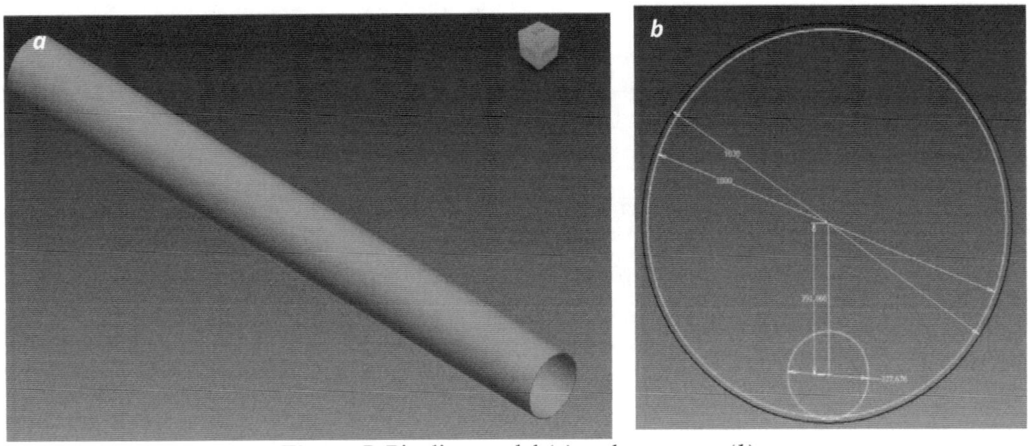

Figure 5. Pipeline model (*a*) and geometry (*b*)

To estimate the suggested model, the stress-strain state analysis is carried out for the pipeline with abrasion that increases during its operation (figure 6).

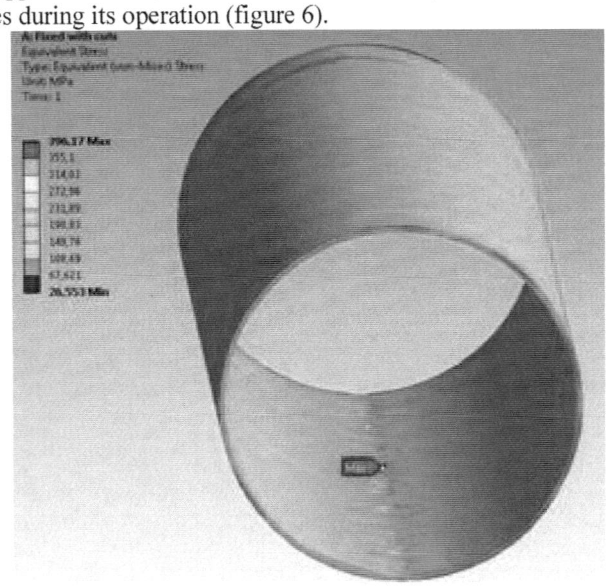

Figure 6. Pipeline equivalent stresses at abrasive wear after 18.1 years. Yield stress is exceeded

The model of abrasive wear of the pipeline, defect geometry, load effects, and the initial data for value calculation allow creating table 2 to represent the results by the tablulated function of maximum stresses in the pipeline wall depending on its lifetime. Also, these results are represented by influence curves in figures 7-9.

Table 2. Pipeline lifetime

Years	0	7.5	15	16	17	18.1
H_{max} **(mm)**	0	2.226215	4.45243	4.749259	5.046087	5.372599
σ_{max} **(MPa)**	230.1	239.63	282.54	332.4	355.72	396.17

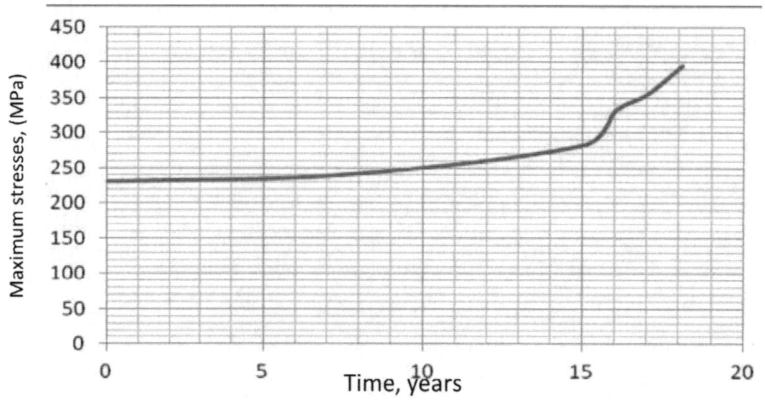

Figure 7. Dependence between maximum stresses and pipeline lifetime

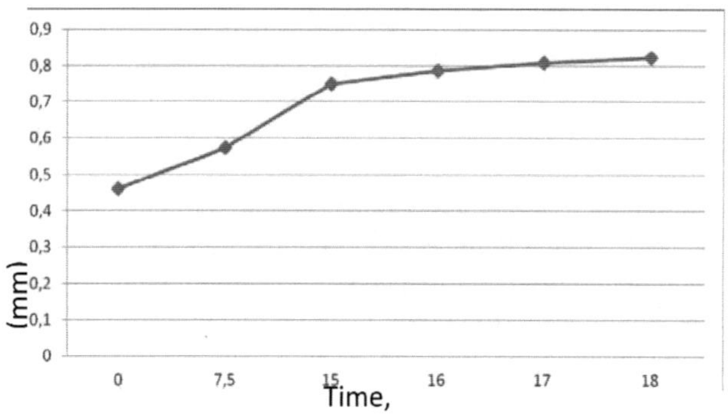

Figure 8. Dependence between displacement (generalized spatial movements) and pipeline lifetime

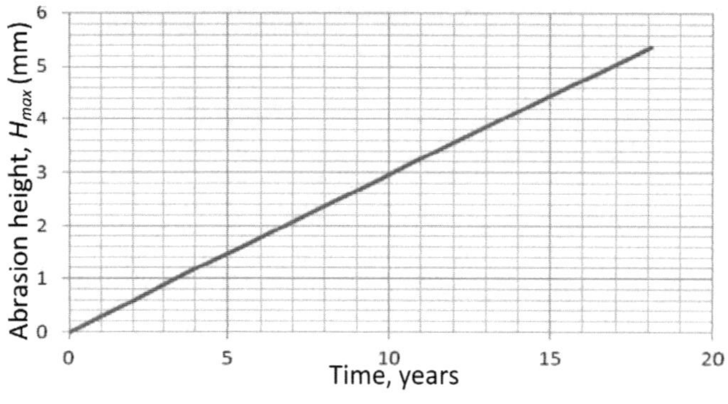

Figure 9. Dependence between abrasion height and pipeline lifetime

These curves show that the stress-strain state of the pipeline is not linear time dependence and changes stepwise. The growth of stresses at the beginning of the pipeline operation is almost insufficient. On the contrary, at the end of its operation stresses increase.

The adequacy of the suggested model is assessed by the comparison of results of numerical calculations and those prescribed by the construction regulations of Russia. For this, let us detect the maximum allowable depth of abrasion of the underground pipeline section. The pipeline section is made of the type 17G2SF steel tubes having 1030 mm diameter and 15 mm thick. The ultimate tensile stress is 372,6 MPa; the pressure of transported product is 5,4 MPa (figure 4).

Circular stresses can be obtained from

$$\sigma_c = \frac{m}{k_1 \cdot k_s} \cdot R_1^n = \frac{0.9}{1.47 \cdot 1} \cdot 372 = 227.8 MPa, \tag{10}$$

where m is the coefficient of operating conditions equaled to 0.9; k_l is the safety factor of material equaled to 1; k_s is the safety factor of the pipeline purpose equaled to 1,47; $R_1{}^n$ is the standard tensile strength of material.

The maximum allowable depth of abrasion is

$$(c) = \delta - \frac{p \cdot D_y}{2 \cdot ((\sigma_c) + p)} = 15 - \frac{5.4 \cdot 1030}{2 \cdot (227.8 + 5.4)} = 3.08m, \tag{11}$$

that is 20,53 % of the pipeline wall thickness. Thus, the error of the suggested model and estimation methodology is not over 10%.

Conclusion

With a view to improve the pipeline serviceability, estimate its mechanical conditions and predict the durability, technical diagnostics should be provided in time as well as the stress-strain state analysis by the finite element method.

The stress-strain state of the pipeline was not linear time dependence and changed stepwise. The growth of stresses at the beginning of the pipeline operation was almost insufficient. On the contrary, at the end of its operation stresses increased.

The allowable circular stresses agreed with the construction regulations equaled to 250 MPa, and stresses will exceed the allowable stresses after 10 years of operation. Computations carried out with ANSYS software package showed that stresses will exceed the allowable stresses after 17 years of operation. Therefore, the construction regulations should be taken into consideration in interpreting results obtained from the application software product based on ASME standards.

References

[1] Liang-Shih Fan, Chao Zhu 1998 Principles of Gas-Solid Flows (Cambridge University Press)
[2] Burkov P V, Tikhonova T V and Kholodov O N 2013 Proc. 9th Int. Conf. (Warsaw) vol 37 (Warsaw: BEL Studio Sp. z o.o) pp 63–66
[3] Putilova I V 2004 Abrazivnyi iznos truboprovodov pnevmotransportnykh ustanovok sistem zoloshlakoudaleniya i pyleprigotovleniya TES [Abrasion wear of hydraulic pulverizing and ash/slag-discharge machinery at thermal power stations] (Moscow)
[4] Tugunov P I 2002 Tipovye raschety pri proektirovanii i ekspluatatsii neftebaz i nefteprovodov [Standard design and operating techniques of tank farms and pipelines] (Ufa: DizainPoligrafServis)
[5] Burkov P V, D Y Chernyavsky, Burkova S P and Konan A 2014 IOP Conf. Ser.: Earth Environ. Sci. vol 21 (Bristol: IOP Publishing Ltd) p 1-5
[6] Burkov P V, Kalmykova K G, Burkova S P and Do T T 2014 IOP Conf. Ser.: Earth Environ. Sci.

vol 21 (Bristol: IOP Publishing Ltd) p 5-7
[7] Huang, S.L., et al. (2011) Oil and Gas Pipeline Deformation Finite Element Analysis of Eddy Current Probe Coil. Journal of Tsinghua University, 51, 390-394.
[8] Huang K, Wu S, Chen L, et al. Stress analysis of oil and gas pipeline parallel laying when traversing tunnels. Journal of Chemical & Pharmaceutical Research, 2014, 6(6).
[9] Wu X, Jiang Y, Lu H, et al. Stress analysis of shallow sea gas pipelines. Res. J. Appl. Sci., Eng. Tech, 2014, 7: 157-160.
[10] Xiaonan W, Yan X, Yuanhai L, et al. Analysis on Gas Pipeline Stress in Tunnel during Pigging .Natural Gas and Oil, 2012, 2: 002.

Computer Simulation of Stress-Strain State of Oil Gathering Pipeline Designed for Ugut Field

P V Burkov[1], **S P Burkova**[2], **V D Samigullin**[2]

[1] Yurga Institute of Technology, 26, Leningradskaya Str., Yurga, Russia
[2] National Research Tomsk Polytechnic University, 30, Lenin Ave., Tomsk, Russia

e-mail: [1]burkovpv@mail.ru

Abstract. The paper presents the stress and strain state modeling of infield pipeline in Ugut oil field. The finite element models of the stress field distribution in the pipeline wall are presented in this paper. The attention is paid to the pipeline reliability under stress conditions induced by the internal pressure and external compressive or tensile loads.

Introduction

One of the main problems that must be solved for the infield pipeline design is the construction of the linear pipeline portion. As the pipeline routing is often laid together with other linear communications, it is necessary to take all possible measures to preserve its integrity. In crossing automobile roads, the pipelines are usually cased in protective steel tubes [1]. Underground pipeline portions laid in severe climatic conditions are especially difficult to maintain. Failures and accidents of such pipelines occur due to their excessive bending accompanied by the instability of the soil-pipe-liquid (gas) system and a non-uniform setting. In order to prevent pipelines from failures and damages induced by the excessive bending, climatic and operating conditions, soils, and potentially dangerous areas should be studied and detected. Along with engineering facilities, the problems of strength and stability are to be solved to discover potentially dangerous areas. Construction regulations that define the design procedure for underground pressure piping, namely: SNiP 2.05.06-85* 'Main pipelines' (item 8.25), SNiP 34-116-97 'Design, construction and reconstruction of infield pipelines' (item 8.1) include the requirements for the combined calculation of the pipeline and the soil mass. The relevance of the stress and strain state design technique for underground pipelines is explained by the fact that at least 1 mm increase of the pipe wall thickness results in a considerable overexpenditure of material. This is because the soil exerts not only the external load application to the pipeline, but also is the medium in which deformations of the linear construction occur. At the same time, construction regulations do not contain directions of how to perform the combined calculations.

The remaining lifetime is estimated as the deterioration storage time until the limit state is reached by the specified failure criterion. The calculation estimation of the remaining lifetime is not usually specified. Calculations are based on formulas used to forecast the structural lifetime by the deterministic or probabilistic schemes. Deterministic schemes imply the use of fixed data on crack

[1]To whom any correspondence should be addressed.

resistance in materials, structural loading, and defects. The remaining lifetime is predicted using factors of safety (crack resistance) and robustness. The remaining lifetime is determined by the non-propagation of the existing or hypothetical cracks (their non-exceedance of safe sizes) out of the realm of safe states of the system deteriorated by the criteria of fracture and robustness mechanics. Probability calculations are carried out using the reliability theory based on statistical data on mechanical properties of materials, loads, effects, and flaw detection. During the operation, the pipeline elements are exposed to external loads leading to the stress-strain state of the material. For example, in crossing points with water channels, automobile roads, pipe blowout and sagging or joints with pump station assemblies, flexural stresses are more intensive. The state of the pipeline material is directly depends on the external loads that induce stresses in it together with the internal pressure.

As is known, the local sections of pipelines subject to the increased stresses undergo degradation processes that affect the functional properties of material due to metal strain age and low-cycle fatigue. In time, this leads to the critical state of material and, as consequence, the decrease of the pipeline efficiency.

Thus, the estimation of metal properties in the pipeline is one of the preventive measures that should be taken to avoid accidental fractures.

Other preventive measures include the reinforcement of the pipeline sections exposed to stresses, cutting off pipeline sections, mechanical stress relief by means of pipeline rectification.

Also, the effect of lumped plasticity in the pipeline elements observed both in manufacturing and operation should be investigated in relation to its reliability under stress conditions induced by the pressure inside the pipeline and external compressive or tensile loads.

Thus, a study of the stress and strain state of such pipeline portions is a relevant problem that must be solved to provide a safe operation of the designed infield pipeline [2-10].

The portion of the infield oil-gathering pipeline being designed for the Ugut field was selected for investigations. PipeSim software was used to carry out hydraulic calculations that allowed the determination of the required diameter of the pipeline, namely 219 mm. Hydraulic calculations were made for the maximum volume values of oil extraction in the Ugut field. As a result, the preliminary pipeline routing was conducted. The standard pressure in the infield pipeline was assumed in accordance with the pressure relieve valve of the measuring equipment established on site of the group of wells and equaled to 4.0 MPa. The type 09GSF steel was accepted for the pipeline construction. This steel possesses higher strength properties. The strength analysis was then conducted in compliance with SNiP 34-116-97. The thickness of the pipeline wall was determined as 8 mm and the minimum pipeline depth from the ground surface to the upper tube was determined as 1.8 m.

The longitudinal section of oil gathering pipeline was constructed based on the accepted design solutions, in which the point of crossing with the automobile road was indicated as shown in figure 1. This section of the pipeline routing is laid in a protective steel casing having 45 m length and 426 mm in diameter. Figure 2 shows the schematic layout of the pipeline across the automobile road.

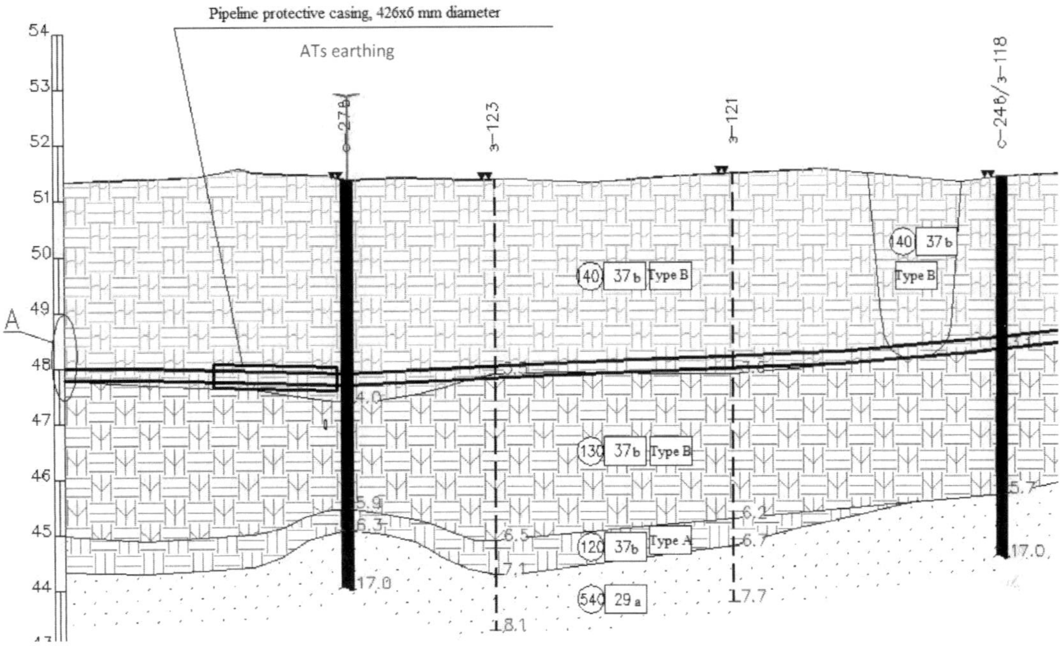

Figure 1. Cross-sectional plan view of the pipeline portion

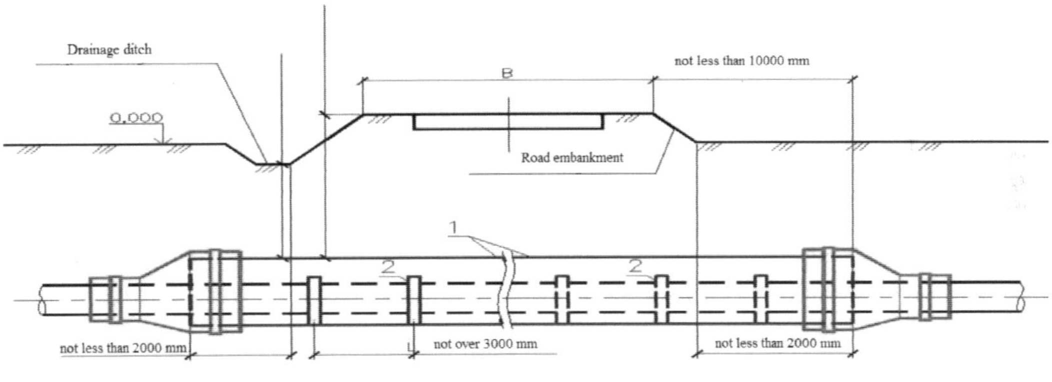

Figure 2. Schematic layout of the pipeline across the automobile road

Results and discussion

The internal changes in metal lead to its fracture and are called a limit stress state. The structural reliability is provided by the comparison of maximum stresses which appear at the most dangerous point and admissible limit values of the given material. The limit stress state of a structure is just the limit beyond which its operation is impermissible. The structural reliability is higher when the level of actual internal stresses is farther from the limit state. Autodesk Inventor software allows the strain-stress computation of the pipeline wall and obtaining output data in the form of three values of primary stresses which are presented by roots of a cubic equation defined by the stress vector components.

The strain-stress computation is based on the calculation methodology of allowable elastic stresses. The stress and strain state of the displacement type defect located on a cylindrical pipe casing is determined by a three-dimensional design model subjected to combined service loads. Since the aim of this work is to study stresses occurring within the defective portion of the pipeline, the design model can be reduced to its defective portion. The internal excessive stress is the service load applied to the pipeline. The dead load of the pipeline portion, wind and snow/ice loads also should be taken into account. In order to provide a static determinability of the model, the boundary conditions include a pin-edge fixing of the pipeline portion to prevent linear motions.

The stress and stain state of the pipeline in protective steel casing was analyzed by the finite element method using Autodesk Inventor software. Once the solid model of the pipeline was invented, the stress and strain state analysis is conducted accounting for all loads that have been calculated at the design stage. The length of the pipeline casing is selected such that to exclude the possible effect of end connections on the estimated area. The pipeline crossing over the road is modeled in conformity with the measurement results obtained during technical diagnostics. Loads distributed across the casing area are applied to its defect area. Welded joints are neglected. The finite element model contains five casing objects unified with each other by the bonded contact, i.e. the complete dependence of displacements is provided in all assemblies. Cartesian system is used for this model. The finite element mesh reproduces the surface curvature and is superimposed automatically. The minimum mesh size was obtained at mash condensation within the defect area and equals to 0.6 mm. Mesh size of 10 mm was accepted for all other structural elements. The non-linear model simulation was carried out using the iteration technique. The values of the equivalent stresses are distributed in the sectional area of casing under the maximum load, i.e. in its outer surface. The finite element models of the stress field distribution in the pipeline wall are presented in figure 5, 6.

Figure 3. Solid model of the pipeline in protective steel casing

The solid model of the pipeline is illustrated in figure 3 and figure 4 demonstrates the longitudinal stress distribution.

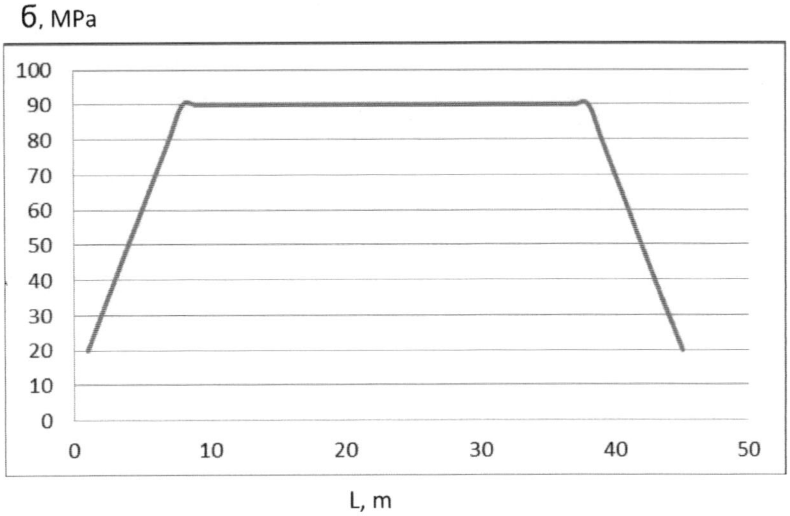

Figure 4. Stress distribution (y) along the pipeline (L)

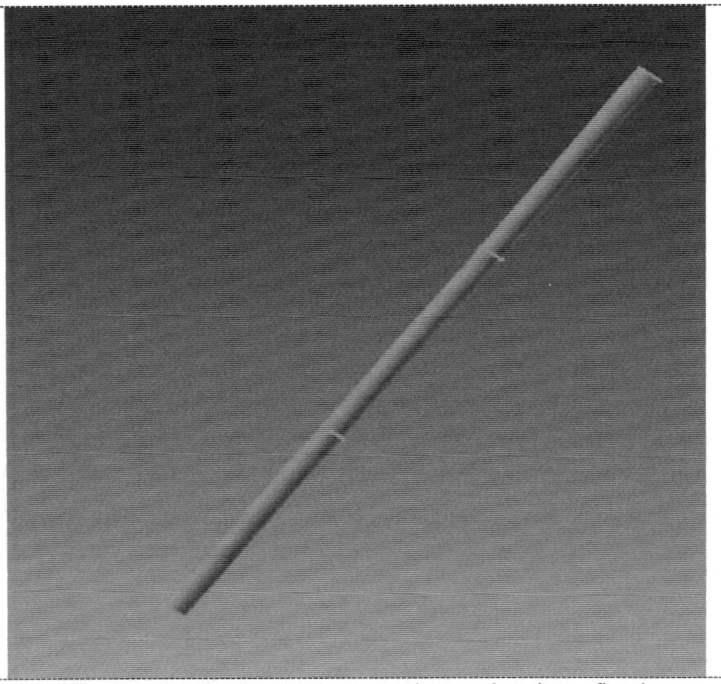

Figure 5. Pipeline section in protective steel casing – fixation

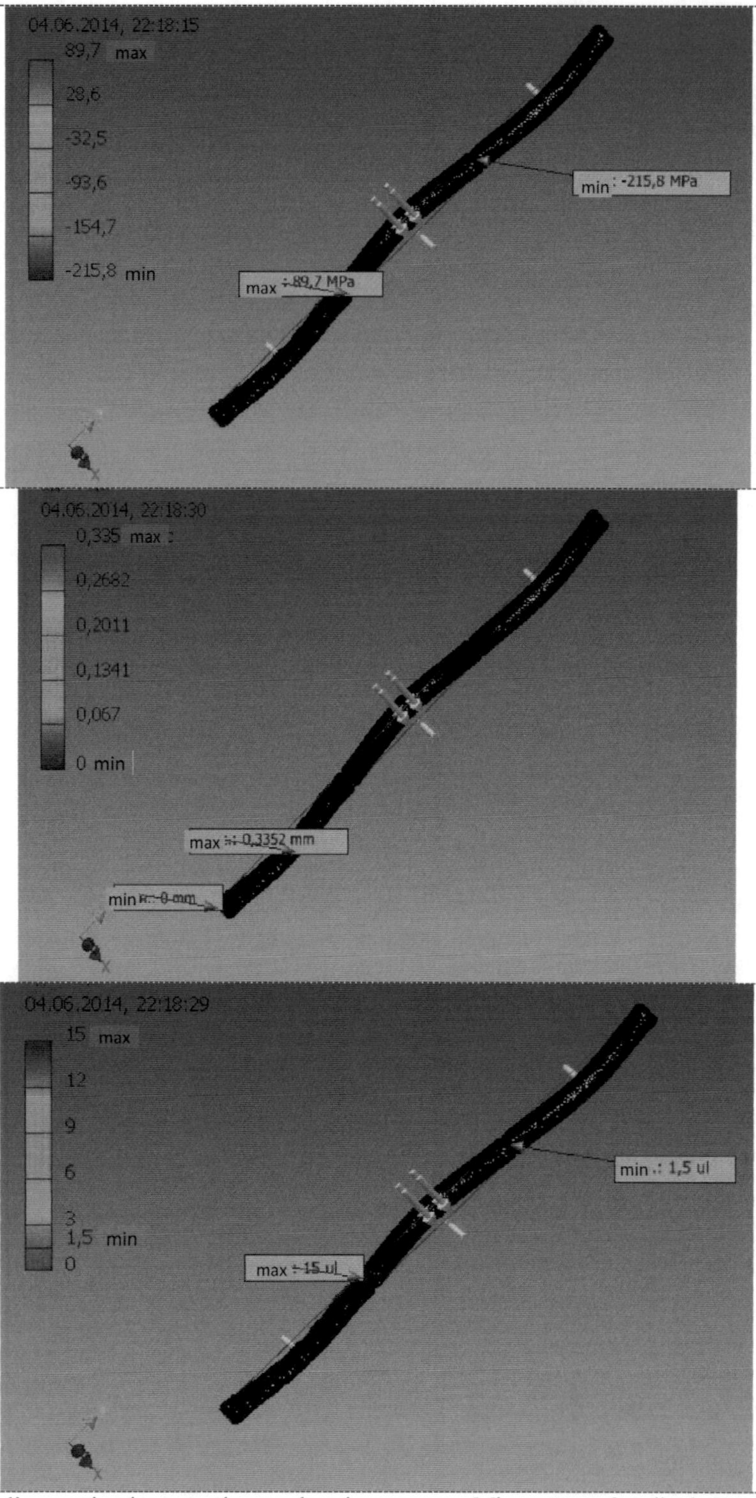

Figure 6. Pipeline section in protective steel casing: *a* – von Mises stress; *b* – displacement; *c* – safety factor

Conclusion

The use of Autodesk Inventor software for the stress and strain state analysis of the pipeline portion in protective steel casing showed that the universal CAD system allows solving a variety of difficult problems. The results obtained in this work showed that all design considerations accepted in the course of this research provide a safe operation of the pipeline protective steel casing in the crossing point with the automobile road, because the maximum stress occurred in the pipeline is less than the yield point of the steel tube. Thus, the strength of the pipeline portion was provided by the design pressure of 4 MPa inside the pipeline having the wall thickness corresponding to the actual value, and the protective casing had no a considerable effect on serviceability of the suggested pipeline

References

[1] Mustafin F M 2007 Tekhnologiya sooruzheniya gazonefteprovodov [Gas oil pipe technology] (Moscow: Nedra) p 632
[2] Burkov P V, Burkova S P, Timofeev V Yu, Ashcheulova A A and Klyus O V 2013 Vestnik KuzSTU, 6 77–79
[3] Burkov P V, D Y Chernyavsky, Burkova S P and Konan A 2014 IOP Conf. Ser.: Earth Environ. Sci. vol 21 (Bristol: IOP Publishing Ltd) p 1-5
[4] Burkov P V, Kalmykova K G, Burkova S P and Do T T 2014 IOP Conf. Ser.: Earth Environ. Sci. vol 21 (Bristol: IOP Publishing Ltd) p 5-7
[5] Xu, J.L. and Hua, B., et al. (2009)The Research Status and Progress of the Technology of Magnetic Memory Testing. Journal of Inspection and Quarantine, 19, 64-66.
[6] Jia, W.L. and Liao, K.X., et al. (2010) Acidic Gas Pipeline Corrosion Defects Detection Technology. Oil and Gas Field Surface Engineering, 29, 1-2.
[7] Huang, S.L., et al. (2011) Oil and Gas Pipeline Deformation Finite Element Analysis of Eddy Current Probe Coil. Journal of Tsinghua University, 51, 390-394.
[8] Huang K, Wu S, Chen L, et al. Stress analysis of oil and gas pipeline parallel laying when traversing tunnels. Journal of Chemical & Pharmaceutical Research, 2014, 6(6).
[9] Wu X, Jiang Y, Lu H, et al. Stress analysis of shallow sea gas pipelines. Res. J. Appl. Sci., Eng. Tech, 2014, 7: 157-160.
[10] Xiaonan W, Yan X, Yuanhai L, et al. Analysis on Gas Pipeline Stress in Tunnel during Pigging . Natural Gas and Oil, 2012, 2: 002

Physicochemical Processes of Metal Lixiviation in the Disintegrator

V I Golik[1], A B Efremenkov[2]

Doctor of Technical Sciences, Professor,
[1]North-Caucasian State University of Technology, Russia, 362021, Vladikavkaz, 44 Nikolaev str.,
[2]Yurga Institute of Technology, TPU affiliate, 26, Leningradskaya Str., 652055, Yurga, Russia

E-mail: vigolik@mail.ru, abe73@rambler.ru

Abstract. The results of lixiviation-based recycling of ore mill tailings in the high-speed disintegrator are presented. The regularities of the change in properties are defined, as well as control directions over them via alteration of variable processing factors are defined. The model to determine the economic effect of the use of solidifying mixes based on mill tailings is proposed. The model of perfectly mixing reactor cascade in continuous operating conditions is developed to describe activation on minerals in the disintegrator.

Introduction

Over the last two centuries, the issue of removing mill tailings of excavated mineral resources from the ground has become urgent on a global level, and made other vital tasks of the mankind less important [1].

Although we are conscious of the necessity to protect our environment from metal and salt pollution – products of natural lixiviation of minerals, physicochemical fundamentals of recycling technologies are developed insufficiently. New technologies to produce and process mill tailings can hardly solve the key problem of wastelessness.

Improperly developed methods for study on materials in order to control them are considered to be the reasons for practice lagging behind the science in this important field.

Technologies of ore dressing are not so developed as those of ore excavating, as the consequence, hundreds of millions tons of tailings are accumulated on the Earth surface.

As the competition in mining industry gets stronger, toughening the requirements for excavated mineral resources, creating artificial masses made of solidifying mixes gets an integral part of underground field mining of commercial minerals. Cement is used as a binder in the basic variant of their preparation, and excavated natural materials – sand and gravel mix as a nonreactive aggregate.

At the industrial enterprises there are accumulated production residues (ashes and slag of thermal power plants, mill tailings of ores, overburdens and off-grade ores etc.), their chemical-mineralogical composition is similar to that of natural resources used for stowing. These man-made materials somehow yield to standard materials in activity or initial reaction rate because of natural or anthropogenic impact [2].

As numerous studies have highlighted, the increase in activity of man-made materials is made possible not only by specific surface when processing in mills, but by growing unsoundness of the lattice, occurring both on the surface of material and in bulk of its elementary particles due to "active lattice" formation while being processed in the disintegrator [3].

Processing in the disintegrator at the impact velocity of particles 160 m/s (12000 revolutions a minute) goes with the growth of specific surface of materials by 35-150 %. The increase in specific surface is conjoined with solubility of substances contained in the additives to the cementing material. Fine grinding of materials in the disintegrator furthers formation of solid homogenous structure of cement stone while their aqueous suspensions are hardening.

Materials, which were considered inert before, e.g. mill tailings of metal ores, become hydraulically active. The texture of hydrated new growths with incorporated hydrosilicates among non-hydrated particles and crystals of new growths is more compact than that of base cement stone. Hydrosilicate fibers overlap pores, which are dimensionally smaller than those of standard cement stone.

The increase in solidity of cement stone produced at the temperature below 30^0C is higher over the first 24 hours in comparison with conventional methods of intensifying processes: the time of hardening goes up to 28 days or the temperature grows up to 75^0C for standard samples. The solidity of samples produced from activated materials is 2-4 folds higher than that of standard samples.

Mechanochemical activation of cement affects the processes of structure formation and the mechanism of gain in strength of cement stone. Simultaneous activation of cement and ash causes the rise of concrete strength by 20-30%, whereas ash or cement activated separately increase the strength in the interval 5 to 10%.

Processing in the disintegrator involves dimensional rearranging of particles. The products of the disintegrator contain small fractions of particles over 125-400 μm and under 5 μm. The concentration of chemically bound water is lower in the samples containing man-made cementing materials, than in those of pure cement, due to partial binding of calcium hydroxide by silica of a finely-grinded additive. There is more chemically bound water in binding material based on slag grinded in a ball mill because of fine fraction loss when processing in the disintegrator.

The scientists share the same view, if complex binding materials cement + additive are used, provided that specific surface is equal, processing of an additive e.g. slag is more efficient in the disintegrator than in a ball mill (Tabl. 1) [4].

Table 1. Activity and chemical composition of hydration product

Type of slag processing	Specific surface, cm^2/g	Ultimate compression strength, MPa		Chemical composition of hydration products in a second after steaming, %		
		1 с.	180 с.	$H_2O_{х.св.}$	Ca $(OH)_2$	$CaCO_3$
Standard variant – cement 100%	3500	14	26	9.6	5.6	5.5
Ball mill	4980	14	29	9,0	1.6	3.1
Ball mill	2100	8	17	4.9	1.7	1.8
Single-phase disintegrator	1500	12	22	7.3	2.5	3.2
Double-phase disintegrator	2100	12	20	7.4	2.5	2.5

Note: ratio cement to slag, %: 60:40.

At present, materials grinded to the specific surface 600-4000 cm^2/g and grains below 10 μm are used in mining industry. Backfills, gel powders, weighting materials, chemical reagents, as well as chemically low-active and difficult-to recycle-products of metallurgical, mining, chemical and energy industries are milled this way. Using aforementioned products as components of binding materials in solidifying mixes requires for their fine dispersion and multi-component mixing, sometimes even making them more chemically active. It is implemented in disintegrators [5].

Intensifying resource saving trends in development of mining industry insist on metal extraction from mill tailings to be recycled as the total cost of lost metals can exceed the cost of extracted ones.

Conventional dressing procedures can't provide extraction of residual metals both from ores and preprocessed mill tailings as 1-2 types of energy impact are used and oriented on extraction of limited amount of mineral components [6].

Materials based on mill tailings of ores are used in mining without extraction of metals from them to fill the used breasts of stopes, e.g. the Korobkovskoye deposit of Kursk Magnetic Anomaly. Here the loss of gold only amounts to 3 tons a year, whereas its concentration in tailings is 0.5 g/t [7]. This practice is considered to be a palliative both in view of economy and environment protection.

An advance of wasteless recycling is in technologies involving metal lixiviation, although chemical lixiviation even though mechanically activated takes a lot of time and does not provide a needed level of extracted metals.

The proposed technology differs from the available mechanochemical one, because basic reagents are not produced from minerals used in the process but fed into the processing chamber of the disintegrator from the outside [8].

In comparison with the standard procedures evolving in recent years mechanochemical technology of recycling mill tailings is distinguished by accelerated metal extraction in the processing chamber of the disintegrator due to the energy resulting from the increase in the speed of rotor revolution and particle impact. Practical value of the phenomenon is in the fact that leaching solutions are not filtrated in the cracks but pressed in them, i.e. metals are extracted alongside with crystal fracturing [9].

In certain conditions reagents – oxidizers extract into the solution merely all metals contained in ores twice as fast as under agitation-based lixiviation. The technology enables to extract metals from mill tailings the way they meet the sanitary requirements, then, secondary mill tailings can be used without restrictions, e.g. for preparing solidifying mixes.

The principle of disintegration is known approximately over 150 years, the phenomenon of lixiviation has been developed since the middle of the last century in uran industry, and the idea of leaching in the disintegrator has been becoming well-known in recent years only [10].

The issue becomes even more topical as environmentally friendly technologies focused on stowing man-made cavities with solidifying mixes have been rapidly developed in mining industry, so mill tailings can be used as raw materials for this purpose.

This paper is aimed at development of complete waste utilization of ore mining and smelting industry via making more precise parameters of physicochemical processes of mechanochemical treatment based on lixiviation in the disintegrator.

Methods of experiment

As the basic method to solve the task experiments were carried out in a laboratory disintegrator, changing state of the substance so it gains a new quality. The parameters of solidifying mixes are studied analytically via comparing traditional technologies of preparing mixes with an alternative procedure based on lixiviation in the disintegrator.

Experimental study on metal extraction based on lixiviation and recycling of mill tailings involve comparing strength factors of conventionally prepared mixes and those ones based on lixiviation in the disintegrator in various conditions of processing, inter alios:

- agitation-based lixiviation in the percolator;
- agitation-based lixiviation in the percolator of preliminary activated dry mill tailings in the disintegrator;
- lixiviation of mill tailings with reagents in the processing chamber of the disintegrator;
- agitation-based lixiviation of mill tailings, which are activated once with reagents in the disintegrator;

- lixiviation of mill tailings, which are activated repeatedly in the processing chamber of the disintegrator.

Variable factors, adjustable when carrying out experiments:

1. Concentration of reagents: sulfuric acid and sodium chloride.
2. Liquid and solid phase ratio (L:S).
3. Time of activation-based lixiviation.
4. The speed of rotor revolution in the disintegrator.
5. The number of processing cycles.

The assessment criterion of samples under consideration is extraction of metals specified in the course of analyzing the samples taken from the disintegrator every hour and at the end of lixiviation process, as well as uniaxial compression strength of solidifying mixes, produced from ferruginous quartzite mill tailings of Lebedinskoye Deposit.

The solidifying mix with constant chemical composition, kg/m^{3}are tested: mill tailings possessing diverse parameters of activation – 1445, water – 380, cement – in dependence on conditions of research.

Results

The strength of conventionally produced solidifying mixes without lixiviation-based extraction of metals and mechanical activation varies monotonically in dependence on the concentration of binding cement (Table 2).

Table 2. The strength of not activated solidifying mixes

Consumption of Portland cement, kg/m^3	30	60	80	100	120	180
Strength, MPa	0.79	0.92	0.101	1.20	1.41	1.80

Mechanical activation of mill tailings in the disintegrator increases the strength of the solidifying mix (Table 3, Point 2).

All modes of lixiviation support metal extraction and reduce the strength of solidifying mix as moisture increases (Table 3, Points 3 – 5), its compensation is possible through raising frequency of processing (Table 3, Point 6).

Table 3. The strength of binding cement mixes relative to activated mill tailings

№	Activation type	Strength, MPa, 24 h.		
		7	14	28
1	Not activated, consumption of Portland cement 100 kg/m^3	1.04	1.11	1.20
2	Reagentless mechanical activation	1.16	1.25	1.32
3	Agitation-based lixiviation without activation	0.52	0.61	0.72
4	Reagentless mechanical activation with agitation-based lixiviation	0.68	0.73	0.88
5	Mechanochemical activation in the disintegrator	0.73	0.77	0.94
6	Three-phase mechanochemical activation in the disintegrator	0.92	1.10	1.22

The strength of the mix under three-phase mechanochemical activation of mill tailings in the disintegrator is similar to that of the basic mix, 100 kg/m^3 ready-made Portland cement is consumed.

Mechanical activation of mill tailings in the disintegrator without reagent lixiviation increases the strength of cementless mixes by 10 and more per cents (Table 4, Point 2). Liquid medium of lixiviation in the disintegrator and agitation vat deteriorates the strength and requires for increase in the frequency of processing to provide the same strength (Table 4, Points 3-6).

Table 4. The effect of activation on strength of cementless mixes

№	Activation type	Strength, MPa, 24 h.		
		7	14	28
1	Not activated	0.64	0.81	1.01
2	Reagentless mechanical activation	0.86	0.95	1.12
3	Agitation-based lixiviation without activation	0.42	0.57	0.62
4	Reagentless mechanical activation with agitation-based lixiviation	0.60	0.69	0.78
5	Mechanochemical activation in the disintegrator	0.63	0.71	0.84
6	Three-phase mechanochemical activation in the disintegrator	0.82	1.00	1.12

Solidifying mixes containing unlixiviated mill tailings activated in the disintegrator and cement (Figure 1).

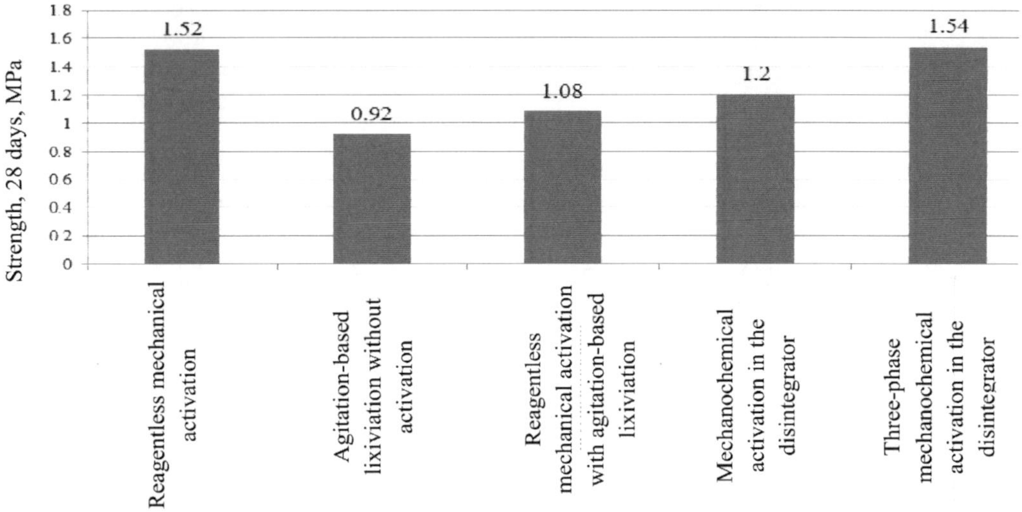

Figure1. The strength of solidifying mix with cement

Activation increases the strength of the mix (Table 5, Point 2). In all conditions of lixiviation the strength gets reduced when one-phase processing and is compensated by its frequency (Table 5, Points 3-6).

The strength of mill tailings activated without cement in the disintegrator is lower by 20-25% (Figure 2), its compensation by frequency of processing is still possible (Table 5, Points 1-6).

Table 5. Parameters of activation

№	Activation type	The strength, 28 days, MPA		
		cement100 kg/m^3		Cementless activation
		Not activated	Activated	
1	Not activated	1.30	-	1.01
2	Reagentless mechanical activation	-	1.52	1.22
3	Agitation-based lixiviation without activation	-	0.92	0.62
4	Reagentless mechanical activation with agitation-based lixiviation	-	1.08	0.78

№	Activation type	The strength, 28 days, MPA		
		cement100 kg/m^3		Cementless activation
		Not activated	Activated	
5	Mechanochemical activation in the disintegrator	-	1.20	0.94
6	Three-phase mechanochemical activation in the disintegrator	-	1.54	1.12

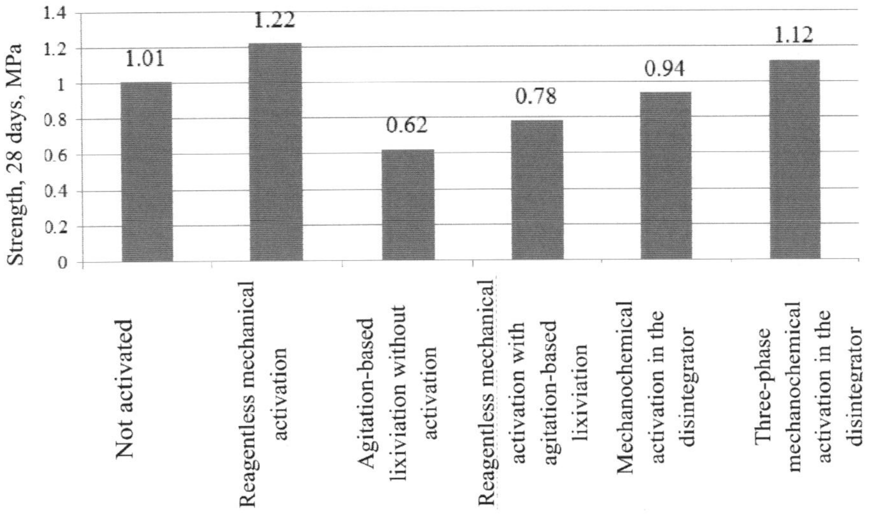

Figure 2. The strength of cementless solidifying mix

The cementless compounds containing activated in the disintegrator mill tailings possessing the strength 0.5-1.0 MPa can be used as solidifying mixes for stowing man-made cavities in corresponding geo-mechanical conditions [11].

The gain coefficient of strength of solidifying mixes based on activated in the disintegrator mill tailings is estimated 20-30%.

The importance of the disintegrator in the outlined above technology consists in the increase of the strength of cementless solidifying mixes containing mill tailings. This task is well-performed by the disintegrator as the strength of the thrice-processed mix is comparable with that of the mix containing ready-made cement [12].

Lixiviation as an integral part of the cycle somehow deteriorates the strength of the cementless compound, nevertheless, enables extraction of residual metals from the mill tailings and makes the recycled tailings safe from the standpoint of sanitary standards and accessible for the wide use [13].

The economic effect of field development using solidifying mixes based on mill tailings over the period of deposit exploitation is:

$$E_{\text{хо}} = (P_{e\text{м}} + P_{\text{хтz}} + P_{\text{хм}}) - (X_{e\text{м}} + X_{\text{тз}})$$

where $E_{\text{хо}}$ – economic effect of the use of mill tailings, rub.;

$P_{e\text{м}}$ – profit from field development, rub.;

$P_{\text{хтz}}$ – profit from tailings used as consolidating stowing, rub.;

$P_{\text{хм}}$ – profit from extraction of metals from tailings, rub.;

expenditures on metal extraction from mill tailings, rub.;

$X_{тз}$ – expenditures on production of consolidating stowing from mill tailings, rub.;

Alongside with mechanochemical activation dispersion and homogenization of multi-component materials are possible in the disintegrator. When processing small mill tailings the function of a high-speed mixer is a foreground one.

The materials of a solidifying mix include low base hydrosilicates with silica. The solubility of particular components of the material can be increased via processing in the disintegrator. The concentration of calcium hydroxide in the liquid phase at the temperature over 50^0C during the first hours depends on the mode of its dry disintegration. When solidifying of not activated mix the residual concentration of lime in the liquid phase of cement solution is 760 mg/l at the end of cementation that is nearly equal to its saturated solution in these conditions [4]. After processing in the disintegrator the residual concentration of lime is 470 mg/l at the rotation frequency of rotors in the disintegrator 6000 revolutions a minute; and 180 mg/l – at 12000 revolutions a minute.

Processing of silica-containing materials in the disintegrator supports their activation and with respect to the mode of activation increases the speed of solidifying in the initial phase of hydration. When consolidating activated compounds thermodynamically stable low based hydrosilicate compounds form in the earlier periods of consolidation as compared with the base variant. In terms of thermodynamics free energy of hydration equals to the difference of free energies of input and output products of the process. The lower free energy of formation of input waterfree binding materials is, the higher is the free energy of hydration.

The disintegrator facilitates the destruction of upper layer of material and reduces the energy of waterfree binding material. The difference in free energies of formation of solids subjected to mechanochemical activation can be similar to free generation energy of salts from oxides

As the result of hydrating the substance in the course of mechanochemical activation hydrates with new thermodynamic properties are formed, key factors of them include dissolubility and diffuse characteristics of ions, inter alios, calcium, and silicon acids formed in the process of hydration and hydrolyze.

The ratio of concentration of calcium and silicon acid ions exceeds the unity that is the reason for high basicity of primary hydrated phases. Calcium ions detach from the crystalline structure of the binding material under action of water polar molecules, whereas necessary for formation of hydrosilicate phases bond rapture in silicon and calcium oxides requires for energy of activation.

The bigger particles of silicon acid are, the more important are the diffuse processes and the more energy is consumed for ion interacting. Dispersion of substance particles conjoining with bond raptures decreases the staging of generation of hydrated compounds and facilitates formation of thermodynamic phases.

The process of treatment in the disintegrator is described by the model of perfectly mixing reactor cascade in conditions of continuous operation:

$$A_к = A_i + V_d \times A_i \times t_к \times n$$

where A_i – the initial activity of material;

$A_к$ – the activity of material in the following phase of processing;

V_d – the speed of changing activity of material;

$t_к$ – the time of material processing in a single cycle;

n – the number of processing cycles.

Conclusions

The efficiency of physical-mechanical processes taking place in the course of mechanochemical technology is specified via comparing strength factors of mixes and extraction of metals in different operating conditions of the disintegrator.

Activation of mill tailings and lixiviation of metals from them in the disintegrator meet general requirements and are to be controlled by changing variable factors of processing.

Low-cement and cementless compounds based on activated in the disintegrator mill tailings possessing the strength 0.5-1.0 MPa can be used as solidifying mixes for stowing man-made cavities in corresponding geo-mechanical conditions. The gain coefficient of strength of solidifying mixes containing mill tailings activated in the disintegrator is estimated 20-30%.

The process of treatment in the disintegrator is described by the model of perfectly mixing reactor cascade in conditions of continuous operation. The economic efficiency of the new technology is specified by the volume of replaced ready-made components by secondary recycled mill tailings subjected to mechanochemical processing and selling simultaneously extracted metals.

References

[1] Rakishev B. R. Integrated utilization of ore at the enterprises of non-ferrous metallurgy of Kazakhstan. Mining journal. 2013. №7. pp.48-53.

[2] Ermolovich E. A., Shok I. A. Man-made wastes in stowage composite materials. Mining journal. 2012. №9. pp.89-95.

[3] Khint I.A. DND - technology: problems and perspectives. Tallinn, 1981. 67 p.

[4] Karimov N.Kh., Agzamov F.A., Izmukhambetov B. S. On hydration of binding materials produced in the disintegrator. Mining Bulletin № 4, 1998. pp. 79-85.

[5] Golik V., Komashchenko V., Morkun V. Feasibility of using the mill tailings for preparation of self-hardening mixtures. Metallurgical and Mining Industry, 2015, №3, p.p. 38-41.

[6] Golik V. I., Hasheva Z. M., Economical Efficiency of Utilization of Allied Mining Enterprises Waste. The Social Sciences 10 (5): 682-686, Medwell Journals, 2015.

[7] Versilov S.O., Frolov A.V., Versilova E.S., Demov V.I. Optimization of litho-technical system of ore-dressing and processing enterprise «underground mine – tailing pit». GIAB. 2014 №8. pp.56-63.

[8] Golik V.I., Stradanchenko S.G, Maslennikov S.A. Experimental Study Of Non-Waste Recycling Tailings Ferruginous Quartzite. Research India Publications. Number 15 (2015), pp 410-416.

[9] Golik V.I., Razorenov Y.I., Polukhin O.N. Metal extraction from ore benefication codas by means of lixiviation in a disintegrator. International Journal of Applied Engineering Research. Volume 10, Number 17 (2015) pp 38105-38109.

[10] Golik V., Komashchenko V., Morkun V. Innovative technologies of metal extraction from the ore processing mill tailings and their integrated use. Metallurgical and Mining Industry, 2015, №3, p.p. 49-52 .

[11] Golik V., Komashchenko V., Morkun V. Geomechanical terms of use of the mill tailings for preparation. Metallurgical and Mining Industry, 2015, №. 4.pp. 321-324.

[12] Golik V.I., Stradanchenko S.G, Maslennikov S.A. Experimental feasibility study of wasteless utilization of ferruginous quartzite mill tailings. Perspective materials. 2015. № 7. pp. 64-71.

[13] Golik V., Doolin A., Komissarova M., Doolin R. Evaluating the Effectiveness of Utilization of Mining Waste. International Business Management 9 (5). 1993-5250. Medwell Journals, 2015.

IOP Conf. Series: Materials Science and Engineering **125** (2016) 012039 doi:10.1088/1757-899X/125/1/012039

Technology of ultrasonic control of gas-shielded welding process

S A Solodsky[1,a], V D Sarychev[2,b] and I S Borisov[1,c]

[1]Cand.Sc, AP. Yurga Institute of Technology (branch) of Tomsk polytechnic university. Russia, Leningradskaya str., 26.
[2]Cand.Sc, AP. Siberian State Industrial University. Russia, Novokuznetsk, Kirova str., 42
[1]student. Yurga Institute of Technology (branch) of Tomsk polytechnic university. Russia, Leningradskaya str., 26.

E-mail: [a]serdgio80@inbox.ru, [b]sarychev_vd@mail.ru, [c]il.borisow2009@yandex.ru

Abstract. A new approach to implementation of electrode metal transfer control under MAG, MIG welding is suggested. The process ensures control of thermal and crystallization processes, stabilizes the time of electrode metal drop formation. The results of the research allow formulating the basic criteria of electrode metal transfer control via ultrasonic exposure, determining the conditions of producing a more equilibrium structure of deposit metal.

1. Relevance

Under rigid and continuously growing requirements to the quality of weld joints designing welding systems capable of controlling electrode metal transfer is currently a priority area of scientific research and practical developments of many companies producing welding equipment for semi-automatic welding [1]. It is evident from significant upsurge in interest to that kind of equipment demonstrated by various industries. The given problem is considered in the works of many authors: Paton B Ye, Dudko D A, Zaruba I I, Potapievsky A G, Dyurgerov N G, Lenivkin V A, Brunov O G, Knyazkov A F, Sarayev Yu N and many others [1, 2, 3].

In the given paper we present a welding system implementing an actual process of controlled transfer. The process is based on exciting resonance high-frequency (10^9 – 10^{12} Hz) mechanic oscillations in the electrode metal drop reducing the size of the drop due to increasing the uniform compression force in the area of the bridge resulting in its faster destruction.

2. Experimental procedure

It is known [4] that magnetic action on ferromagnetic materials makes them change their linear dimensions due to lattice distortion, i.e. magnetostriction effect is observed. Thus, the electromagnetic field may change the linear dimensions of ferromagnetic material placed into it. Subsequently, if we expose ferromagnetic material to variable magnetic field it will induce energy impulses in the material creating surface acoustic waves. In ferro- and ferrimagnetics (Fe, Ni, Co, Gd, Tb and other, some alloys, ferrites) magnetostriction reaches significant magnitude (relative deformation $\Delta l/l \approx 10^{-6} - 10^{-2}$ m). [5].

There are only two types of ultrasonic waves that can propagate in the uniform isotropic indefinite solid medium – longitudinal and shear waves. For the longitudinal waves the movement of particles is parallel to the direction of wave propagation and deformation is a combination of uniform compression (tension) and pure shear. For the shear waves movement of particles is perpendicular to the direction of wave propagation and deformation is pure shear. On the semi-infinite solid-fluid interface Rayleigh surface waves can propagate being the combination of inhomogeneous longitudinal and shear waves with exponentially diminishing amplitudes as they move away from the interface.

The electrode wire applied for machine arc gas-shielded welding, being a ferromagnetic, demonstrates the magnetostriction effect and it is possible to induce the required longitudinal oscillations with specified frequency in it. For this purpose a magnetostriction transducer is applied. It transforms the energy of the magnetic field into mechanic ultrasonic energy.

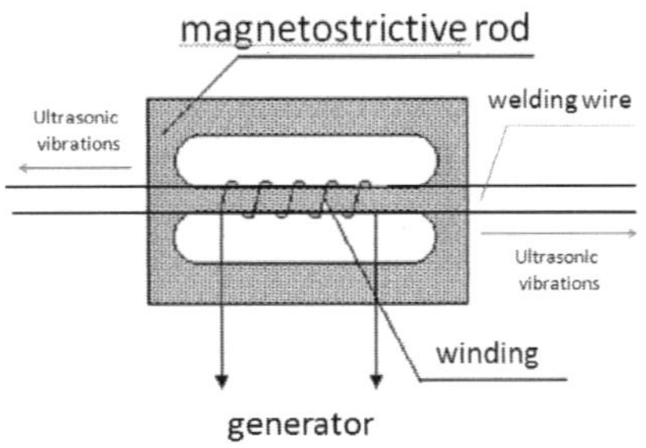

Figure 2.3 Magnetostriction transducer

When considering metal as elastic anisotropic medium we should notice that, under point energy deposition, two types of volume waves: longitudinal waves where particle displacement occurs along the direction of the wave propagation and transverse waves where the particles are displaced in the direction perpendicular to the direction of the wave propagation. It is possible to say that longitudinal and transverse waves are volume natural oscillations.

In the case of the pile-up of the pulse upon the electrode wire the transverse waves may not be taken into consideration as the point-like pulse is piled-up across the whole diameter and, thus, the transverse waves being excited in every point are mutually absorbed. Let us set longitudinal waves velocity as (c_n). Then the velocity of longitudinal waves propagation in elastic materials with known constants is expressed as follows [1];

$$c_n = \sqrt{\frac{E(1-\nu)}{\rho(1+\nu)(1-2\nu)}}, \quad (1)$$

where ρ – density of material; ν – Poisson's ratio; E – Young modulus.

So, surface drift is propagated as an elastic wave along axis x and exponentially damping as it moves inward the wire. This surface wave is called Rayleigh wave R(c) and its velocity equals:

$$R(c) = (2 - m^2)^2 - 4\sqrt{1 - m^2} = 0, \quad (2)$$

where $m = c/c_n$, where c – sound velocity in a material.

Subsequently, if we find the velocity of Rayleigh wave propagation we can obtain the required frequency of the electrode wire as it is known that, under velocities close to R(c), but not exceeding its value the wave becomes concave and under c_n>R(c) it becomes convex.

If we apply electrode wire as ferromagnetic material the piled-up electromagnetic pulses induce a surface wave on it, the wave which propagates in both directions from the center of electromagnetic pulse application. Under machine gas-shielded welding such electromagnetic pulses are applied to the wire moving towards the welding arc zone. The travel speed of the wire is not taken into account during the calculations as velocity of Rayleigh surface waves propagation is an order of magnitude faster.

Consequently, Rayleigh waves travel by the electrode wire surface towards its end, i.e. up to the bridge of electrode metal drop and on the solid-fluid interface the waves (Love-waves) are excited, they have vertical polarization and their structure and velocity are completely different from those of Rayleigh waves. They consist of a weak inhomogeneous wave in the fluid which amplitude slowly diminishes as the wave moves away from the interface of two strongly inhomogeneous waves in the solid (longitudinal and transverse). Due to this phenomenon the wave energy and motion of the particles are mainly localized in the fluid, not in the solid [6].

These waves continuously radiate energy into the fluid as they propagate forming in the fluid an inhomogeneous wave moving away from the interface. The phase velocity of the given surface acoustic wave equals c_n and the attenuation coefficient for λ wave length is $\sim$ 0.1. Beside the attenuating wave, on the solid-fluid interface there is always a continual wave running along the interface with phase velocity.

Consequently, energy of high-frequency electromagnetic field is transmitted through the wire to the molten drop due to Rayleigh and Love waves. This energy increases the activity of the surface of electrode metal drop, thus, increasing the uniform compression force effecting the bridge and contributing to its faster destruction. Subsequently, the size of the drop able to detach from the electrode wire is reduced.

3. Implementation of the process

In the process of continuous feed welding elastic longitudinal oscillations are piled-up at the wire. Piling-up of elastic longitudinal oscillations is implemented due to magnetostriction properties of electrode wire which allow its application as generator of mechanical longitudinal impulses. Elastic longitudinal impulses appear when longitudinal magnetic field with ultrasonic frequency of $10^9 - 10^{12}$ Hz is piled-up on electrode wire. longitudinal oscillations induce elastic waves in the electrode wire. In the process of welding elastic waves transmit elastic longitudinal oscillations to the drop of molten metal which is at the butt end of electrode wire. When the frequency of longitudinal oscillations of the wire coincide with the frequency of molten metal drop oscillations, it creates resonance of oscillations which causes dramatic increase of forced oscillations amplitude. The effect of mechanical detachment of electrode metal drop is created. of Electrode metal is transferred into the weld pool due to the resonance phenomenon [6].

The work and the method of the welding process implementation is as follows (Figure 1)

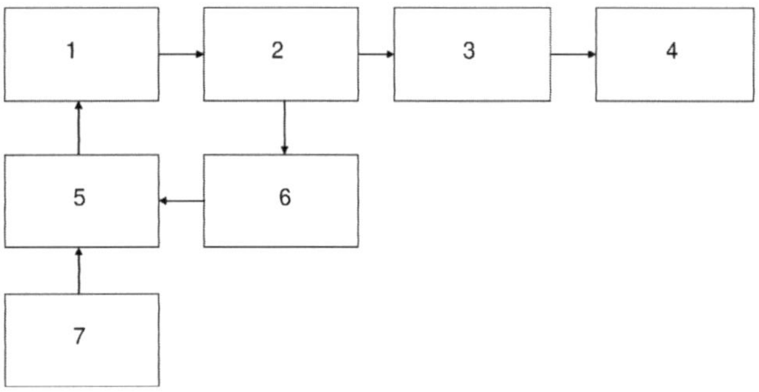

Figure 1 – Flow chart of the welding process implementation: 1 – ultrasonic generator; 2 – inertia-free solenoid; 3 – weld pool; 4 – molten metal drop; 5 – comparing and correcting unit; 6 – resonance indicator; 7 – unit of intended resonance frequency [7]

Ultrasonic generator 1 with variable frequency from 10^9 Hz to 10^{12} Hz feeds pulses of current to inertia-free solenoid 2 inside which electrode wire 3 is passed at a uniform rate. Inertia-free solenoid 2 produces variable longitudinal magnetic field under the effect of pulse current. Elastic longitudinal oscillations with the amplitude of $(4\text{-}8)\cdot 10^{-9}$ m and frequency equaling that of the generator are induced in the electrode wire 3 due to magnetostriction effect under the action of longitudinal magnetic field. Longitudinal oscillations move along electrode wire 3 according to the law of elastic waves propagation and Rayleigh waves induced at the butt end of electrode wire are transmitted to electrode metal drop 4. When frequency of Rayleigh waves and longitudinal oscillations of the wire coincide with the natural oscillations of molten metal drop we observe resonance and sharp rise of the amplitude of induced oscillations. The drop of electrode metal 4 is mechanically detached from electrode metal 3. To stabilize the process the control unit includes resonance indicator 6 registering sharp increase of oscillations amplitude. The signal from resonance indicator 6 arrives at comparing and correcting unit 5 where the signal is compared to the signal from resonance indicator 6 and that of the unit of intended resonance frequency 7. If the signals do not coincide the comparing and correcting unit 5 corrects the frequency of mechanical oscillations of the drop changing the frequency of ultrasonic generator 1. The ultrasonic generator is presented in Figure 2.

4. Material and methods of research

The research conducted at the unit allowed experimental determining one of possible resonance frequencies ν_R within megahertz range. The frequency depends on welding conditions and electrode wire material. The current on magnetostriction transducer under the given frequency ν_R is 21 A and it corresponds to field density creating maximum magnetostriction effect for the iron.

Studies of the results of welding with resonance control in CO_2 atmosphere were conducted on the samples produced from sheet metal. To study the welding processes in CO_2 atmosphere with resonance control we applied a welding unit which comprises a magnetostriction transducer with control system designed and produced in Yurga Institute of Technology of Tomsk Polytechnic University under the supervision of AP of Welding Production department, Cand.Sc. Brunov O G, arc welder – welding rectifier VS-300 B UZ, digiscope N338 which two channels were used to obtain oscillograms of voltage and current of the arc. To film the process of metal transfer the unit was placed on the automatic welding head GSP – 2.

For high-speed filming of the welding process and physic models processing we applied high-speed camera "VideoSprint" with camera lens Nikon (maximal filming speed 8000 frames per second). Macroslices were produced by the central laboratory of Yurga Engineering Plant.

On Figure 2 a, b, c the oscillograms of the process are presented. Analysis of the oscillograms showed that under the effect of off-resonance frequencies the process of transfer does not differ significantly from that with classic electrode wire feed (Figures 3 a, b). The oscillograms of the welding process with the frequency of wire oscillations $18\cdot 10^5$ Hz, i.e. the transfer process is stabilized at the resonance frequency and the size of the transferred drop is reduced.

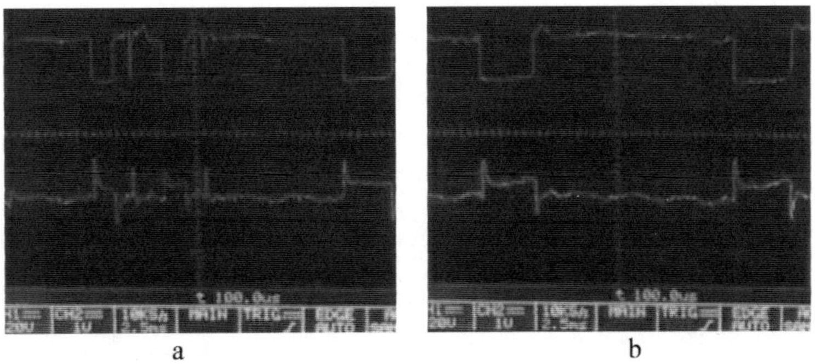

a b

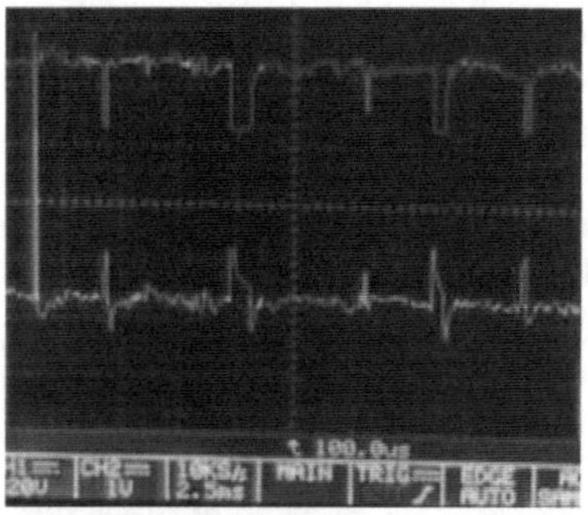

c

Figure 2: a – Oscillograms of the welding process with wire oscillation frequency $7\cdot10^4$ Hz. b - Oscillograms of the welding process with wire oscillation frequency $22\cdot10^6$ Hz. Frequency of drop transfer is 42 Hz, c – Oscillograms of the welding process with the obtained wire oscillation frequency ν_p

Frames of high-speed filming of the welding process are presented in (Figure 4) (Uд = 22 B, Iд= 180 A).

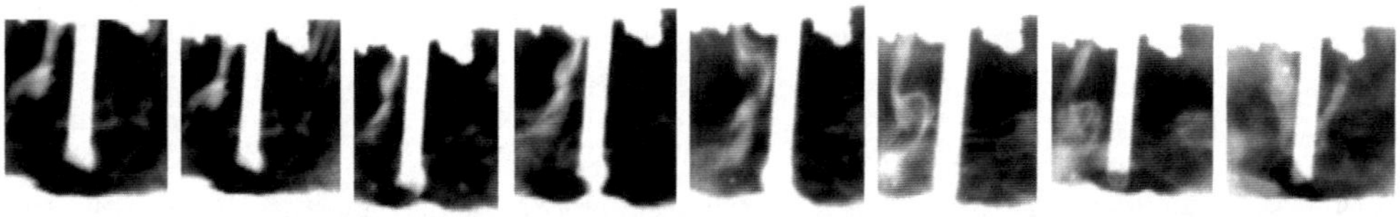

Figure 4. Frames of high-speed filming of the studied process on the resonance frequency (1500 frames per second)

Analysis of the records [8, 9] showed that application of resonance allows increasing stability of the welding process due to introduction of additional force pushing the drop towards the weld pool. The given approach allows reducing oscillation of arc spots, thus, improving stability of the drop, reducing its size, increasing frequency of transfer. All that leads to reduction of spattering and alloying elements burning, improvement of weld formation without application of expensive pulse control systems.

5. Conclusion

The developed welding method is characterized by the following advantages:

a) transfer of electrode metal drop is realized due to resonance causing mechanical detachment of electrode metal drop from the electrode wire;

b) control of electrode metal transfer happens without periodic alteration of arc-welding current;

c) the suggested method of welding for realizing controlled transfer does not require a switched-mode power supply and special means of pulsed wire feed.

References

[1] Saraev Y N Adaptive pulse-arc welding methods for construction and repair of the main pipelines. Proceedings of The 2nd South-East European IIW International Congress "Welding – HIGH-TECH Technology in 21st century". Sofia, Bulgaria, October 21st-24th 2010, p. 174 – 177.

[2] Dyurgerov N G and Lenivkin V A Stability of pulsed-arc consumable electrode welding. Welding Internationa 2016.

[3] Chinakhov D A Study of thermal cycle and cooling rate of steel 30ХГСА single-pass weld joints

// Applied Mechanics and Materials. – Vols. 52–54. – 2011. – p. 442-447. - Mode of access: http://www.scientific.net/AMM.52-54.442

[4] Vonsovsky S V, Magnetism, M. 1971; Belov K P, Elastic, thermal and electric phenomena in ferromagnetic, 2nd ed., M. – L., 1957.

[5] Ultrasonic transducers, transl. from English, under the editorship of I P Golyamina, M., 1972.

[6] Viktorov I A Acoustic surface waves in solids. Publishing house: Nauka, M–1981.

[7] Patent for invention № 2481931 (RF). Method of consumable machine welding with piling-up mechanic nanoimpulses over electrode wire feed. Solodsky S A, Brunov O G, Zelenkovsky A A, Vasilyev D V I priority of invention 4th of May 2012 V23K 9/173. Published 20.03.2014. Bul.№8.

[8] Chinakhov D A and Agrenich E P Computer simulation of thermo-mechanical processes at fusion welding of alloyed steels // Materials Science Forum. – Vols. 575-578 (2008). – Pp. 833-836.

[9] Valuev D V and Danilov V I Reasons for Negative Formation of Structures in Carbon Steel Processing of Pressure // 7th International Forum on Strategic Technology (IFOST – 2012): Proceedings: in 2 vol., Tomsk, September 18–21, 2012. – Tomsk: TPU Press, 2012 – Vol. 2 – p. 151-154.

AUTHOR INDEX

AUTHOR INDEX